T0192519

# *U*-Statistics

# STATISTICS: Textbooks and Monographs

A Series Edited by

## D. B. Owen, Coordinating Editor
*Department of Statistics*
*Southern Methodist University*
*Dallas, Texas*

R. G. Cornell, Associate Editor
for Biostatistics
*University of Michigan*

W. J. Kennedy, Associate Editor
for Statistical Computing
*Iowa State University*

A. M. Kshirsagar, Associate Editor
for Multivariate Analysis and
Experimental Design
*University of Michigan*

E. G. Schilling, Associate Editor
for Statistical Quality Control
*Rochester Institute of Technology*

## ADDITIONAL VOLUMES IN PREPARATION

# *U*-Statistics

## Theory and Practice

### A. J. Lee

University of Auckland
Auckland, New Zealand

**CRC Press**
Taylor & Francis Group
Boca Raton  London  New York

CRC Press is an imprint of the
Taylor & Francis Group, an **informa** business

Published in 1990 by CRC Press
Taylor & Francis Group
6000 Broken Sound Parkway NW, Suite 300
Boca Raton, FL 33487-2742

First issued in paperback 2020

© 1990 Taylor & Francis Group, LLC
CRC Press is an imprint of Taylor & Francis Group, an Informa business

No claim to original U.S. Government works

ISBN 13: 978-0-367-58015-5 (pbk)
ISBN 13: 978-0-8247-8253-5 (hbk)

**Visit the Taylor & Francis Web site at**
**http://www.taylorandfrancis.com**

**and the CRC Press Web site at**
**http://www.crcpress.com**

Library of Congress catalog number: 90-3458

---

**Library of Congress Cataloging-in-Publication Data**

---

Catalog record is available from the Library of Congress

---

*To LBL and the memory of LDL*

# PREFACE

Over forty years have elapsed since P. R. Halmos and Wassily Hoeffding introduced the class of $U$-statistics into statistical practice. Since that time a great many periodical articles and parts of several books have extended and applied the theory, and research interest in the subject seems to be accelerating.

The class of $U$-statistics is important for at least three reasons. First, a great many statistics in common use are in fact members of this class, so that the theory provides a unified paradigm for the study of the distributional properties of many well-known test statistics and estimators, particularly in the field of non-parametric statistics. Second, the simple structure of $U$-statistics makes them ideal for studying general estimation processes such as bootstrapping and jackknifing, and for generalising those parts of asymptotic theory concerned with the behaviour of sequences of sample means. Third, application of the theory often generates new statistics useful in practical estimation problems.

It thus seems appropriate to attempt a monograph describing in a reasonably comprehensive way the accumulated theory of the last forty years, and to detail some of the more interesting applications of this theory. While portions of several textbooks have dealt with $U$-statistics, these accounts have necessarily been incomplete; for example Serfling (1980) deals with asymptotic aspects but is not concerned with applications, and Randles and Wolfe (1979) treat applications to non-parametric statistics but do not deal with some of the more abstruse asymptotics. The books by Puri and Sen (1971) and Sen (1981) concentrate on applications to multivariate non-parametrics and sequential nonparametrics respectively. The present work aims to survey the literature in English and to present a blend of theory and practical applications, although in view of the existence of the works just referred to we have not attempted a systematic exposition of classical nonparametric theory from a $U$-statistic viewpoint. Instead, we give an account of the basic theory, with a selection of advanced topics determined by the author's own tastes, and illustrate the theory by means of examples and applications scattered throughout the text. Some more complex applications are collected in a final chapter.

The book is organised as follows: Chapter 1 introduces the basic statistics based on i.i.d. sequences, discusses the optimal properties of $U$-statistics, and explains how to calculate variances. The $H$-decomposition of a $U$-statistic into uncorrelated components of decreasing order in the sample size is fundamental to the asymptotic theory, and this decomposition is discussed next. Chapter 2 deals with various generalisations of this basic theme such as generalised $U$-statistics, weighted and trimmed $U$-statistics and generalised $L$-statistics. Relaxations of the i.i.d. assumptions are also explored.

Chapter 3 is the heart of the theoretical part of the book, and covers the asymptotic theory. We treat asymptotic distributions, strong consistency, Berry-Esseen rates, invariance principles and the law of the iterated logarithm. A general theme is how the $H$-decomposition coupled with the corresponding result for sample means yields $U$-statistic variants of the classical theorems of probability theory.

Chapter 4 is devoted to the study of several related classes of statistics. One such is the class of symmetric statistics, which contains $U$-statistics as a special case, and many of the properties of $U$-statistics carry over to this more general class. Another related class of statistics is that of von Mises statistics or $V$-statistics, which also may be expressed in terms of $U$-statistics.

Computation of $U$-statistics can involve averaging over large numbers of terms, so it is natural to consider statistics that average over only a subset of these terms. We are thus led to the idea of incomplete $U$-statistics, and these are the subject of the final part of Chapter 4.

The problem of estimating the standard errors of $U$-statistics is considered in Chapter 5, where the emphasis is on methods based on resampling.

Our final chapter offers a selection of applications of the theory described in the previous chapters.

Finally, warm thanks are due to Marilyn Talamaivao, who typed a difficult manuscript with great efficiency, and Donald Knuth for providing the tools. Most of all, thanks are due to Nick Fisher, who kindled my interest in the subject and even volunteered to read the manuscript!

*A. J. Lee*

# CONTENTS

# CHAPTER ONE

## Basics

### 1.1 Origins

Consider a functional $\theta$ defined on a set $\mathcal{F}$ of distribution functions on $\mathbb{R}$ :

$$\theta = \theta(F), \quad F \in \mathcal{F}. \tag{1}$$

Suppose we wish to estimate $\theta(F)$ on the basis of a sample $X_1, ..., X_n$ of random variables, which until further notice are assumed to be independently and identically distributed with d.f. $F$. We assume that $F$ is an unknown member of $\mathcal{F}$, but $\mathcal{F}$ is known. The following questions were first raised in a fundemental 1946 paper by P. R. Halmos, which may be regarded as representing the beginnings of our subject:

(a) Does there exist an estimator of $\theta$ that will be unbiased whatever the distribution function $F$ may be? Can we characterise the sets $\mathcal{F}$ and the functionals $\theta$ for which the answer is yes?

(b) If such an estimator exists, what is it? If several exist, which is the best?

The first part of question (a) is easily disposed of.

Let $\mathcal{F}$ be any subset of the set of distribution functions on $\mathbb{R}$, and let $\theta(F)$ be a functional defined on $\mathcal{F}$. Suppose that for each sufficiently large integer $n$, there is a function $f_n(X_1, \ldots, X_n)$ of $n$ variables such that

$$E\{f_n(X_1, \ldots, X_n)\} = \theta(F) \tag{2}$$

for all $F$ in $\mathcal{F}$, where $X_1, \ldots, X_n$ is a sequence of independent random variables distributed as $F$. Such a functional $\theta$ is said to admit an unbiased estimator, and our first theorem, due to Halmos, characterises such functionals.

**Theorem 1.** *A functional $\theta$ defined on a set $\mathcal{F}$ of distribution functions admits an unbiased estimator if and only if there is a function $\psi$ of $k$ variables such that*

$$\theta(F) = \int_{-\infty}^{\infty} \cdots \int_{-\infty}^{\infty} \psi(x_1, x_2, \ldots, x_k) \, dF(x_1) \ldots dF(x_k) \tag{3}$$

1

*for all $F$ in $\mathcal{F}$.*

**Proof.** Suppose that $\theta$ is of the form (3). Then the unbiased estimator

$$f_n(X_1, \ldots, X_n) = \psi(X_1, \ldots, X_k)$$

satisfies (2) for $n \geq k$ and so $\theta$ admits an unbiased estimator. Conversely, if (2) holds then (3) holds with $k = n$, so that $\theta$ is of the desired form.

A functional satisfying (3) for some function $\psi$ is called a *regular statistical functional* of degree $k$, and the function $\psi$ is called the *kernel* of the functional. The estimates in the proof of Theorem 1 are obviously unsatisfactory since they use only the information from $k$ of the observations in the sample, but an intuitively reasonable estimator will be one based on a symmetric function $\psi$ of all $n$ observations, since the random variables $X_1, \ldots, X_n$ are independent and identically distributed. Thus we restrict ourselves to estimators of the form $\hat{\theta} = f_n(x_1, \ldots, x_n)$ where $f_n$ is a symmetric function satisfying (2). (A symmetric function is one invariant under permutations of its arguments.)

Let us regard estimators as being identical if they agree on some Borel set $E$. The choice of $E$ depends on the set $\mathcal{F}$ under consideration : for example if $\mathcal{F}$ consists of all distributions on the set $\{0, 1\}$ then $E$ could be $\{0, 1\}$. Alternatively, if $\mathcal{F}$ is the set of all distributions whose means exist, it would be natural to take $E$ to be $\mathbb{R}$, the set of real numbers.

It turns out that if $\mathcal{F}$ is sufficiently large, then there is only one symmetric unbiased estimator (up to equality on $E$). The concept of "sufficiently large" can be made precise in a variety of ways; we will assume first that $\mathcal{F}$ is large enough to include all distributions with finite support in $E$; i.e. all distributions whose distribution functions are step functions whose (finitely many) points of increase are in $E$. We can then prove

**Theorem 2.** *Let $\mathcal{F}$ contain all distributions with finite support in $E$, and let $\theta$ be a regular functional satisfying (3). Then up to equality on $E$, there is a unique symmetric unbiased estimator of $\theta$.*

**Proof.** Let $\psi^{[n]}(x_1, \ldots, x_n) = \{(n-k)!/n!\} \sum \psi(x_{i_1}, \ldots, x_{i_k})$ where the sum extends over all $n!/(n-k)!$ permutations $(i_1, \ldots, i_k)$ of distinct integers

2

chosen from $\{1, 2, \ldots, n\}$. Then $\psi^{[n]}(X_1, \ldots, X_n)$ is unbiased, since

$$\int \cdots \int_{\mathbb{R}_n} \psi^{[n]}(x_1, \ldots, x_n) \prod_{i=1}^{n} dF(x_i) = \int \cdots \int_{\mathbb{R}_k} \psi(x_1, \ldots, x_k) \prod_{i=1}^{k} dF(x_i)$$
$$= \theta(F).$$

Now let $f$ be any other symmetric unbiased estimator. Then by applying Lemma A below to the function $f - \psi^{[n]}$ we see that $\psi^{[n]}$ is unique.

**Lemma A.** *Let $\mathcal{F}$ contain all distributions with finite support in $E$, and let $f$ be a symmetric function of $n$ variables with*

$$\int \cdots \int_{\mathbb{R}_n} f(x_1, \ldots, x_n) \prod_{i=1}^{n} dF(x_i) = 0 \quad for \ all \quad F \in \mathcal{F}.$$

*Then $f(x_1, \ldots, x_n) = 0$ whenever $x_i \in E, \quad i = 1, 2, \ldots, n$.*

**Proof.** For $i = 1, 2, \ldots, n$, let $x_i$ be a point in $E$, and let $F$ be a distribution with points of increase at $x_1, \ldots, x_n$, and jumps $p_1, \ldots, p_n$ at these points. Then

$$\int \cdots \int f(x_1, \ldots, x_n) \prod_{i=1}^{n} dF(x_i) = \sum_{i_1=1}^{n} \cdots \sum_{i_n=1}^{n} f(x_{i_1} \ldots, x_{i_n}) p_{i_1} \cdots p_{i_n} = 0$$

and so the integral is a homogeneous polynomial in $p_1, \ldots, p_n$ vanishing identically on the simplex $\sum p_i = 1, p_i \geq 0$. It follows that the polynomial vanishes identically; in particular so does the coefficient of $p_1 \ldots p_n$, which is given by $\sum f(x_{i_1}, \ldots, x_{i_n})$ where the sum is taken over all permutations $(i_1, \ldots, i_n)$ of $\{1, 2, \ldots n\}$. But $f$ is symmetric in its arguments so that this implies that $f(x_1, \ldots, x_n) = 0$.

Further, in the case when $E = \mathbb{R}$, the essentially unique symmetric estimate $\psi^{[n]}$ is also the one with minimum variance:

**Theorem 3.** *Let $\theta$ be a regular functional of degree $k$ defined by (3) on a set $\mathcal{F}$ of distribution functions containing all distributions having finite*

3

support. Let $f$ be an unbiased estimate of $\theta$ based on a sample of size $n$, so that $f$ satisfies (2). Then $Var\, f \geq Var\, \psi^{[n]}$ for all $F$ in $\mathcal{F}$.

**Proof.** Define $f^{[n]}(x_1, \ldots, x_n) = (n!)^{-1} \sum_{(n)} f(x_{i_1}, \ldots, x_{i_n})$ where here and in the sequel the sum $\sum_{(n)}$ is taken over all permutations $(i_1, \ldots, i_n)$ of $\{1, 2, \ldots, n\}$. Then $f^{[n]}$ is a symmetric unbiased estimator, and so by Theorem 2 agrees with $\psi^{[n]}$ on $\mathbb{R}$.

Hence by the Cauchy-Schwartz inequality

$$(\psi^{[n]})^2 = \left( \sum_{(n)} \frac{1}{n!} f(x_{i_1}, \ldots, x_{i_n}) \right)^2$$

$$\leq \sum_{(n)} \left( \frac{1}{n!} \right)^2 \sum_{(n)} f^2(x_{i_1}, \ldots, x_{i_n})$$

$$= \frac{1}{n!} \sum_{(n)} f^2(x_{i_1}, \ldots, x_{i_n})$$

where the sums are taken over all permutations, and so

$$E(\psi^{[n]})^2 \leq \frac{1}{n!} \sum_{(n)} E\{f^2(X_{i_1}, \ldots, X_{i_n})\}$$

$$= \frac{1}{n!} \sum_{(n)} E\{f^2(X_1, \ldots, X_n)\}$$

$$= E\{f^2(X_1, \ldots, X_n)\}$$

which, since $E(\psi^{[n]}) = E(f) = \theta$, proves the result.

Theorems 2 and 3 suffice for many of the examples in the next section, but sometimes we want to consider functionals defined on the class of all absolutely continuous distribution functions, rather than a class containing all finitely supported distribution functions. This occurs particularly in nonparametric statistics. A theorem to cover this case is due to Fraser (1954), (1957). For the purposes of this theorem, we call estimators $f(X_1, \ldots, X_n)$ and $g(X_1, \ldots, X_n)$ *identical* if $f = g$ a.e with respect to Lebesgue measure on $\mathbb{R}_n$.

4

**Theorem 4.** *(i) Let $\theta$ be a regular statistical functional of degree $k$ with kernel $\psi$, defined on a set $\mathcal{F}$ of distribution functions containing all absolutely continuous d.f.s. Then $\psi^{[n]}$ is the unique symmetric unbiased estimator of $\theta$.*

*(ii) The estimator $\psi^{[n]}$ has minimum variance in the class of all unbiased estimators of $\theta$.*

**Proof.** The proofs of Theorems 2 and 3 can be applied mutatis mutandis to the present case once Lemma B below is proved.

**Lemma B.** *Let $\theta$ be a regular statistical functional having a symmetric kernel $\psi$ of degree $k$ defined on the set $\mathcal{F}$ of absolutely continuous distribution functions, and suppose that $\theta(F) = 0$ for all $F \in \mathcal{F}$. Then $\psi = 0$ a.e. on $\mathbb{R}_k$.*

**Proof** For all densities (and hence for all nonnegative integrable functions $f$ on $\mathbb{R}$ ), we have

$$\int \cdots \int \psi(x_1, \ldots, x_k) f(x_1) \ldots f(x_k) dx_1 \ldots dx_k = 0. \tag{5}$$

From (5), it follows that if $A_1, \ldots, A_k$ are half open-half closed bounded intervals of $\mathbb{R}$, $I_{A_i}$ is the indicator function of the set $A_i$ and if $c_1, \ldots, c_k$ are non non-negative numbers, then

$$0 = \int \cdots \int \psi(x_1, \ldots, x_k) \prod_{j=1}^{k} \sum_{i=1}^{k} c_i I_{A_i}(x_j) dx_j$$

$$= \sum_{i_1=1}^{k} \cdots \sum_{i_k=1}^{k} c_{i_1} \ldots c_{i_k} \int_{A_{i_1} \times \cdots \times A_{i_k}} \psi(x_1, \ldots, x_k) dx_1 \ldots dx_k \tag{6}$$

for all $c_i \geq 0, i = 1, 2, \ldots, k$.

As in Lemma A, (6) is a homogeneous polynomial of degree $k$ vanishing identically for $c_i \geq 0$, and hence everywhere. The coefficient of $c_1 \ldots c_k$ is thus zero, and so by the symmetry of $\psi$ we obtain

$$\int_{A_1 \times \ldots \times A_k} \psi(x_1, \ldots, x_k) dx_1 \ldots dx_k = 0$$

5

for arbitrary intervals $A_1, \ldots, A_k$. It follows that

$$\int_E \psi(x_1, \ldots, x_k) dx_1 \ldots dx_k = 0$$

for all $k$-dimensional Borel sets $E$ and hence that $\psi = 0$ a.e.

The conditions of Lemma $A$ and $B$ are related to the concept of *completeness* which is described in e.g. Fraser (1957) p23 and Lehmann (1983) p46. A statistic $T(X_1, \ldots, X_k)$ based on a random sample $X_1, \ldots, X_k$ is said to be *complete* with respect to a family $\mathcal{F}$ of distribution functions if

$$\int h(T(x_1, \ldots, x_k)) \, dF(x_1) \ldots dF(x_k) = 0$$

for all $F \in \mathcal{F}$ implies $h = 0$ a.e. $(F)$. Now let $X_{(1)} < \cdots < X_{(k)}$ be the *order statistics* of the sample, i.e. the sample arranged in ascending order. The order statistics are a $(k$-dimensional) statistic, and in this case

$$\int h(T(x_1, \ldots, x_k)) \, dF(x_1) \ldots dF(x_k)$$
$$= \int h^{[n]}(x_1, \ldots, x_k) \, dF(x_1) \ldots dF(x_k).$$

Thus we see that the completeness of the order statistics relative to a class $\mathcal{F}$ is exactly equivalent to the uniqueness of symmetric estimators unbiased for all $F \in \mathcal{F}$. This unique estimator will then be the minimum variance estimator.

These results have been extended by various authors. Bell, Blackwell and Breiman (1960) consider them in the setting of general probability spaces. Hoeffding (1977) and Fisher (1982) deal with the case when the family of distributions $\mathcal{F}$ is subject to certain restrictions of the form

$$\int u_i(x) \, dF(x) = c_i \text{ or } \int\int u_i(x_1, x_2) \, dF(x_1) \, dF(x_2) = c_i, \quad F \in \mathcal{F},$$

for known functions $u_i$ and constants $c_i$. (For example, the distributions in $\mathcal{F}$ might be required to have certain moments.) They conclude that symmetric estimators that are unbiased for all $F \in \mathcal{F}$ are no longer always unique, and characterise such estimators.

6

Yamato and Maesono (1986) consider families $\mathcal{F}$ whose members are invariant under the action of finite groups of transformations, and in particular those $\mathcal{F}$ consisting of symmetric distributions. They show that the usual $U$-statistic is no longer the unique unbiased estimator, but a related "invariant" $U$-statistic, which exploits the invariance, is in fact the unbiased estimator having minimum variance. The efficiencies of the ordinary and "invariant" $U$-statistics are compared in Yamato and Maesono (1989).

For families $\mathcal{F}$ containing all finitely supported or all absolutely continuous distributions, Theorems 3 and 4 justify restricting consideration to symmetric unbiased estimators, and so we choose as an estimate of the regular functional $\theta$ the essentially unique estimator $\hat{\theta} = \psi^{[n]}$. Define $\psi^{[k]}(x_1, \ldots, x_k) = (1/k!) \sum \psi(x_{i_1}, \ldots, x_{i_k})$ where the sum is taken over all permutations $(i_1, \ldots, i_k)$ of $\{1, 2, \ldots, k\}$. Then we can write

$$\hat{\theta} = \binom{n}{k}^{-1} \sum_{(n,k)} \psi^{[k]}(X_{i_1}, \ldots, X_{i_k}) \tag{4}$$

where the sum $\sum_{(n,k)}$ is taken over all subsets $1 \leq i_1 < \cdots < i_k \leq n$ of $\{1, 2, \ldots, n\}$. We will use the notation $\sum_{(n,k)}$ repeatedly in the sequel. We will also use the notation $\mathcal{S}_{n,k}$ to denote the set of $k$-subsets of $\{1, \ldots, n\}$.

Note that

$$\int \cdots \int_{\mathbb{R}_k} \psi(x_1, \ldots, x_k) \prod_{i=1}^{k} dF(x_i) = \int \cdots \int_{\mathbb{R}_k} \psi^{[k]}(x_1, \ldots, x_k) \prod_{i=1}^{k} dF(x_i)$$

so that without loss of generality we may take the functions $\psi$ defining regular functionals $\theta$ as in (2) to be symmetric. The unique symmetric unbiased estimators are then of the form

$$\hat{\theta} = \binom{n}{k}^{-1} \sum_{(n,k)} \psi(X_{i_1}, \ldots, X_{i_k})$$

and are called $U$-statistics.

In the above discussion, we have assumed for the sake of simplicity that the random variables $X_1, \ldots, X_n$ take values in $\mathbb{R}$. However, there is nothing in the above theory that requires this, and in fact they may take values in any suitable space.

7

## 1.2 *U*-statistics

We saw in Section 1.1 that statistics of the form

$$U_n = \binom{n}{k}^{-1} \sum_{(n,k)} \psi(X_{i_1}, \dots X_{i_k}) \tag{1}$$

have desirable properties as estimators of regular functionals. Such statistics are known as *U*-statistics, due to their unbiasedness, and were so named by Hoeffding in his seminal 1948 paper (Hoeffding (1948a)) which began the systematic study of this class of statistics. We begin by considering some elementary examples.

### Example 1. Sample mean.

Let $\mathcal{F}$ be the set of all distributions whose means exist, so that $\mathcal{F}$ contains all distributions having finite support on $\mathbb{R}$. Then the mean functional is

$$\theta(F) = \int_{-\infty}^{\infty} x \, dF(x)$$

and the *U*-statistic that estimates $\theta(F)$ is just the sample mean $\overline{X}_n = \frac{1}{n} \sum_{i=1}^{n} X_i$.

### Example 2. Sample variance.

Let $\mathcal{F}$ be the set of all distributions with second moment finite:

$$\mathcal{F} = \left\{ F : \int |x|^2 dF(x) < \infty \right\}.$$

Then we can define the variance functional on $\mathcal{F}$ by

$$Var\, F = \int_{-\infty}^{\infty} \int_{-\infty}^{\infty} \tfrac{1}{2}(x_1 - x_2)^2 dF(x_1) \, dF(x_2)$$

which is estimated by the sample variance $s_n^2 = \binom{n}{2}^{-1} \sum_{1 \le i < j \le n} \tfrac{1}{2}(X_i - X_j)^2$.

### Example 3. Sample covariance.

Let $\mathcal{F}$ be all bivariate distributions for which $\int_{\mathbb{R}_2} |xy| \, dF(x, y)$ is finite. Then the covariance functional is

$$Cov(x, y) = \int_{\mathbb{R}_2} \int_{\mathbb{R}_2} \tfrac{1}{2}(x_1 - x_2)(y_1 - y_2) \, dF(x_1, y_1) \, dF(x_2, y_2)$$

and the corresponding $U$-statistic is

$$U_n = \binom{n}{2}^{-1} \sum_{1 \le i < j \le n} \frac{1}{2}(X_i - X_j)(Y_i - Y_j),$$

the sample covariance.

**Example 4.   Probability weighted moments.**

For any random variable $X$ having a distribution function $F$ whose first absolute moment is finite, define *probability weighted moments* $\beta_r$ by

$$\beta_r = E[X\{F(X)\}^r] = \int x\{F(x)\}^r \, dF(x), \qquad r = 0, 1, \ldots$$

The constants $\{\beta_r\}$ characterise the distribution, in the sense that different distributions have different sequences $\{\beta_r\}$. Note that $\beta_{r-1}$ is related to the maximum of $X_1, \ldots X_r$ :

$$E \frac{1}{r} \max(X_1, \ldots, X_r) = \beta_{r-1}$$

so that $\beta_{r-1}$ is a regular functional defined on the set of all d.f.s with finite first absolute moments by

$$\beta_{r-1} = \int \cdots \int_{\mathbb{R}_r} \frac{1}{r} \max(x_1, \ldots, x_r) dF(x_1) \ldots dF(x_r).$$

This follows because the density of $\max(X_1, \ldots, X_r)$ is $rF(x)^{r-1} dF(x)$. A $U$-statistic to estimate this functional is

$$U_n = \binom{n}{r}^{-1} \sum_{(n,r)} \frac{1}{r} \max(X_{i_1}, \ldots, X_{i_r}).$$

**Example 5.   Kendall's Tau.**

Two points $P_1$ and $P_2$ on the plane are said to be   *concordant* if the line joining them has positive slope, and *discordant* if the slope is negative. If $\mathcal{F}$ is the set of distribution functions of all absolutely continuous bivariate random vectors (X,Y), then a measure of association between X and Y is the functional $\tau$ defined on $\mathcal{F}$ by

$$\tau = Pr(P_1 \text{ and } P_2 \text{ are concordant}) - Pr(P_1 \text{ and } P_2 \text{ are discordant})$$

9

where $P_1$ and $P_2$ are two independent points distributed as $(X, Y)$. The functional $\tau$ is called Kendall's coefficient of concordance, or Kendall's Tau, and satisfies all the usual properties of a correlation, in that it takes values in the interval $[-1,1]$, is zero when $X$ and $Y$ are independent, and is equal to $\pm 1$ whenever $Y = f(X)$ for some monotone function $f$.

If we define a kernel $t$ by

$$t(P_1, P_2) = \begin{cases} 1 & \text{if } P_1 \text{ and } P_2 \text{ are concordant;} \\ -1 & \text{if } P_1 \text{ and } P_2 \text{ are discordant;} \end{cases}$$

then $t(P_1, P_2) = \text{sgn}(X_1 - X_2)(Y_1 - Y_2)$, $\tau = E\, t(P_1, P_2)$ and a $U$-statistic estimator of $\tau$ is

$$t_n = \binom{n}{2}^{-1} \sum_{(n,2)} t(P_i, P_j)$$

which is the proportion of concordant pairs of points in a sample $P_1, \ldots, P_n$ minus the proportion of pairs of points that are discordant.

## 1.3 The variance of a $U$-statistic

The variance of a $U$-statistic based on i.i.d. random variables can usefully be expressed in terms of certain conditional expectations. Define for $c = 1, 2, \ldots, k$ the conditional expectations

$$\psi_c(x_1, \ldots, x_c) = E\{\psi(x_1, \ldots, x_c, X_{c+1}, \ldots, X_k)\} \tag{1}$$

and their variances

$$\sigma_c^2 = Var\{\psi_c(X_1, \ldots, X_c)\}. \tag{2}$$

Define also

$$\sigma_0^2 = 0.$$

Then

**Theorem 1.** *The functions $\psi_c$ defined in (1) above have the properties*
*(i) $\psi_c(x_1, \ldots, x_c) = E\{\psi_d\ (x_1, \ldots, x_c, X_{c+1}, \ldots, X_d)\}$ for $1 \le c < d \le k$,*

(ii) $E\{\psi_c(X_1, \ldots, X_c)\} = E\psi(X_1, \ldots, X_k)$.

**Proof.**

(i) $E\{\psi_d(x_1, \ldots, x_c, X_{c+1}, \ldots, X_d)\}$

$$= \int \cdots \int \psi_d(x_1, \ldots, x_c, x_{c+1}, \ldots, x_d) \prod_{i=c+1}^{d} dF(x_i)$$

$$= \int \cdots \int \left\{ \int \cdots \int \psi(x_1, \ldots, x_d, \ldots, x_k) \prod_{i=d+1}^{k} dF(x_i) \right\} \prod_{i=c+1}^{d} dF(x_i)$$

$$= \int \cdots \int \psi(x_1, \ldots, x_k) \prod_{i=c+1}^{k} dF(x_i)$$

$$= \psi_c(x_1, \ldots, x_c).$$

(ii) $E\{\psi_c(X_1, \ldots, X_c)\} = \int \cdots \int \psi_c(x_1, \ldots, x_c) \prod_{i=1}^{c} dF(x_i)$

$$= \int \cdots \int \left\{ \int \cdots \int \psi(x_1, \ldots, x_k) \prod_{i=c+1}^{k} dF(x_i) \right\} \prod_{i=1}^{c} dF(x_i)$$

$$= \int \cdots \int \psi(x_1, \ldots, x_k) \prod_{i=1}^{k} dF(x_i)$$

$$= E\{\psi(X_1, \ldots, X_k)\}.$$

The variance $\sigma_c^2$ of the conditional expectation has an interpretation as a covariance, as shown in the next theorem. We use the notation $\psi(S)$ to denote $\psi(X_{i_1}, \ldots, X_{i_k})$, where $S = \{i_1, \ldots, i_k\}$.

**Theorem 2.** *An alternative expression for $\sigma_c^2$ is*

$$\sigma_c^2 = Cov(\psi(S_1), \psi(S_2))$$

*where $S_1, S_2$ are two k-subsets of $\{1, 2, \ldots, n\}$ with c elements in common.*

**Proof.** By (ii) of Theorem 1, it suffices to show that

$$E\{\psi(X_1, \ldots, X_k)\psi(X_1, \ldots, X_c, \ X_{k+1}, \ldots, X_{2k-c})\} = E\{\psi_c(X_1, \ldots, X_c)\}^2$$

11

which follows from

$$E\{\psi(X_1,\ldots,X_k)\psi(X_1,\ldots,X_c,X_{k+1},\ldots,X_{2k-c})\}$$

$$= \int \cdots \int \psi(x_1,\ldots,x_k)\psi(x_1,\ldots,x_c,x_{k+1},\ldots,x_{2k-c}) \prod_{i=1}^{2k-c} dF(x_i)$$

$$= \int \cdots \int \left\{ \int \cdots \int \psi(x_1,\ldots,x_k) \prod_{i=c+1}^{k} dF(x_i) \right\} \times$$

$$\left\{ \int \cdots \int \psi(x_1,\ldots,x_c,x_{k+1},\ldots,x_{2k-c}) \prod_{i=k+1}^{2k-c} dF(x_i) \right\} \prod_{i=1}^{c} dF(x_i)$$

$$= \int \cdots \int \psi_c^2(x_1,\ldots,x_c) \prod_{i=1}^{c} dF(x_i)$$

$$= E\{\psi_c^2(X_1,\ldots,X_c)\}.$$

A useful expression for the variance of a $U$-statistic can be developed in terms of the quantities $\sigma_c^2$:

**Theorem 3.** *Let $U_n$ be a $U$-statistic with a kernel $\psi$ of degree $k$. Then*

$$Var\, U_n = \binom{n}{k}^{-1} \sum_{c=1}^{k} \binom{k}{c}\binom{n-k}{k-c}\sigma_c^2. \tag{3}$$

**Proof.** We have

$$Var\, U_n = Var\left\{ \binom{n}{k}^{-1} \sum_{(n,k)} \psi(X_{i_1},\ldots,X_{i_k}) \right\}$$

$$= Cov\left\{ \binom{n}{k}^{-1} \sum_{(n,k)} \psi(X_{i_1},\ldots,X_{i_k}),\; \binom{n}{k}^{-1} \sum_{(n,k)} \psi(X_{j_1},\ldots,X_{j_n}) \right\}$$

$$= \binom{n}{k}^{-2} \sum_{(n,k)} \sum_{(n,k)} Cov\,(\psi(X_{i_1},\ldots,X_{i_k}),\; \psi(X_{j_1},\ldots X_{j_k})) \tag{4}$$

where the sum is taken over all pairs $\{i_1,\ldots,i_k\}$ and $\{j_1,\ldots,j_k\}$ of $k$-subsets of the set $\{1,2,\ldots,n\}$. We need to enumerate how many of these

12

$\binom{n}{k}^2$ pairs have one element in common, two elements in common and so on, in order to apply Theorem 2.

Consider the number of ways of choosing a pair of $k$-subsets that have $c$ elements in common. Clearly the first member of the pair can be chosen $\binom{n}{k}$ ways. The $c$ elements of the second set that are common with the first can be chosen in $\binom{k}{c}$ ways, and the $k - c$ elements distinct from these can then be chosen in $\binom{n-k}{k-c}$ ways. Thus the number of pairs having $c$ elements in common, and hence having

$$Cov\{\psi(X_{i_1},\ldots,X_{i_k}),\ \psi(X_{j_1},\ldots X_{j_k})\} = \sigma_c^2$$

is $\binom{n}{k}\binom{k}{c}\binom{n-k}{k-c}$, and so (4) is equal to

$$\binom{n}{k}^{-1} \sum_{c=1}^{k} \binom{k}{c}\binom{n-k}{k-c}\sigma_c^2,$$

proving the theorem.

Some examples using Theorem 3 follow:

**Example 1. Sample mean.**
We have $\psi_1(x) = x$, and so $\sigma_1^2 = \sigma^2 = Var\, X_1$. Hence $Var\, \overline{X} = \sigma^2/n$ from (3).

**Example 2. Sample variance.**
In the case of the sample variance, the degree of the kernel is $k = 2$, and $\psi_1(x) = E\left\{\frac{1}{2}(x - X_2)^2\right\} = \frac{1}{2}(\sigma^2 + (x - \mu)^2)$. Thus, writing $\mu_\nu$ for $E(X_1 - \mu)^\nu$, we have

$$\begin{aligned}
\sigma_1^2 &= Var\left\{\frac{1}{2}(\sigma^2 + (X_1 - \mu)^2)\right\} \\
&= \frac{1}{4}\left\{E(X_1 - \mu)^4 - [E\{(X_1 - \mu)^2\}]^2\right\} \\
&= \frac{1}{4}(\mu_4 - \sigma^4).
\end{aligned}$$

Also $\psi_2(x_1, x_2) = \frac{1}{2}(x_1 - x_2)^2$ so that we get $\sigma_2^2 = Var\left\{\frac{1}{2}(X_1 - X_2)^2\right\} = \frac{1}{2}(\mu_4 + \sigma^4)$ and hence, denoting the sample variance by $s_n^2$, we get

$$\begin{aligned}
Var\, s_n^2 &= \binom{n}{2}^{-1}(2(n-2)\sigma_1^2 + \sigma_2^2) \\
&= \frac{\mu_4}{n} - \frac{(n-3)\sigma^4}{n(n-1)}
\end{aligned}$$

upon substituting for $\sigma_1^2$ and $\sigma_2^2$ in terms of $\mu_4$ and $\sigma^2$.

## Example 3. Sample covariance.

Once again we have $k = 2$. The conditional expectations are

$$\psi_1\{(x,y)\} = \tfrac{1}{2}E(x - X_2)(y - Y_2) = \sigma_{XY} + (x - \mu_X)(y - \mu_Y)$$

where $\mu_X = E(X_1)$, $\mu_Y = E(Y_1)$, $\sigma_{XY} = Cov(X_1, Y_1)$ and

$$\psi_2\{(x_1, y_1), (x_2, y_2)\} = \tfrac{1}{2}(x_1 - x_2)(y_1 - y_2).$$

The variances of the conditional expectations are

$$\sigma_1^2 = \tfrac{1}{4}Var\{(X_1 - \mu_X)(Y_1 - \mu_Y)\} = \tfrac{1}{4}(\mu_{2,2} - \sigma_{XY}^2)$$

and

$$\sigma_2^2 = \tfrac{1}{4}Var\{(X_1 - X_2)(Y_1 - Y_2)\} = \tfrac{1}{2}(\mu_{2,2} + \sigma_X^2 \sigma_Y^2)$$

writing $\mu_{2,2}$ for $E\{(X_1 - \mu_X)^2(Y_1 - \mu_Y)^2\}$ and $\sigma_X^2, \sigma_Y^2$ for $Var\,X$ and $Var\,Y$. Hence using (3) we obtain

$$Var\,U_n = \frac{\mu_{2,2}}{n} - \frac{\{(n-2)\sigma_{XY}^2 - \sigma_X^2\sigma_Y^2\}}{n(n-1)}.$$

## Example 4. Kendall's Tau.

The estimator $t_n$ of Example 5 of Section 1.2 is obviously unbiased and by Theorem 3 its variance is given by

$$Var\,t_n = \binom{n}{2}^{-1}\{2(n-2)\sigma_1^2 + \sigma_2^2\}$$

where $\sigma_1^2 = Var\,t_1(X_1, Y_1)$ and $\sigma_2^2 = 1 - \tau^2$.

The conditional expectation $t_1(x_1, y_1)$ is given by

$$
\begin{aligned}
t_1(x,y) &= E\{t((x,y),(X_1,Y_1))\} \\
&= Pr\big((x - X_1)(y - Y_1) > 0\big) - Pr\big((x - X_1)(y - Y_1) < 0\big) \\
&= Pr\big((X_1 > x \text{ and } Y_1 > y) \text{ or } (X_1 < x \text{ and } Y_1 < y)\big) \\
&\qquad - Pr\big((X_1 > x \text{ and } Y_1 < y) \text{ or } (X_1 < x \text{ and } Y_1 > y)\big) \\
&= 1 - 2F(x,\infty) - 2F(\infty,y) + 4F(x,y) \\
&= (1 - 2F_1(x))(1 - 2F_2(y)) + 4(F(x,y) - F_1(x)F_2(y))
\end{aligned}
$$

14

where $F_1$ and $F_2$ are the marginal d.f.s of $X_1$ and $Y_1$.

Under independence of $X_1$ and $Y_1$, $F(x,y) = F_1(x)F_2(y)$ and so $t_{(1)}(x,y) = (1 - 2F_1(x)) \times (1 - 2F_2(y))$. The random variables $U$ and $V$ given by $U = 1 - 2F_1(X)$ and $V = 1 - 2F_2(Y)$ are independent uniform r.v.s on [-1,1] so that

$$Var\, t_1(X,Y) = Var\, UV = E(U^2)E(V^2) - (EU)^2(EV)^2$$
$$= \left( \frac{1}{2} \int_{-1}^{-1} u^2\, du \right)^2$$
$$= \frac{1}{9}.$$

Thus under independence

$$Var\, t_n = \binom{n}{2}^{-1} (2(n-2)/9 + 1)$$
$$= 2(2n+5)/9n(n+1).$$

Many results concerning the conditional variances can be proved. A sample follows:

**Theorem 4.** For $0 \le c \le d \le k$

$$\frac{\sigma_c^2}{c} \le \frac{\sigma_d^2}{d}.$$

**Proof.** Using the facts that the quantities $\delta_d^2 = \sum_{c=0}^{d}(-1)^c \binom{d}{c}\sigma_{d-c}^2$ are positive and that $\sigma_c^2 = \sum_{d=1}^{c} \binom{c}{d}\delta_d^2$ (see Section 1.6) we see that

$$c\sigma_d^2 - d\sigma_c^2 = c\sum_{j=1}^{d} \binom{d}{j}\delta_j^2 - d\sum_{j=1}^{c} \binom{c}{j}\delta_j^2$$
$$= \sum_{j=1}^{c} \left( c\binom{d}{j} - d\binom{c}{j} \right) \delta_j^2 + \sum_{j=c+1}^{d} c\binom{d}{j}\delta_j^2$$

Since $c\binom{d}{j} - d\binom{c}{j} \ge 0$ for $d \ge c \ge j \ge 1$, the first term in the above sum is positive, and so $c\sigma_d^2 - d\sigma_c^2 \ge 0$.

15

**Theorem 5.** *The function $Var\, nU_n$ is decreasing in $n$.*

**Proof.** Using Theorem 3 we may write

$$Var\, nU_n - Var\{(n+1)U_{n+1}\} = \sum_{c=1}^{k} d_{n,c}\sigma_c^2$$

where

$$d_{n,c} = n^2 \binom{n}{k}^{-1}\binom{k}{c}\binom{n-k}{k-c} - (n+1)^2\binom{n+1}{k}^{-1}\binom{k}{c}\binom{n+1-k}{k-c}$$

$$= \binom{n}{k}^{-1}\binom{n+1-k}{k-c}\binom{k}{c}(n+1-k)^{-1}\{n(c-1)-(k-1)^2\}. \quad (5)$$

Let $[x]$ denote the greatest integer less than or equal to $x$. Then if $c_0 = 1 + \left[\frac{(k-1)^2}{n}\right]$ we see that $d_{n,c} > 0$ for $c > c_0$, and $d_{n,c} \le 0$ for $c \le c_0$. Using Theorem 4 for $c = 1, \ldots, k$ we obtain

$$cd_{n,c}\frac{\sigma_{c_0}^2}{c_0} \le \sigma_c^2 d_{n,c} \quad (6)$$

and summing (6) over $c$ from 1 to $k$ yields

$$Var\, nU_n - Var(n+1)U_{n+1} \ge \frac{\sigma_{c_0}^2}{c_0}\sum_{c=1}^{k} cd_{n,c}.$$

Finally, using the identity

$$\binom{n}{k}^{-1}\sum_{c=1}^{k}c\binom{k}{c}\binom{n-k}{k-c} = \frac{k^2}{n}$$

we see from (5) that $\sum_{c=1}^{k} cd_{n,c} = 0$, which proves the theorem.

Futher results of this type may be found in Karlin and Rinott (1982).

### 1.4. The covariance of two $U$-statistics

Let $U_n^{(1)}$ and $U_n^{(2)}$ be two $U$-statistics, both based on a common sample $X_1, \ldots, X_n$ but having different kernels $\psi$ and $\phi$ of degrees $k_1$ and $k_2$ respectively, with $k_1 \le k_2$. We can develop results similar to those of the previous section for the covariance between $U_n^{(1)}$ and $U_n^{(2)}$.

16

Define $\sigma_{c,d}^2$ to be the covariance between the conditional expectations $\psi_c(X_1, \ldots, X_c)$ and $\phi_d(X_1, \ldots, X_d)$, and if $S$ is a set, let $|S|$ denote the number of elements in $S$. Then we have

**Theorem 1.** *Suppose that $c \leq d$. If $S_1$ is in $S_{n,k_1}$ and $S_2$ in $S_{n,k_2}$ with $|S_1 \cap S_2| = c$, then*

$$\sigma_{c,d}^2 = Cov(\psi(S_1), \phi(S_2)).$$

**Proof.** The proof is almost identical to that of Theorem 2 of Section 1.3 and hence is omitted.

Note that as a consequence of Theorem 1, $\sigma_{c,c}^2 = \sigma_{c,c+1}^2 = \ldots = \sigma_{c,k_2}^2$ for $c = 1, 2, \ldots, k_1$.

Theorem 1 can be used to obtain a formula for the covariance of $U_n^{(1)}$ and $U_n^{(2)}$:

**Theorem 2.** *Let $U_n^{(1)}$ and $U_n^{(2)}$ be as above. Then*

$$Cov\left(U_n^{(1)}, U_n^{(2)}\right) = \binom{n}{k_1}^{-1} \sum_{c=1}^{k_1} \binom{k_2}{c}\binom{n-k_2}{k_1-c}\sigma_{c,c}^2.$$

**Proof.**

$$Cov\left(U_n^{(1)}, U_n^{(2)}\right) = \binom{n}{k_1}^{-1}\binom{n}{k_2}^{-1} \sum_{(n,k_1)}\sum_{(n,k_2)} Cov\left(\psi(S_1), \phi(S_2)\right)$$

$$= \binom{n}{k_1}^{-1}\binom{n}{k_2}^{-1} \sum_{c=1}^{k_1}\sum_{|S_1 \cap S_2|=c} Cov\left(\psi(S_1), \phi(S_2)\right)$$

$$= \binom{n}{k_1}^{-1} \sum_{c=1}^{k} \binom{k_2}{c}\binom{n-k_2}{k_1-c}\sigma_{c,c}^2$$

since there are exactly $\binom{n}{k_2}\binom{k_2}{c}\binom{n-k_2}{k_1-c}$ pairs of sets $(S_1, S_2)$ with $S_1$ in $S_{n,k_1}$, $S_2$ in $S_{n,k_2}$ and $|S_1 \cap S_2| = c$.

We apply Theorem 2 to calculate the variance of a well-known non-parametric statistic.

17

**Example 1. The Wilcoxon one-sample statistic.**

Let $X_1, \ldots, X_n$ denote a random sample from an absolutely continuous distribution having distribution function $F$ and density $f$, and let $R_i$ be the rank of $|X_i|$, $i = 1, 2, \ldots, n$. (That is, $R_i$ denotes the position of $|X_i|$ when the random variables $|X_1|, \ldots, |X_n|$ are arranged in ascending order).

A statistic in common use for testing if the distribution $F$ is symmetric about zero is the Wilcoxon one sample rank statistic $T^+$, which is computed by summing the quantities $R_i$ corresponding to the positive $X_i$. The statistic $T^+$ is not a $U$-statistic, but can be written as a linear combination of $U$-statistics by introducing the so-called *Walsh average* $\frac{1}{2}(X_i + X_j)$ for $1 \leq i \leq j \leq n$. It is clear that a Walsh average for $i \neq j$ is positive either (a) if $X_i > 0$ and $|X_j| < X_i$; or (b) if $X_j > 0$ and $|X_i| < X_j$, so that for $i < j$,

$$I\{X_i + X_j > 0\}$$
$$= I\{X_i > 0\} \; I\{|X_j| < X_i\} + I\{X_j > 0\} \; I\{|X_i| < X_j\} \qquad (1)$$

where the *indicator function* $I\{\ \}$ takes on values zero or one according as the condition in the braces is false or not. Summing over $1 \leq i \leq j \leq n$ and using (1) yields

$$\sum_{1 \leq i \leq j \leq n} I\left\{\frac{1}{2}(X_i + X_j) > 0\right\} = \sum_{(n,2)} I\{X_i > 0\}I\{|X_j| < X_i\}$$

$$+ \sum_{(n,2)} I\{X_j > 0\}I\{|X_i| < X_j\} + \sum_{i=1}^{n} I\{X_i > 0\}$$

$$= \sum_{i=1}^{n}\sum_{j=1}^{n} I\{X_i > 0\}I\{|X_j| \leq X_i\}$$

$$= \sum_{i=1}^{n} I\{X_i > 0\}R_i$$

$$= T^+, \qquad (3)$$

so $T^+$ is just the number of positive Walsh averages.

Hence if the kernels $\psi(x_1, x_2)$ and $\phi(x)$ are defined by

$$\psi(x_1, x_2) = \begin{cases} 1, & \text{if } x_1 + x_2 > 0, \\ 0, & \text{otherwise,} \end{cases}$$

18

and

$$\phi(x) = \begin{cases} 1 & \text{if } x > 0, \\ 0 & \text{otherwise,} \end{cases}$$

then

$$T^+ = \sum_{(n,2)} \psi(X_i, X_j) + \sum_{i=1}^{n} \phi(X_i)$$

$$= \binom{n}{2} U_n^{(1)} + n U_n^{(2)}$$

say. Now define

$$p_1 = Pr(X_1 > 0),$$
$$p_2 = Pr(X_1 + X_2 > 0),$$
$$p_3 = Pr(X_1 + X_2 > 0, X_1 > 0),$$

and

$$p_4 = Pr(X_1 + X_2 > 0, X_2 + X_3 > 0).$$

Then

$$Cov(\psi(X_1, X_2), \ \psi(X_2, X_3)) = p_4 - p_2^2,$$
$$Var \ \psi(X_1, X_2) = p_2(1 - p_2),$$
$$Var \ \phi(X_1) = p_1(1 - p_1)$$

and

$$Cov(\psi(X_1, X_2), \ \phi(X_1)) = p_3 - p_1 p_2$$

so we get

$$Var \ U_n^{(1)} = \binom{n}{2}^{-1} \{(n-2)(p_4 - p_2^2) + p_2(1 - p_2)\}$$

and

$$Var \ U_n^{(2)} = p_1(1 - p_1)/n.$$

Using Theorem 2, it follows that

$$Cov(U_n^{(1)}, U_n^{(2)}) = 2(p_3 - p_1 p_2)/n$$

19

and so

$$ET^+ = \binom{n}{2} p_2 + np_1 \tag{2}$$

and

$$Var\, T^+ = Var \left\{ \binom{n}{2} U_n^{(1)} + nU_n^{(2)} \right\}$$

$$= \binom{n}{2} \{(n-2)(p_4 - p_2^2) + p_2(1-p_2) + 4(p_3 - p_1 p_2)\} + np_1(1-p_1). \tag{3}$$

If the distribution is symmetric about zero, $p_1 = \frac{1}{2}, p_2 = \frac{1}{2}, p_3 = \frac{3}{8}$ and $p_4 = \frac{1}{3}$ so that (2) and (3) reduce to

$$ET^+ = n(n+1)/4 \tag{4}$$

and

$$Var\, T^+ = n(n+1)(2n+1)/24. \tag{5}$$

To see that the probabilities $p_3$ and $p_4$ do have the values claimed, let $f$ and $F$ be the (symmetric) density and distribution function of the $X's$. Then

$$Pr(X_1 + X_2 > 0, X_1 > 0) = \int_0^\infty f(v) \int_0^\infty f(u-v)\, du\, dv$$

$$= \int_0^\infty f(v)F(v)\, dv$$

$$= \frac{3}{8}$$

upon integrating by parts, and

$$Pr(X_1 + X_2 > 0, X_2 + X_3 > 0)$$

$$= \int_0^\infty \int_0^\infty \int_{-\infty}^\infty f(w)f(v-w)f(u-v+w)\, dw\, dv\, du$$

$$= \int_{-\infty}^\infty f(w) \int_{-\infty}^\infty f(v-w)F(v-w)\, dv\, dw$$

$$= \int_{-\infty}^\infty f(w)\tfrac{1}{2}(1 - F^2(-w))\, dw$$

$$= \tfrac{1}{2}E(1 - F^2(X_1))$$

once again integrating by parts. Since $F(X_1)$ is uniformly distributed on $[0,1]$, the result now follows by noting that $E(U^2) = \frac{1}{3}$ for a variate $U$ uniformly distributed on $[0,1]$.

We can also consider the covariance between $U$-statistics based on different numbers of observations. Let $U_n$ be based on kernel $\psi$. Then using the methods of Theorem 3 of Section 1.3 we can compute the covariance between $U_n$ and $U_m$:

**Theorem 3.** *Let $U_n$ and $U_m$ be $U$-statistics based on the same kernel $\psi$ of degree $k$ but on different numbers $n$ and $m$ of observations. Then if $m < n$*

$$Cov(U_n, U_m) = \binom{n}{k}^{-1} \sum_{c=1}^{k} \binom{k}{c}\binom{n-k}{k-c}\sigma_c^2$$

$$= Var\, U_n.$$

**Proof.** Of the $\binom{m}{k}\binom{n}{k}$ terms in the sum

$$\sum_{(m,k)}\sum_{(n,k)} Cov(\psi(S_1), \psi(S_2))$$

exactly $\binom{m}{k}\binom{k}{c}\binom{n-k}{k-c}$ have $c$ variables in common. The result now follows as in Theorem 3 of Section 1.3.

## 1.5. Higher moments of $U$-statistics

Expressions for the higher moments of $U$-statistics can in principle be computed using the techniques of Theorem 3 of Section 1.3 but the results would hardly assume a simple form. For the asymptotics of Chapter 3 simple bounds are sufficient, and we now establish these. Our first theorem is due to Grams and Serfling (1973).

**Theorem 1.** *Suppose that $E|\psi(X_1, \ldots, X_k)|^r < \infty, where\ r \geq 2$. Then*

$$E|U_n - \theta|^r = O(n^{-\frac{r}{2}}).$$

**Proof.** Let $p = \left[\frac{n}{k}\right]$, the greatest integer $\leq \frac{n}{k}$, and define

$$W(x_1, \ldots, x_n) = p^{-1}\{\psi(x_1, \ldots, x_k) + \psi(x_{k+1}, \ldots, x_{2k})$$

$$+ \ldots + \psi(x_{(p-1)k+1}, \ldots, x_{pk})\}.$$

Then

$$\sum_{(n)} W(x_{\nu_1}, \ldots, x_{\nu_n}) = k!(n-k)! \sum_{(n,k)} \psi(x_{i_1}, \ldots, x_{i_k})$$

(recall that $\sum\limits_{(n)}$ denotes summation over all $n!$ permutations $(\nu_1, \ldots, \nu_n)$ of $\{1, 2, \ldots, n\}$) and so

$$U_n = \frac{1}{n!} \sum_{(n)} W(X_{\nu_1}, \ldots, X_{\nu_n}).$$

Hence

$$U_n - \theta = \frac{1}{n!} \sum_{(n)} (W(X_{\nu_1}, \ldots, X_{\nu_n}) - \theta)$$

and thus by Minkowski's inequality

$$E|U_n - \theta|^r \le E|W(X_1, \ldots, X_n) - \theta|^r.$$

But $W(X_1, \ldots, X_n) - \theta$ is an average of $p$ i.i.d random variables, and so the result follows by Lemma A below.

**Lemma A.** Let $X_1, \ldots, X_n$ be a sequence of identically and independently distributed zero mean random variables satisfying $E|X_n|^r < \infty$ and let $S_n = X_1 + \cdots + X_n$. If $r \ge 2$ then

$$E\left(|n^{-1}S_n|^r\right) = O(n^{-\frac{r}{2}}).$$

**Proof.** By a theorem of Marcinkiewicz and Zygmund (see e.g. Chow and Teicher (1978) p356) there exists a constant $A$ not depending on the $X's$ with

$$E\{|n^{-1}S_n|^r\} \le n^{-r} A E\left\{(\sum_{j=1}^{n} X_j^2)^{\frac{r}{2}}\right\}. \tag{1}$$

For $r \ge 2$, the inequality

$$\left(\sum_{j=1}^{n} x_j^2\right)^{\frac{r}{2}} \le n^{\frac{r}{2}-1} \left(\sum_{j=1}^{n} |x_j|^r\right)$$

22

follows from Hölder's inequality, and hence

$$E\left\{(\sum_{j=1}^{n} X_j^2)^{\frac{r}{2}}\right\} \le n^{\frac{r}{2}} E(|X_1|^r)$$

which in conjunction with (1) establishes the lemma.

The bound of Theorem 1 can be improved a bit when the first few $\sigma_c^2$ are zero. The next result is by Serfling (1980):

**Theorem 2.** *Suppose that* $0 = \sigma_1 = \cdots = \sigma_{d-1}^2 < \sigma_d^2$. *Let* $r$ *be an integer* $\ge 2$. *Then*

$$E(U_n - \theta)^r = O\left(n^{-[\frac{rd+1}{2}]}\right)$$

*where* $[x]$ *denotes the greatest integer* $\le x$.

**Proof.** Set $N = \binom{n}{k}$ and let $S_1, \ldots, S_N$ be the $N$ $k$-subsets of $\{1, 2, \ldots, n\}$. Further, denote $\psi(X_{i_1}, \ldots, X_{i_k}) - \theta$ by $\phi(S)$, where $S = \{i_1, \ldots, i_k\}$. Then

$$E(U_n - \theta)^r = \binom{n}{k}^{-r} E\left\{\sum_{i=1}^{N} \phi(S_i)\right\}^r$$

$$= \binom{n}{k}^{-r} \sum_{i_1=1}^{N} \cdots \sum_{i_r=1}^{N} E\{\phi(S_{i_1}) \ldots, \phi(S_{i_r})\} \qquad (2)$$

where $S_{i_1}, \ldots, S_{i_r}$ are $r$ not necessarily distinct subsets. Consider a typical term of this sum, with fixed but arbitrary sets $S_{i_1}, \ldots, S_{i_r}$. Let $q$ be the number of elements that belong to at least two of the sets $S_{i_j}$, and let $p_j$ be the number of elements in $S_{i_j}$ that appear in at least one other set. Then by Lemma B we obtain

$$2q \le \sum_{j=1}^{r} p_j.$$

For fixed $q, p_1, \ldots, p_j$, the number of sets $S_{i_1}, \ldots, S_{i_r}$ satisfying the above conditions is of order $n^{q+(k-p_1)+\cdots+(k-p_r)} = n^{q+rk-\sum p_j}$ since the common indices in $S_{i_1}$ unique to that set can be chosen in order $n^{k-p_1}$ ways and similarly for $S_{i_1}, \ldots, S_{i_r}$. Moreover we need consider only those combinations for which $p_1, \ldots, p_r$ are greater than $d$. To see this, consider

$$E\{\phi(S_{i_1}) \ldots \phi(S_{i_r})\} = E\left[E\{\phi(S_{i_1}) \ldots \phi(S_{i_r}) | X_{j_1} \ldots X_{j_{k-p_1}}\}\right]$$

$$= E\left[\phi(S_{i_2}) \ldots \phi(S_{i_r}) E\{\phi(S_{i_1}) | X_{j_1} \ldots X_{j_{k-p_1}}\}\right]$$

23

where $j_1, \ldots, j_{k-p_1}$ are the elements in $S_{i_1}$ not appearing on the other sets. But since $\sigma_1 = \cdots = \sigma_{d-1}^2 = 0$,

$$E\{\phi(S_{i_1})|X_{j_1}, \ldots, X_{j_{k-p_1}}\} = 0$$

for $p_1 \le d - 1$, and similarly for the other indices.

Thus we may assume that $\sum_{j=1}^r p_j \ge rd$ and hence

$$q + rk - \sum_{j=1}^r p_j \le rk - \left(\sum_{j=1}^r p_j - \left[\frac{1}{2}\sum_{j=1}^r p_j\right]\right)$$

$$= rk - \left[\frac{1}{2}\left(\sum_{j=1}^r p_j + 1\right)\right]$$

$$\le rk - \left[\frac{1}{2}(rd+1)\right],$$

using the fact that for any integer $x$, $x - \left[\frac{x}{2}\right] = \left[\frac{1}{2}(x+1)\right]$.

Hence the sum of terms corresponding to fixed values of $q, p_1, \ldots, p_r$ is $O\left(n^{rk-[\frac{1}{2}rd+1]}\right)$ and since the number of such sets of values is bounded as $n$ increases, the sum in (2) is $O\left(n^{rk-[\frac{1}{2}rd+1]}\right)$. The result follows upon noting that $\binom{n}{k}^{-r}$ is $O(n^{-rk})$.

**Lemma B.** *Let $S_1, \ldots, S_n$ be finite sets, and define $T_i = \cup_{j\ne i} S_j$. Denote the number of elements in a set $S$ by $|S|$. Then*

$$2|\cup_{i\ne j}(S_i \cap S_j)| \le \sum_{i=1}^n |S_i \cap T_i| \tag{3}$$

**Proof.** The proof is by induction on the number of elements in $\cup_{i\ne j} S_i \cap S_j$. First assume that $|\cup_{i\ne j} S_i \cap S_j| = 1$, so that there is an element $x$ with $x \in S_i$ and $x \in S_j$ for two indices $i$ and $j$. Then $x \in S_i \cap T_i$ and $x \in S_j \cap T_j$ so that (1) is true.

Now assume that $|\cup_{i\ne j} S_i \cap S_j| = m + 1$, and suppose that the result is true for sets for which $|\cup_{i\ne j} S_i \cap S_j| = m$. Let $x \in \cup_{i\ne j} S_i \cap S_j$, so that we can choose indices $i'$ and $j'$ with $x \in S_{i'} \cap S_{j'}$. Define $S_i^* = S_i - \{x\}$

24

and $T_i^* = \bigcup_{j \neq i} S_j^*$ for $i = 1, 2, \ldots, n$. Then $S_i \cap T_i - \{x\} = S_i^* \cap T_i$ and $\bigcup_{i \neq j}(S_i \cap S_j) - \{x\} = \bigcup_{i \neq j}(S_i^* \cap S_j^*)$ so that

$$2|\bigcup_{i \neq j}(S_i \cap S_j)| = 2|\{x\} \cup \bigcup_{i \neq j}(S_i^* \cap S_j^*)|$$

$$\leq 2 + 2|\bigcup_{i \neq j}(S_i^* \cap S_j^*)| \tag{4}$$

$$\leq 2 + \sum_{i=1}^{n}|S_i^* \cap T_i^*|$$

by the induction hypothesis. Further,

$$\{x\} \cup (S_{i'}^* \cap T_{i'}^*) = S_{i'} \cap T_{i'}$$

and $\{x\} \cap (S_{i'}^* \cap T_{i'}^*)$ is empty, so that

$$1 + |S_{i'}^* \cap T_{i'}^*| = |S_{i'} \cap T_{i'}| \tag{5}$$

and similarly for index $j$, so that from (4) and (5) we get

$$2|\bigcup_{i \neq j} S_i \cap S_j| \leq \sum_{i=1}^{n}|S_i \cap T_i|$$

proving the lemma.

## 1.6. The $H$-decomposition

This section treats a representation of $U$-statistics of degree $k$ in terms of sums of uncorrelated $U$-statistics of degree $1, 2, \ldots, k$ which will be used frequently in the sequel. The decomposition in the form presented here is due to Hoeffding (1961), but has analogues in many parts of statistics, most notably in the analysis of variance in balanced experimental designs. It is a generalisation of the widely used projection technique used in many parts of nonparametric statistics. The geometric basis of the decomposition is discussed in the next section.

To describe the decomposition, we introduce kernels $h^{(1)}, h^{(2)}, \ldots, h^{(k)}$ of degrees $1, 2, \ldots, k$ which are defined recursively by the equations

$$h^{(1)}(x_1) = \psi_1(x_1) - \theta$$

and

$$h^{(c)}(x_1, \ldots, x_c) = \psi_c(x_1, \ldots, x_c) - \sum_{j=1}^{c-1} \sum_{(c,j)} h^{(j)}(x_{i_1}, \ldots, x_{i_j}) - \theta \qquad (1)$$

for $c = 2, 3, \ldots, k$.

Let $S_j\{i_1, \ldots, i_k\}$ denote the sum $\sum h^{(j)}(x_{\nu_1}, \ldots, x_{\nu_j})$ summed over all $j$-subsets $\{\nu_1, \ldots, \nu_j\}$ of $\{i_1, \ldots, i_k\}$.

Then using the relationship

$$\sum_{(n,k)} S_j\{i_1, \ldots, i_k\} = \binom{n-j}{k-j} \sum_{(n,j)} h^{(j)}(x_{\nu_1}, \ldots, x_{\nu_j}),$$

the identity

$$\binom{n}{k}^{-1} \binom{n-j}{k-j} = \binom{k}{j}\binom{n}{j}^{-1}$$

and the relationship (1) for $c = k$, we can write

$$U_n = \binom{n}{k}^{-1} \sum_{(n,k)} \psi(x_{i_1}, \ldots, x_{i_k})$$

$$= \binom{n}{k}^{-1} \sum_{(n,k)} \left( \sum_{j=1}^{k} S_j\{i_1, \ldots, i_k\} + \theta \right)$$

$$= \theta + \binom{n}{k}^{-1} \sum_{j=1}^{k} \binom{n-j}{k-j} \sum_{(n,j)} h^{(j)}(x_{\nu_1}, \ldots, x_{\nu_j})$$

$$= \theta + \sum_{j=1}^{k} \binom{k}{j} H_n^{(j)}$$

where $H_n^{(j)}$ is the $U$-statistic of degree $j$ based on kernel $h^{(j)}$. We state this as a theorem:

**Theorem 1.** For $j = 1, 2, \ldots, k$, let $H_n^{(j)}$ be the $U$-statistic based on the kernel $h^{(j)}$ defined by (1). Then

$$U_n = \theta + \sum_{j=1}^{k} \binom{k}{j} H_n^{(j)}. \qquad (2)$$

26

The decomposition (2) is the $H$-decomposition, named after its inventor Hoeffding. Its usefulness lies in the fact that the terms $H_n^{(j)}$ are uncorrelated, with variances of decreasing order in $n$. We establish these facts in the next few theorems. Note that the terms of the decomposition (2) can also be written as $U$-statistics of order $k$: we have using (1) for $c = k$

$$\binom{k}{j} H_n^{(j)} = \binom{k}{j} \binom{n}{j}^{-1} \sum_{(n,j)} h^{(j)}(X_{\nu_1}, \ldots, X_{\nu_j})$$

$$= \binom{n}{k}^{-1} \sum_{(n,k)} S_j\{i_1, \ldots, i_k\},$$

a $U$-statistic of degree $k$ with kernel $g_j(x_1, \ldots, x_k) = S_j\{1, 2, \ldots, k\}$. Also, if $R_n^{(c)}$ is the remainder obtained by truncating the $H$-decomposition after $c$ terms, we get

$$U_n = \theta + \sum_{j=1}^{c} \binom{k}{j} H_n^{(j)} + R_n^{(c)} \tag{3}$$

where $R_n^{(c)}$ is a $U$-statistic with kernel $\sum_{j=c+1}^{k} g_j$.

An alternative representation of the functions $h^{(c)}$ can be given in terms of integrals. Let $G_x$ denote the distribution function of a single point mass at $x$. Then we can write

$$h^{(1)}(x_1) = \psi_1(x_1) - \theta$$

$$= \int \cdots \int \psi(x_1, u_2, \ldots, u_k) \prod_{i=2}^{k} dF(u_i)$$

$$- \int \cdots \int \psi(u_1, \ldots, u_k) \prod_{i=1}^{k} dF(u_i)$$

$$= \int \cdots \int \psi(u_1, \ldots, u_k)(dG_{x_1}(u_1) - dF(u_1)) \prod_{i=2}^{k} dF(u_i).$$

Similarly,

$$h^{(2)}(x_1, x_2) = \int \cdots \int \psi(u_1, \ldots, u_k)(dG_{x_1}(u_1) - dF(u_1))$$

$$\times (dG_{x_2}(u_2) - dF(u_2)) \prod_{i=3}^{k} dF(u_i)$$

27

and in general

$$h^{(j)}(x_1, \ldots, x_j)$$
$$= \int \cdots \int \psi(u_1, \ldots, u_k) \prod_{i=1}^{j} (dG_{x_i}(u_i) - dF(u_i)) \prod_{i=j+1}^{k} dF(u_i). \quad (4)$$

To verify (4), we need only show that the functions in (4) satisfy (1). To this end, note that the identity

$$\prod_{i=1}^{k} dG_{x_i}(u_i) = \prod_{i=1}^{k} \{(dG_{x_i}(u_i) - dF(u_i)) + dF(u_i)\}$$

leads to

$$\prod_{i=1}^{k} dG_{x_i}(u_i) = \sum_{c=0}^{k} \sum_{(k,c)} \prod_{j=1}^{c} (dG_{x_{i_j}}(u_{i_j}) - dF(u_{i_j})) \prod_{j=c+1}^{k} (dF(u_{i'_j})) \quad (5)$$

where the inner sum is taken over all $c$-subsets $S = \{i_1, \ldots, i_c\}$ of $\{1, \ldots, k\}$ and the corresponding complements $S' = \{i'_{c+1}, \ldots, i'_k\}$. Multiplying (5) by $\psi(u_1, \ldots, u_k)$ and integrating, we obtain (1).

This integral representation (4) is useful in establishing properties of the functions $h^{(j)}$. We begin with studying the means and conditional variances. Note that, using the notational convention (1) of Section 1.3, we denote the $c$th conditional expectation of $h^{(j)}$ by $h_c^{(j)}$.

**Theorem 2.**

(i) For $c = 1, 2, \ldots, j-1$, and $j = 1, 2, \ldots, k$,  $h_c^{(j)}(x_1, \ldots, x_c) = 0$;

(ii) $E\{h^{(j)}(X_1, \ldots, X_j)\} = 0$.

**Proof.** From the definitions (1) of Section 1.3, we can write

$$h_{j-1}^{(j)} = E\{h^{(j)}(x_1, \ldots, x_{j-1}, X_j)\}$$

28

and using the integral representation (4) we obtain

$$h^{(j)}(x_1,\ldots,x_j)$$

$$= \int \cdots \int \psi(u_1,\ldots,u_k) \prod_{i=1}^{j}(dG_{x_i}(u_i) - dF(u_i)) \prod_{i=j+1}^{k} dF(u_i)$$

$$= \int \cdots \int \psi(u_1,\ldots,x_j,\ldots,u_k) \prod_{i=1}^{j-1}(dG_{x_i}(u_i) - dF(u_i)) \prod_{i=j+1}^{k} dF(u_i)$$

$$- \int \cdots \int \psi(u_1,\ldots,u_k) \prod_{i=1}^{j-1}(dG_{x_i}(u_i) - dF(u_i)) \prod_{i=j}^{k} dF(u_i)$$

$$= \int \cdots \int \psi(u_1,\ldots,u_k) \prod_{i=1}^{j-1}(dG_{x_i}(u_i) - dF(u_i)) \prod_{i=j+1}^{k} dF(u_i)$$

$$- h^{(j-1)}(x_1,\ldots,x_{j-1}).$$

Integrating both sides with respect to $u_j$ with measure $dF(u_j)$ yields

$$E\{h^{(j)}(x_1,\ldots,x_{j-1},X_j)\} = h^{(j-1)}(x_1,\ldots,x_{j-1}) - h^{(j-1)}(x_1,\ldots,x_{j-1})$$

$$= 0$$

and so $h_{j-1}^{(j)}(x_1,\ldots,x_{j-1}) = 0$.

It thus follows from Theorem 1 (i) of Section 1.3 that

$$h_c^{(j)}(x_1,\ldots,x_c) = E\{h_{j-1}^{(j)}(x_1,\ldots,x_c,X_{c+1},\ldots,X_{j-1})\}$$

$$= 0, \tag{6}$$

proving (i).

From Theorem 1 (ii) of Section 1.3 it follows that

$$E\{h^{(j)}(X_1,\ldots,X_j)\} = E\{h_c^{(j)}(X_1,\ldots,X_c)\}$$

$$= 0$$

by applying (6).

The variances of the conditional expectations of the $h^{(j)}$ are thus all zero, except the $j$th. Define

$$\delta_j^2 = Var\{h_j^{(j)}(X_1,\ldots,X_j)\}$$

$$= Var\{h^{(j)}(X_1,\ldots,X_j)\}.$$

Then by Theorem 3 of Section 1.3

$$Var\, H_n^{(j)} = \binom{n}{j}^{-1} \sum_{c=1}^{j} \binom{j}{c}\binom{n-j}{j-c} Var\{h_c^{(j)}(X_1,\ldots,X_c)\}$$

$$= \binom{n}{j}^{-1} \delta_j^2. \qquad (7)$$

The usefulness of the $H$-representation lies in the fact that the variances of the component $U$-statistics $H_n^{(j)}$ are of order $n^{-j}$ and are uncorrelated. To establish the second assertion, we prove our next theorem.

**Theorem 3.**

(i) Let $j < j'$ and let $S_1$ and $S_2$ be a $j$-subset of $\{1,2,\ldots,n\}$ and a $j'$-subset of $\{1,2,\ldots,n\}$ respectively. Then

$$Cov(h^{(j)}(S_1), h^{(j')}(S_2)) = 0$$

and so

$$Cov(H_n^{(j)}, H_n^{(j')}) = 0.$$

(ii) Let $S_1 \neq S_2$ be two distinct $j$-subsets of $\{1,2,\ldots,n\}$ .Then

$$Cov(h^{(j)}(S_1), h^{(j)}(S_2)) = 0.$$

**Proof.** (i) Because of Theorem 2 (ii), we need only prove that

$$E\{h^{(j)}(S_1)h^{(j')}(S_2)\} = 0.$$

Since $j < j'$, there is an element of $S_2$ that is not in $S_1$ and hence there is a r.v. $X_{j'}$ say, that appears in $h^{(j')}(S_2)$ but not in $h^{(j)}(S_1)$. Thus we can write

$$Eh^{(j)}(S_1)h^{(j')}(S_2) = E\{E(h^{(j)}(S_1)h^{(j')}(S_2)|X_{j'})\}$$
$$= E\{h^{(j)}(S_1)E(h^{(j')}(S_2)|X_{j'})\}$$

since $h^{(j)}(S_1)$ and $X_{j'}$ are independent. But $E(h^{(j')}(S_2)|X_{j'}) = h_1^{(j')}(X_{j'})$ $= 0$ by Theorem 2 (i) and so $Eh^{(j)}(S_1)h^{(j')}(S_2) = 0$, proving the first part. For the second part,

$$Cov(H_n^{(j)}, H_n^{(j')}) = \binom{n}{j}^{-1}\binom{n}{j'}^{-1} \sum_{(n,j)}\sum_{(n,j')} Cov(h^{(j)}(S_1)h^{(j')}(S_2))$$

$$= 0.$$

30

(ii) If $S_1 \cap S_2$ is empty, the result follows by independence. Otherwise, suppose that $S_2 - S_1 = \{i_1, \ldots i_c\}$ with $c < j$. (If $S_2 \subseteq S_1$ then consider $S_1 - S_2$). Then

$$
\begin{aligned}
E(h^{(j)}(S_1)h^{(j)}(S_2)) &= E\{E(h^{(j)}(S_1)h^{(j)}(S_2)|X_{i_1}, \ldots, X_{i_c})\} \\
&= E\{h^{(j)}(S_1)E(h^{(j)}(S_2)|X_{i_1}, \ldots, X_{i_c})\} \\
&= 0
\end{aligned}
$$

since $h_c^{(j)}(x_1, \ldots, x_c) = 0$ by Theorem 2(i).

An expression for $Var\, U_n$ in terms of the variances $\delta_j$ of the $h^{(j)}$ now follows easily.

**Theorem 4.** *The variance of $U_n$ is given by*

$$
Var\, U_n = \sum_{j=1}^{k} \binom{k}{j}^2 \binom{n}{j}^{-1} \delta_j^2.
$$

**Proof.**

$$
\begin{aligned}
Var\, U_n &= Var\left\{ \theta + \sum_{j=1}^{k} \binom{k}{j} H_n^{(j)} \right\} \\
&= \sum_{j=1}^{k} \binom{k}{j}^2 Var\, H_n^{(j)} \\
&= \sum_{j=1}^{k} \binom{k}{j}^2 \binom{n}{j}^{-1} \delta_j^2
\end{aligned}
$$

using (2) and (7).

Relations describing $\sigma_c^2$ in terms of the $\delta_j^2$ and vice versa can now also be derived; these fill a gap left in the proof of Theorem 4 of Section 1.3. From (1) we have

$$
\psi_c(X_1, \ldots, X_c) = \sum_{j=1}^{c} \sum_{(c,j)} h^{(j)}(X_{i_1}, \ldots, X_{i_j}) + \theta
$$

31

and so taking variances of each side, and using Theorem 3 and (7) we get

$$\sigma_c^2 = \sum_{j=1}^{c} Var\{\sum_{(c,j)} h^{(j)}(X_{i_1}, \ldots, X_{i_j})\}$$

$$= \sum_{j=1}^{c} \binom{c}{j}^2 Var(H_c^{(j)})$$

$$= \sum_{j=1}^{c} \binom{c}{j} \delta_j^2. \tag{8}$$

We can invert this last expression with the aid of standard identities involving binomial coefficients. Multiplying (8) on both sides by $(-1)^{d-c}\binom{d}{c}$ and summing from $c = 1$ to $d$ yields

$$\sum_{c=1}^{d} \binom{d}{c}(-1)^{d-c}\sigma_c^2 = \sum_{c=1}^{d}\sum_{j=1}^{c} \binom{c}{j}\binom{d}{c}(-1)^{d-c}\delta_j^2$$

$$= \sum_{j=1}^{d} \left\{ \sum_{c=0}^{d-j} \binom{d-j}{c}(-1)^c \right\} (-1)^{d-j}\binom{d}{j}\delta_j^2.$$

Since the term in braces is zero except when $j = d$; we obtain

$$\delta_d^2 = \sum_{c=1}^{d}(-1)^{d-c}\binom{d}{c}\sigma_c^2 \tag{9}$$

so that the $\delta_d^2$'s are the same as the quantities defined in Theorem 4 of Section 1.3. We have thus filled the gap in that theorem as promised.

It follows from the relationship (9) that if a $U$-statistic is degenerate of order $d$ (i.e. if $0 = \sigma_1^2 = \ldots = \sigma_d^2 < \sigma_{d+1}^2$) then the first $d$ terms of the $H$-decomposition vanish, since in this case $\delta_1^2 = \ldots = \delta_d^2 = 0$. In particular, such a statistic has a variance of order $n^{-(d+1)}$. We will see in Chapter 3 that the order of the degeneracy determines the asymptotic distribution of $U$-statistics.

The integral representation (4) also gives a formula for the function $h^{(c)}$ in terms of the conditional expectations $\psi_j$. Write

$$\prod_{i=1}^{c}(dG_{x_i}(u_i) - dF(u_i))$$

$$= 1 + \sum_{d=1}^{c}(-1)^d \sum_{(c,d)} \prod_{j=1}^{d} dG_{x_{i_j}}(u_{i_j}) \prod_{j=d+1}^{c} dF(u_{i_{l_j}}) \tag{10}$$

32

where the inner sum is taken over all $d$-subsets $\{i_1, \ldots, i_d\}$ of $\{1, \ldots, c\}$ and $\{i\prime_{d+1}, \ldots, i\prime_c\}$ is the complement of $\{i_1, \ldots, i_d\}$. Substituting (10) in (4) yields

$$h^{(c)}(x_1, \ldots, x_c) = \psi_c(x_1, \ldots, x_c)$$
$$+ \sum_{d=1}^{c-1} (-1)^d \sum_{(c,c-d)} \psi_{c-d}(x_{i_1}, \ldots, x_{i_{c-d}}) + (-1)^c \theta. \quad (11)$$

An example of the $H$- decomposition is given below:

**Example 1. The sample variance.**

Since $\psi_1(x_1) = \frac{1}{2}(\sigma^2 + (x_1 - \mu)^2)$ and $\psi_2(x_1, x_2) = \frac{1}{2}(x_1 - x_2)^2$ we have using (11)

$$h^{(1)}(x_1) = \psi_1(x_1) - \sigma^2 = \frac{1}{2}((x_1 - \mu)^2 - \sigma^2)$$

and

$$\begin{aligned} h^{(2)}(x_1, x_2) &= \psi_2(x_1, x_2) - \psi_1(x_1) - \psi_1(x_2) + \theta \\ &= \frac{1}{2}(x_1 - x_2)^2 - \frac{1}{2}(\sigma^2 + (x_1 - \mu)^2) \\ &\quad - \frac{1}{2}(\sigma^2 + (x_2 - \mu)^2) + \sigma^2 \\ &= \frac{1}{2}\{(x_1 - x_2)^2 - (x_1 - \mu)^2 - (x_2 - \mu)^2\} \\ &= -(x_1 - \mu)(x_2 - \mu). \end{aligned}$$

Thus

$$\begin{aligned} s^2 &= \frac{1}{2} \sum_{1 \le i < j \le n} (X_i - X_j)^2 \\ &= \sigma^2 + \left\{ \frac{1}{n} \sum_{i=1}^{n} (X_i - \mu)^2 - \sigma^2 \right\} \\ &\quad - \binom{n}{2}^{-1} \sum_{i<j} (X_i - \mu)(X_j - \mu). \end{aligned}$$

Also from (9) we obtain

$$\delta_1^2 = \sigma_1^2 + \frac{1}{4}(\mu_4 - \sigma^4)$$

and

$$\delta_2^2 = \sigma_2^2 - 2\sigma_1^2 = \frac{1}{2}(\mu_4 + \sigma^4) - \frac{2}{4}(\mu_4 - \sigma^4)\sigma^4.$$

33

## 1.7 A geometric perspective on the $H$-decomposition

The notion of *projection* is useful in many areas of statistics. Typically, a statistic of interest is regarded as a member of vector space, and is projected orthogonally onto some subspace. The projection may be of interest in its own right (as in the case of analysis of variance) or useful as an approximation (as in the case of rank statistics). In analysis of variance, means are projected onto subspaces to obtain main effects and interactions. In the theory of rank statistics, the projection is onto the subspace of *linear rank statistics*.

In the case of $U$-statistics, the basic space is the set $\mathcal{L}_2$ of all random variables having variances, which can be equipped with an inner product $E\,UV$. Consider the subspace $\mathcal{L}_2^{(1)}$ of all r.v.s of the form $\sum_{i=1}^{n} \psi(X_i)$ when $X_1, \ldots, X_n$ is a fixed sequence of i.i.d. random variables, and $E\psi^2(X_1)$ is finite. Clearly the first element $H_n^{(1)}$ of the $H$-decomposition is of this form, and (3) of Section 1.6 with $c = 1$ expresses $U_n - \theta$ as the sum of an element in $\mathcal{L}_2^{(1)}$ and an element orthogonal to $\mathcal{L}_2^{(1)}$, since by Theorem 3 of Section 1.6 we can write

$$U_n - \theta = kH_n^{(1)} + R_n^{(1)} \tag{1}$$

with $R_n^{(1)}$ orthogonal to $H_n^{(1)}$.

It follows that $kH_n^{(1)}$ is the projection onto $\mathcal{L}_2^{(1)}$. The equation (1) splits up $\mathcal{L}_2$ into the direct sum of two orthogonal subspaces

$$\mathcal{L}_2 = \mathcal{L}_2^{(1)} \oplus (\mathcal{L}_2^{(1)})^{\perp}.$$

If we define $\mathcal{L}_2^{(j)}$ to be the Hilbert space of all $U$-statistics of degree $j$ with square integrable kernels, then the $H$-decomposition is just the representation of $U_n - \theta$ (a member of $\mathcal{L}_2^{(k)}$) as the sum of its component projections onto the spaces $\mathcal{M}_j = \mathcal{L}_2^{(j)} \cap (\mathcal{L}_2^{(j-1)})^{\perp}$, for $j = 1, \ldots, k$.

To see this, it is enough to prove that $H_n^{(j)}$ is in $\mathcal{L}_2^{(j)}$ and orthogonal to $\mathcal{L}_2^{(j-1)}$. The first assertion is obvious from the definition. To prove the second, we must show that $E(H_n^{(j)} \sum_{(n,j-1)} f(X_{i_1}, \ldots, X_{i_{j-1}})) = 0$, which in turn will follow by symmetry if $E\{h^{(j)}(X_{i_1}, \ldots, X_{i_j})f(X_1, \ldots, X_{j-1})\} = 0$ for any set of indices $\{i_1, \ldots, i_j\}$.

There must be an index, $i$ say, that is in the first set but not the second. Then $f$ is independent of $X_i$, and

$$E(h^{(j)}f) = E\{E(h^{(j)}f|X_i)\}$$
$$= E\{fE(h^{(j)}|X_i)\}$$
$$= 0$$

since $h_1^{(j)} = 0$ by Theorem 2(i) of Section 1.6.

## 1.8 Bibliographic details

What are now termed $U$-statistics were first identified as minimum-variance unbiased estimators by Halmos (1946) and the material in Section 1.1 is taken from his paper, and also from Fraser (1954), who proves Lemma B of Section 1.1. See also Bell, Blackwell and Breiman (1960) for generalisations of these ideas. The term "$U$-statistic" is due to Hoeffding, who showed how to compute variances and demonstrated the asymptotic normality of this class of statistics in his fundamental paper (Hoeffding (1948a)). Sections 1.2–1.4 are based on this reference. The books by Fraser (1957), Puri and Sen (1971), Randles and Wolfe (1979) and Serfling (1980) all have basic material on $U$-statistics.

The moment results of Section 1.5 are taken from Serfling (1980) and Grams and Serfling (1973). The material on the H-decomposition is adapted from Hoeffding's unpublished 1961 technical report, and also appears in many other places, for example in Sproule (1974). A similar decomposition, valid for any symmetric statistic, is due to Rubin and Vitale (1980) and is described in Chapter 4. For a more general result, see Efron and Stein (1981).

# CHAPTER TWO

## Variations

### 2.1 Introduction

Many generalisations of the basic theory outlined in the previous chapter can be made. For example, we may allow the kernel to be a function of more than one sample, and consider so-called *generalised U-statistics*, which are the subject of Section 2.2 below. Generalising in another direction, we may relax the assumptions of independence and identity of distribution, as is done in Sections 2.3–2.5. We may also allow the form of the kernel to depend on the actual set of variables chosen, perhaps by introducing a set-dependent weighting factor, or perhaps by trimming. Such generalisations are explored in Sections 2.6–2.7. Brief bibliographic comments are in Section 2.8.

### 2.2 Generalised $U$-statistics

For $j = 1, 2, \ldots, m$, let $X_{j1}, \ldots, X_{jn_j}$ be independently and identically distributed with distribution function $F_j$, with $X_{ij}$ independent of $X_{i'j'}$ for $j \neq j'$, and let $\psi$ be a function of $k_1 + \ldots + k_m$ arguments

$$\psi(x_{11}, \ldots, x_{1k_1}; x_{21}, \ldots, x_{2k_2}; \ldots; x_{m1}, \ldots, x_{mk_m})$$

which is symmetric in each set of arguments $x_{j1}, \ldots, x_{jk_j}$. Using an obvious extension of the notation of the previous section, the *generalised U-statistic* based on $\psi$ is a statistic of the form

$$\binom{n_1}{k_1}^{-1} \cdots \binom{n_m}{k_m}^{-1} \sum_{(n_1, k_1)} \cdots \sum_{(n_m, k_m)} \psi(S_1; \ldots; S_m). \qquad (1)$$

and is an unbiased estimator of $E\,\psi(X_{11}, \ldots, X_{mk_m})$. Typically in applications the r.v.s $X_{j1}, \ldots, X_{jn_j}$ are a sample from the $j$th out of a set of $m$ populations. For notational simplicity, we confine ourselves to the case $m = 2$, but the next few results carry over to the general case in an obvious manner.

As in Section 1.3, define

$$\psi_{c,d}(x_1, \ldots, x_c; y_1, \ldots, y_d)$$
$$= E\{\psi(x_1, \ldots, x_c, X_{c+1}, \ldots, X_{k_1}; y_1, \ldots, y_d, Y_{d+1}, \ldots, Y_{k_2})\}$$

and

$$\sigma^2_{c,d} = Var \, \psi_{c,d}(X_1, \ldots, X_c; Y_1, \ldots, Y_d).$$

Then, in a now familiar vein, we have

**Theorem 1.** *Let $S_1$ and $S_2$ be two sets in $\mathcal{S}_{n,k_1}$ with c indices in common and $T_1, T_2$ be two sets in $\mathcal{S}_{n,k_2}$ with d indices in common. Then*

$$\sigma^2_{c,d} = Cov(\psi(S_1, T_1), \psi(S_2, T_2)).$$

**Proof.** Same as Theorem 1 of Section 1.3 and consequently omitted.

Using Theorem 1, we obtain

**Theorem 2.** *Let $U_{n_1,n_2}$ be a generalised U-statistic of the form (1) with $m = 2$ based on a kernel $\psi$ having degrees $k_1$ and $k_2$. Then*

$$Var \, U_{n_1,n_2} = \sum_{c=0}^{k_1} \sum_{d=0}^{k_2} \frac{\binom{k_1}{c}\binom{k_2}{d}\binom{n_1-k_1}{k_1-c}\binom{n_2-k_2}{k_2-d}}{\binom{n_1}{k_1}\binom{n_2}{k_2}} \sigma^2_{c,d}. \tag{2}$$

**Proof.** There are exactly $\binom{n_1}{k_1}\binom{k_1}{c}\binom{n_1-k_1}{k_1-c} \times \binom{n_2}{k_2}\binom{k_2}{d}\binom{n_2-k_2}{k_2-d}$ pairs of pairs of sets $(S_1, T_1)$ and $(S_2, T_2)$ with $|S_1 \cap S_2| = c$ and $|T_1 \cap T_2| = d$. The result follows as usual via Theorem 1.

Many examples of generalised U-statistics are to be found in the theory of nonparametric statistics. We mention one in connection with the two-sample problem.

**Example 1. The two-sample Wilcoxon (Mann-Whitney) statistic.**

Let $X_1, \ldots, X_{n1}$ and $Y_1, \ldots, Y_{n_2}$ be two independent samples with $n_1 \geq n_2$ from absolutely continuous distributions $F$ and $G$, and let $R_j$ denote the

rank of $Y_j$ in the combined sample. Then the Wilcoxon rank sum statistic
is

$$W = \sum_{j=1}^{n_2} R_j.$$

If we define

$$\phi(x, y) = \begin{cases} 1 & \text{if } x < y; \\ 0 & \text{otherwise;} \end{cases}$$

and

$$U = \sum_{i=1}^{n_1} \sum_{j=1}^{n_2} \phi(X_i, Y_j)$$

then in the absence of ties it can be shown that

$$W = U + n_2(n_2 + 1)/2.$$

The statistic $U$ is the Mann-Whitney $U$-statistic, and the related statistic
$U_{n_1,n_2} = (n_1 n_2)^{-1} U$ is clearly a generalised $U$-statistic. The mean of $U_{n_1,n_2}$
is just $E\phi(X_1, Y_1) = Pr(X_1 < Y_1)$, while by Theorem 2 the variance is

$$Var\, U_{n_1,n_2} = (n_1 n_2)^{-1} \{(n_1 - 1)\sigma_{0,1}^2 + (n_2 - 1)\sigma_{1,0}^2 + \sigma_{1,1}^2\}$$

where

$$\sigma_{0,1}^2 = Pr(X_1 < Y_1, X_2 < Y_1) - Pr^2(X_1 < Y_1),$$
$$\sigma_{1,0}^2 = Pr(X_1 < Y_1, X_1 < Y_2) - Pr^2(X_1 < Y_1)$$

and

$$\sigma_{1,1}^2 = Pr(X_1 < Y_1)\{1 - Pr(X_1 < Y_1)\}.$$

If $F$ and $G$ are the distribution functions of $X$ and $Y$ with corresponding
densities $f$ and $g$, then the joint density of $U = Y_1 - X_1$ and $V = Y_1 - X_2$
is given by the equation

$$h(u, v) = \int_{-\infty}^{\infty} f(w - u)f(w - v)g(w)dw$$

and so

$$Pr(X_1 < Y_1, X_2 < Y_1) = Pr(0 < U, 0 < V)$$
$$= \int_0^{\infty} \int_0^{\infty} \int_{-\infty}^{\infty} f(w - u)f(w - u)g(w)\, dw\, du\, dv$$
$$= \int_{-\infty}^{\infty} (1 - F(w))^2 g(w)\, dw.$$

39

Similarly $Pr(X_1 < Y_1, X_1 < Y_2) = \int_{-\infty}^{\infty}(1 - G(w))^2 f(w)dw$. In the case when $F = G$ these both reduce to $\frac{1}{3}$ so that when $F = G, \sigma_{0,1}^2 = \sigma_{1,0}^2 = \frac{1}{12}, \sigma_{1,1}^2 = \frac{1}{4}$ and

$$Var\, U_{n_1,n_2} = (n_1 n_2)^{-1}\{(n_1 - 1)/12 + (n_2 - 1)/12 + \tfrac{1}{4}\}$$
$$= (n_1 n_2)^{-1}(N + 1)/12$$

where $N = n_1 + n_2$.

Thus in the case $F = G$, it follows that $EU = n_1 n_2/2$, $Var\, U = n_1 n_2(N + 1)/12$, $EW = n_2(N + 1)/2$ and $Var\, W = n_1 n_2(N + 1)/12$.

The $H$-decomposition for generalised $U$-statistics can be developed in a manner similar to that for ordinary $U$-statistics: If $G_x$ again denotes the d.f. of a single point mass at $x$, and if we define

$$h^{(c,d)}(x_1,\ldots,x_c;y_1,\ldots,y_d)$$

$$= \int \cdots \int \psi(u_1,\ldots,u_{k_1};v_1,\ldots,v_{k_2}) \prod_{i=1}^{c}(dG_{x_i}(u_i) - dF_1(u_i))$$

$$\times \prod_{i=c+1}^{k_1} dF_1(u_i) \prod_{j=1}^{d}(dG_{y_j}(v_j) - dF_2(v_j)) \prod_{j=d+1}^{k_2} dF_2(v_j)$$

then it follows just as in Section 1.6 that

$$\psi(x_1,\ldots,x_{k_1};y_1,\ldots,y_{k_2})$$

$$= \sum_{c=0}^{k_1}\sum_{d=0}^{k_2}\sum_{(k_1,c)}\sum_{(k_2,d)} h^{(c,d)}(x_{i_1},\ldots,x_{i_c};y_{j_1},\ldots,y_{j_d}). \quad (3)$$

The following may be proved in a manner analogous to Theorems 1, 2 and 3 of Section 1.6:

**Theorem 3.** *The generalised $U$-statistic admits the representation*

$$U_{n_1,n_2} = \sum_{c=0}^{k_1}\sum_{d=0}^{k_2}\binom{k_1}{c}\binom{k_2}{d}H_{n_1,n_2}^{(c,d)}$$

*where $H_{n_1,n_2}^{(c,d)}$ is the generalised $U$-statistic based on the kernel $h^{(c,d)}$ and is given by*

$$H_{n_1,n_2}^{(c,d)} = \binom{n_1}{c}^{-1}\binom{n_2}{d}^{-1}\sum_{(n_1,c)}\sum_{(n_2,d)} h^{(c,d)}(X_{i_1},\ldots,X_{i_c};Y_{j_1},\ldots,Y_{j_d}).$$

40

The functions $h^{(c,d)}$ implicitly defined by (3) satisfy

(i) $E\{h^{(c,d)}(X_1,\ldots,X_c;Y_1,\ldots,Y_d)\} = 0$ ;

and

(ii) $Cov\left(h^{(c,d)}(S_1,S_2), h^{(c',d')}(S_2',S_2')\right) = 0$ for all integers $c,d,c',d'$ and sets $S_1, S_2, S_1', S_2'$ unless $c = c', d = d'$ and $S_1 = S_1'$ and $S_2 = S_2'$.

The generalised $U$-statistics $H_{n_1,n_2}^{(c,d)}$ are thus all uncorrelated. Their variances are given by

$$Var\, H_{n_1,n_2}^{(c,d)} = \binom{n_1}{c}^{-1} \binom{n_2}{d}^{-1} \delta_{c,d}^2 \qquad (4)$$

where $\delta_{c,d}^2 = Var\, h^{(c,d)}(X_1,\ldots,X_c;Y_1,\ldots,Y_d)$.

## Example 2. The Mann-Whitney statistic (continued).

Recall that this statistic is related to the generalised $U$-statistic

$$U_{n_1 n_2} = (n_1 n_2)^{-1} \sum_{i=1}^{n_1} \sum_{j=1}^{n_2} \phi(X_i, Y_j)$$

where the function $\phi$ is defined in Example 1. Direct calculation yields

$$h^{(0,0)} = Pr(X_1 < Y_1),$$
$$h^{(1,0)}(x) = 1 - G(x) - h^{(0,0)},$$
$$h^{(0,1)}(y) = F(y) - h^{(0,0)}$$

and

$$h^{(1,1)}(x,y) = \phi(x,y) - F(y) - (1 - G(x)) + h^{(0,0)},$$

where $F$ and $G$ denote the d.f.'s of $X_1,\ldots,X_{n_1}$, and $Y_1,\ldots,Y_{n_2}$. If $F = G$, and is absolutely continuous, then $Var\, h^{(1,0)}(X_1) = Var\, h^{(0,1)}(Y_1) = \frac{1}{12}$ since $F(Y_1)$ and $G(X_1)$ are then uniformly distributed. Also $\phi(X_1, Y_1) = \phi(F(X_1), F(Y_1))$ so that $Var\, h^{(1,1)} = Var\{\phi(U,V) - U - (1-V)\}$ where $U$ and $V$ are uniformly distributed on [0,1], and a simple calculation yields $Var\, h^{(1,1)} = \frac{1}{12}$. Thus

$$Var\, U_{n_1,n_2} = Var\{H_{n_1,n_2}^{(1,0)} + H_{n_1,n_2}^{(0,1)} + H_{n_1,n_2}^{(1,1)}\}$$
$$= Var\, H_{n_1,n_2}^{(1,0)} + Var\, H_{n_1,n_2}^{(0,1)} + Var\, H_{n_1,n_2}^{(1,1)}$$
$$= (12n_1)^{-1} + (12n_2)^{-1} + (12n_1 n_2)^{-1}$$
$$= (n_1 + n_2 + 1)/(12 n_1 n_2)$$

41

by the above arguments and (4).

## 2.3 Dropping the identically distributed assumption

Suppose that $X_1, \ldots, X_n$ are independent but no longer identically distributed, and denote the distribution function of $X_i$ by $F_i$.

The $U$-statistic (1) of Section 1.2 now has expected value

$$EU_n = \binom{n}{k}^{-1} \sum_{(n,k)} \theta\{i_1, \ldots, i_k\}$$

where $\theta\{i_1, \ldots, i_k\} = E\psi(X_{i_1}, \ldots, X_{i_k})$. The variance is correspondingly complicated. Let $S$ and $T$ be two $k$-subsets of $\{1, \ldots, n\}$ and denote by $\psi(S)$ and $\psi(T)$ the kernels evaluated for the corresponding variables. Let $\sigma^2(S, T)$ be the covariance of $\psi(S)$ and $\psi(T)$. Then

$$
\begin{aligned}
Var \, U_n &= \binom{n}{k}^{-2} \sum_{(n,k)} \sum_{(n,k)} \sigma^2(S, T) \\
&= \binom{n}{k}^{-1} \sum_{c=1}^{k} \binom{k}{c}\binom{n-k}{k-c} \bar{\sigma}_{c,n}^2
\end{aligned}
\tag{1}
$$

where $\bar{\sigma}_{c,n}^2$ is the averaged covariance of all terms having $c$ variables in common:

$$\bar{\sigma}_{c,n}^2 = \left\{ \binom{n}{k}\binom{k}{c}\binom{n-k}{k-c} \right\}^{-1} \sum_{|S \cap T|=c} \sigma^2(S, T). \tag{2}$$

Thus with this new interpretation for $\sigma_c^2$, equation (3) of Section 1.3 is still valid.

For an asymptotic formula for $Var \, U_n$, suppose that there is a constant $A$ such that $Var \, \psi(X_{i_1}, \ldots, X_{i_k}) < A$ for all sets $\{i_1, \ldots, i_k\}$. Then $\sigma^2(S, T) < A$ by the Cauchy-Schwartz inequality, and hence by virtue of (2), $\bar{\sigma}_{c,n}^2$ is uniformly bounded by $A$. Thus it follows from (1) that

$$Var \, U_n = k^2 \bar{\sigma}_{1,n}^2 n^{-1} + o(n^{-1}). \tag{3}$$

## 2.4 $U$-statistics based on stationary random sequences.

Let $X_1, X_2, \ldots$ be a strictly stationary random sequence, i.e. a sequence of dependent random variables having the property that the joint distribution of $X_{t_1+s}, \ldots, X_{t_m+s}$ does not depend on $s$ for any set of positive integers $\{t_1, \ldots, t_m\}$ and any integer $s$ for which $t_j + s \geq 1$, $j = 1, 2, \ldots, m$. In particular, the distributions of each $X_t$ are identical with distribution function $F$ say, so that the mean $\mu = E\,X_t$ is constant for all $t$. We take $\mu$ to be zero without loss of generality. Also, the joint distribution of any two random variables $X_s$ and $X_t$ depends only on the difference $|s - t|$, so we may denote the joint d.f. of $X_s$ and $X_{s+t}$ by $F_t$.

Let $\psi$ be a symmetric function of two variables. (For ease of exposition we confine attention in this section to the case $k = 2$, but details of the general case may be found in Sen (1963).) Define

$$Y_{s,t} = \psi(X_s, X_t)$$

and let $\theta_t = EY_{t+s,s}$. The $U$-statistic $U_n$ based on $X_1, \ldots, X_n$ and kernel $\psi$ has expected value

$$EU_n = \binom{n}{2}^{-1} \sum_{(n,2)} \theta_{|i-j|}$$

$$= \binom{n}{2}^{-1} \sum_{j=1}^{n-1} (n-j)\theta_j. \qquad (1)$$

The covariance of $U_n$ depends in a complicated way on the covariance structure of the underlying process $X_t$, and general formulae are too complex to be of much use. However, useful asymptotic formulae can be derived in the case of *weakly dependent* sequences, for which random variables far apart in the sequence are only slightly associated. This vague idea can be made precise in several ways, including the concepts of *m-dependence* and *strong mixing*, which we now describe in more detail.

### 2.4.1 M-dependent stationary sequences

The stationary sequence $X_1, X_2, \ldots$ is said to be *m-dependent* if $X_s$ and $X_t$ are independent whenever $|s - t| > m$. If $U_n$ is based on such a sequence, then $\theta_t = EY_{t+s,s}$ is constant for $t > m$ and equal to $\theta$ say. Then

$$EU_n = \binom{n}{2}^{-1} \sum_{j=1}^{m} (n-j)\theta_j + \binom{n}{2}^{-1} \sum_{j=m+1}^{n-1} (n-j)\theta$$
$$= \theta + O(n^{-1})$$

and so $U_n$ is asymptotically unbiased for $\theta$.

To study the variance of $U_n$, it is convenient to introduce a reduced form of the statistic, namely

$$U_n^* = \binom{n-m}{2}^{-1} \sum_{\substack{1 \leq i < j \leq n \\ j-i > m}} \psi(X_i, X_j) \tag{1}$$

where the sum is now taken over all pairs of indices $(i,j)$ with $i < j$ and $j - i > m$. There are $\binom{n-m}{2}$ such pairs, and for each such pair $E\psi(X_i, X_j) = \theta$, so that $U_n^*$ is exactly (rather than asymptotically) unbiased for $\theta$.

To develop an asymptotic formula for $Var\,U_n$, we approximate $U_n$ by $U_n^*$ and consider the variance of $U_n^*$. We need the quantities

$$\psi_1(x) = E\psi(x, X_t),$$
$$\sigma_1^2 = Var\,\psi_1(X_t)$$
$$\sigma_{1,t}^2 = Cov(\psi_1(X_{t+s}), \psi_1(X_s))$$

and

$$\sigma_{2,t}^2 = Var\,\psi(X_{t+s}, X_s).$$

Note that $\sigma_{2,t}^2$ is independent of $t$ for $t > m$, so we write $\sigma_{2,t}^2 = \sigma_2^2$ for $t > m$. Also note that $\sigma_{1,t}^2 = 0$ for $t > m$. We can now prove

**Theorem 1.** Define $\sigma^2 = \sigma_1^2 + 2\sum_{h=1}^{m} \sigma_{1,h}^2$. Then

$$Var\,U_n^* = 4\sigma^2 n^{-1} + O(n^{-2}) \tag{2}$$

and

$$Var\,U_n = 4\sigma^2 n^{-1} + O(n^{-2}). \tag{3}$$

**Proof.** To prove (2) we need to be able to evaluate

$$Cov(\psi(X_{s_1}, X_{s_2}), \psi(X_{t_1}, X_{t_2})) \tag{4}$$

44

where $s_2 - s_1 > m$ and $t_2 - t_1 > m$. There are various cases to consider.

(a) If $|s_i - t_j| > m$ for $i, j = 1, 2$ then all four random variables are independent, and the covariance is zero. There are six possible ordered pairs of unordered pairs $(\{s_1, s_2\}, \{t_1, t_2\})$ that can be made from four integers $1 \leq i < j < k < l \leq n$, and these 6 pairs of pairs satisfy the condition $|s_i - t_j| > m$ if and only if $j - i > m, k - j > m$ and $l - k > m$. Hence the number of terms (4) that are zero is $6\binom{n-3m}{4}$ by Lemma A below.

(b) Consider terms for which $0 < |s_i - t_j| = h \leq m$ for exactly one of the four possible differences $|s_i - t_j|$, and the other differences are all greater than $m$. We claim that there an exactly $12\binom{n-2m-h}{3}$ such terms, and each is equal to $\sigma_{1,h}^2$. To see this, note that there are twelve possible arrangements of four fixed integers $s_1, s_2, t_1, t_2$ with $s_2 - s_1 > m, t_2 - t_1 > m$ with exactly one of the four quantities $0 < |s_i - t_j| = h \leq m$ and the rest $> m$; these are

$$s_1 < s_2 < t_1 < t_2 \quad \text{with} \quad t_1 - s_2 = h;$$
$$t_1 < t_2 < s_1 < s_2 \quad \text{with} \quad s_1 - t_2 = h;$$
$$s_1 < t_1 < s_2 < t_2 \quad \text{with} \quad t_1 - s_1 = h;$$
$$t_1 < s_1 < t_2 < s_2 \quad \text{with} \quad s_1 - t_1 = h;$$
$$s_1 < t_1 < s_2 < t_2 \quad \text{with} \quad s_2 - t_1 = h;$$
$$t_1 < s_1 < t_2 < s_2 \quad \text{with} \quad t_2 - s_1 = h;$$
$$s_1 < t_1 < s_2 < t_2 \quad \text{with} \quad t_2 - s_2 = h;$$
$$t_1 < s_1 < t_2 < s_2 \quad \text{with} \quad s_2 - t_2 = h;$$
$$s_1 < t_1 < t_2 < s_2 \quad \text{with} \quad t_1 - s_1 = h;$$
$$t_1 < s_1 < s_2 < t_2 \quad \text{with} \quad s_1 - t_1 = h;$$
$$s_1 < t_1 < t_2 < s_2 \quad \text{with} \quad s_2 - t_2 = h;$$
$$t_1 < s_1 < s_2 < t_2 \quad \text{with} \quad t_2 - s_2 = h;$$

and for each possible arrangement, the number of ways of choosing the four integers satisfying the appropriate constraint is the same as the number of ways of choosing three integers $i, j, k$ with $j - i > m, k - j > m$ and $k \leq n - h$. For example, every arrangement $s_1 < s_2 < t_1 < t_2$ with $t_1 - s_2 = h$ is equivalent to choosing $s_1 = i, s_2 = j, t_1 = j + h$ and $t_2 = k + h$

with $i < j < k, k - j > m, j - i > m$ and $k \leq n - h$. Similar considerations hold for all the other combinations, and so the number of terms of type (b) is $12\binom{n-2m-h}{3}$. To see that (4) takes the value $\sigma_{1,h}^2$ for each arrangement, we consider the 12 cases separately. For example if $s_1 < s_2 < t_1 < t_2$ with $t_1 - s_2 = h$, then (4) equals

$$\int (\psi(x_1, x_2) - \theta)(\psi(x_3, x_4) - \theta)dF(x_1)dF_h(x_2, x_3)dF(x_4)$$
$$= \int (\psi_1(x_2) - \theta)(\psi_1(x_3) - \theta)dF_h(x_2, x_3)$$
$$= Cov(\psi_1(X_1), \psi_1(X_{1+h}))$$
$$= \sigma_{1,h}^2;$$

the other cases are similar.

(c) Now consider the case when exactly one of the differences $|s_i - t_j|$ is zero and the rest are greater than $m$. There are now six terms corresponding to a fixed choice of integers $s_1, s_2, t_1, t_2$ with exactly one difference zero and the rest greater than $m$, since the twelve cases considered in (b) are now identical in pairs. Once again the number of integers satisfying one of the six constraints is equal to the number of triples $1 \leq i < j < k \leq n$ with $k - j > m$, and $j - i > m$. Hence by Lemma A, the number of terms (4) of type (c) is $6\binom{n-2m}{3}$. An argument similar to part (b) shows that each of the six configurations leads to a value of $\sigma_1^2$ for (4).

(d) Finally consider all other terms. From parts (a), (b) and (c) there are $\binom{n-m}{2} - 6\binom{n-3m}{4} - 6\binom{n-2m}{3} - 12\sum_{h=1}^m \binom{n-2m-h}{3} = O(n^2)$ such.

Combining results (a), (b), (c) and (d), we see that

$$Var\, U_n^* = \binom{n-m}{2}^2 \sum_{\substack{1 \leq s_1 < s_2 \leq n \\ s_2 - s_1 > m}} \sum_{\substack{1 \leq t_1 < t_2 \leq n \\ t_2 - t_1 > m}} Cov(\psi(X_{s_1}, X_{s_2}), \psi(X_{t_1}, X_{t_2}))$$

$$= \binom{n-m}{2}^{-2} \Big\{ \sum_{\text{pairs of type (a)}} Cov(\psi(X_{s_1}, X_{s_2}), \psi(X_{t_1}, X_{t_2}))$$

$$+ \sum_{\text{pairs of type (b)}} Cov(\psi(X_{s_1}, X_{s_2}), \psi(X_{t_1}, X_{t_2}))$$

$$+ \sum_{\text{pairs of type (c)}} Cov(\psi(X_{s_1}, X_{s_2}), \psi(X_{t_1}, X_{t_2})) + O(n^2) \Big\}$$

$$= \binom{n-m}{2}^{-2} \left[ \left\{ \sum_{h=1}^{m} 12 \binom{n-2m-h}{3} \sigma_{1,h}^2 \right\} + 6 \binom{n-2m}{3} \sigma_1^2 + O(n^2) \right]$$

$$= 4 \left( \sigma_1^2 + 2 \sum_{h=1}^{m} \sigma_{1,h}^2 \right) n^{-1} + O(n^{-2})$$

$$= 4\sigma^2 n^{-1} + O(n^{-2}).$$

To prove (3), we need only prove that $U_n$ and $U_n^*$ are close. We can write

$$Var \, U_n = Var \, U_n^* + Var(U_n - U_n^*) + 2Cov(U_n^*, U_n - U_n^*) \qquad (5)$$

and

$$Var \, U_n = \binom{n}{2}^{-1} \sum_{(n,2)} Cov(U_n, \psi(X_i, X_j)). \qquad (6)$$

Now $Cov(U_n, \psi(X_i, X_j))$ depends only on $|i - j|$, so it is legitimate to write $C_{h,n} = Cov(U_n, \psi(X_i, X_j))$ when $|i - j| = h$. For $h > m$, $C_{h,n} \, (= C_n)$ say, is independent of $h$ and is equal to $Cov(U_n, U_n^*)$ since

$$Cov(U_n, U_n^*) = \binom{n-m}{2}^{-1} \sum_{\substack{1 \leq i < j \leq n \\ j-i > m}} Cov(U_n, \psi(X_i, X_j)).$$

Hence from (6) we get

$$Var \, U_n = \binom{n}{2}^{-1} \left\{ (n-1)C_{1,n} + \cdots + (n-m)C_{m,n} + \binom{n-m}{2} C_n \right\}. \quad (7)$$

Now

$$C_{h,n} = Cov(U_n, \psi(X_1, X_{1+h}))$$

$$= \binom{n}{2}^{-1} \sum_{1 \leq i < j \leq n} Cov(\psi(X_i, X_j), \psi(X_1, X_{1+h}))$$

and the number of pairs for which $\psi(X_i, X_j)$ and $\psi(X_1, X_{1+h})$ are *not* independent is $O(n)$, so that $C_{h,n}$ is $O(n^{-1})$, and hence (7) entails

$$Var \, U_n = Cov(U_n, U_n^*) + O(n^{-2}). \qquad (8)$$

Thus from (8) and (5) we get

$$Var \, U_n = Var \, U_n^* - Var(U_n - U_n^*) + O(n^{-2})$$

and the result will follow if we can show that $Var(U_n - U_n^*) = O(n^{-2})$. Since $EU_n = EU_n^* + O(n^{-1})$, we need in fact only show that $E(U_n - U_n^*)^2$ is $O(n^{-2})$. Now

$$U_n - U_n^* = \left(U_n - \binom{n}{2}^{-1}\binom{n-m}{2}U_n^*\right) + \left(\binom{n}{2}^{-1}\binom{n-m}{2} - 1\right)U_n^*$$

$$= V_1 + V_2 \quad \text{say.}$$

The r.v. $V_1$ can be written

$$V_1 = \binom{n}{2}^{-1}\sum_{\substack{1 \le i < j \le n \\ j - i \le m}}\psi(X_i, X_j),$$

and

$$E(V_1^2) = \binom{n}{2}^{-2}\sum_{\substack{1 \le i < j \le m \\ j - i \le m}}\;\sum_{\substack{1 \le i' < j' \le m \\ j' - i' \le m}}Cov(\psi(X_i, X_j), \psi(X_{i'}, X_{j'})).$$

Since there are only $O(n^2)$ terms in the double sum, $E(V_1^2) = O(n^{-2})$. Also $\binom{n}{2}^{-1}\binom{n-m}{2} - 1 = O(n^{-1})$ and hence $E(V_2^2) = O(n^{-2})E\{U_n^{*2}\} = O(n^{-2})$. Hence by Minkowski's inequality

$$\{E(V_1 + V_2)^2\}^{\frac{1}{2}} \le \{E(V_1^2)\}^{\frac{1}{2}} + \{E(V_2^2)\}^{\frac{1}{2}}$$

$$= O(n^{-1}) + O(n^{-1})$$

$$= O(n^{-1})$$

and so $E(U_n - U_n^*)^2 = E(V_1 + V_2)^2 = O(n^{-2})$, which proves the theorem.

It remains only to dispose of the combinatorial lemma used in the proof of Theorem 1.

**Lemma A.** *Let $k \ge 2$ Then the number of k-tuples of integers $1 \le i_1 < \cdots < i_k \le n$ that satisfy $i_j - i_{j-1} > m$ for $j = 2, 3, \ldots, k$ is $\binom{n-(k-1)m}{k}$.*

**Proof.** The proof is by induction on $k$. For $k = 2$ the proof is trivial. Assuming the result is true for $k - 1$, we may choose a $k$-tuple of integers satisfying the desired conditions by first choosing $i_1$ from $1, 2, \cdots,$

48

$n - (k-1)(m+1)$ and then choosing $i_2 < \cdots < i_k$ from $i_1 + m + 1, \ldots, n$ in such a way that $i_j - i_{j-1} > m$ for $j = 3, 4, \ldots, k$. By the induction hypothesis this can be done in $\binom{n-m-i-(k-2)m}{k-1}$ ways, so that the total number of ways the $k$ integers can be chosen is

$$\sum_{i=1}^{n-(k-1)(m+1)} \binom{n - i - (k-1)m}{k-1}.$$

The result follows by making the appropriate substitutions in the identity

$$\sum_{j=1}^{N} \binom{k + N - 1 - j}{k-1} = \binom{k - 1 + N}{k}$$

which follows for example from (12.16) in Feller (1968) p65.

### 2.4.2 Weakly dependent stationary sequences

In general, if the $X$'s form a stationary sequence, the remainder $R_n$ in the $H$-decomposition

$$U_n - \theta = k\ H_n^{(1)} + R_n$$

will not be uncorrelated with $H_n^{(1)}$, but progress is still possible if it is asymptotically negliglible in the sense that

$$Var\, U_n = k^2 Var\, H_n^{(1)} + o(n^{-1}). \tag{1}$$

The variance of $H_n^{(1)}$ is given by

$$
\begin{aligned}
Var\, H_n^{(1)} &= n^{-2} \sum_{i=1}^{n} \sum_{j=1}^{n} Cov(h^{(1)}(X_i), h^{(1)}(X_j)) \\
&= n^{-2} \{ nVar(h^{(1)}(X_1)) + 2(n-1)Cov(h^{(1)}(X_1), h^{(1)}(X_2)) \\
&\quad + \cdots + 2Cov(h^{(1)}(X_1), h^{(1)}(X_n)) \}.
\end{aligned}
$$

Since $h^{(1)}(x) = \psi_1(x) - \theta$ we may use the notation of the last section and write

$$Var\, H_n^{(1)} = n^{-1}(\sigma_1^2 + 2 \sum_{t=1}^{n-1} \sigma_{1,t}^2) - 2n^{-2} \sum_{t=1}^{n-1} t\sigma_{1,t}^2$$

49

The second term in the above equation will converge sufficiently quickly to zero as $n \to \infty$ if $\sigma_{1t}^2 \to 0$ sufficiently fast as $t \to \infty$, and if (1) is true we will then have

$$Var\, U_n = k^2 \sigma^2 n^{-1} + o(n^{-1}) \qquad (2)$$

where $\sigma^2 = \sigma_1^2 + 2\sum_{t=1}^{\infty} \sigma_{1,t}^2$.

Hence we need to impose conditions on the underlying process that ensure that $\sigma_{1,t}^2$ converges to zero suitably quickly. These conditions will impose some form of *weak dependence* on the sequence $X_n$ in the sense that observations sufficiently far apart are almost independent. There are several ways of making this idea precise; for a general discussion see e.g. Ibragimov and Linnik (1971) Chapter 17 or Ibragimov and Rosanov (1978) Chapter 4. We impose here the condition of *absolute regularity*. (The discussion of this concept involves the introduction of measure-theoretic ideas, so that the mathematical level of this section is a little higher than that of the rest of this chapter.)

Assume that the random variables $X_t$ constituting our stationary process are all defined on the same probability space $(\Omega, \mathcal{F}, P)$, and let $\mathcal{M}(a, b)$ be the $\sigma$-field generated by events of the form $\{w : (X_{t_1}(w), \ldots, X_{t_k}(w)) \epsilon B\}$ where $k$ is an arbitrary positive integer, $a \le t_1 < \ldots < t_k \le b$ and $B$ is a $k$-dimensional Borel set. The stationary process is then said to be *absolutely regular* if the sequence $\beta(n)$ defined by

$$\beta(n) = E \sup_{A \in \mathcal{M}(t+n,\infty)} |P(A|\mathcal{M}(-\infty,t)) - P(A)|$$

converges to zero as $n \to \infty$. Intuitively speaking, the condition requires that the conditional distribution of the r.vs $X_s, s \ge t + n$, given the past values of the process up to time $t$, is almost (in an average sense) the *unconditional* distribution. The coefficient $\beta(n)$ is called the *coefficient of absolute regularity*. Note that it follows directly from the definition that $\beta(n)$ is decreasing in $n$. The condition of absolute regularity requires that it decrease to zero.

The coefficient $\beta(n)$ has an interpretation as the variation of certain signed measures, (or equivalently as the variation of certain functions of

50

bounded variation) which is central to the verification of (2), so we digress briefly to explain this. For a fuller discussion than that presented here, see Halmos (1950) Chapter 6, or Burrill (1972) Chapter 5.

If $(\Omega, \mathcal{F})$ is a measurable space (i.e. a set $\Omega$ and a $\sigma$-field $\mathcal{F}$ of subsets of $\Omega$) a *finite signed measure* $\mu$ on $(\Omega, \mathcal{F})$ is an additive finite set function on $\mathcal{F}$, and the *variation* of $\mu$ is defined by

$$|\mu| = \sup \sum_{i=1}^{n} |\nu(E_i)| \tag{3}$$

where the supremum is taken over all finite collections $E_1, \ldots, E_n$ of disjoint sets in $\mathcal{F}$. An equivalent formulation is to introduce the *upper and lower variations* of $\mu$, given by

$$\mu^+(E) = \sup_{A \subseteq E} \mu(A)$$

and

$$\mu^-(E) = \inf_{A \subseteq E} \mu(A).$$

The set functions $\mu^+$ and $\mu^-$ are finite measures, and $\mu(E) = \mu^+(E) - \mu^-(E)$ for all $E$ in $F$. The measure $|\mu|$ defined by $|\mu|(E) = \mu^+(E) + \mu^-(E)$ is called the *total variation* of $\mu$ and the variation $|\mu|$ defined in (3) is just $|\mu|(\Omega)$. This measure satisfies

$$|\int f d\mu| \leq \int |f| d|\mu| \leq \sup |f| |\mu| \tag{4}$$

for all functions $f$ integrable with respect to $\mu$.

The connection between $\beta(n)$ and the variation of a signed measure is given in the next lemma, which like the lemma and theorem that follow is due to Yoshihara (1976).

**Lemma A.** Let $t_1 < \cdots < t_k$ be integers, let $F, G_j$ and $H_j$ be the distribution functions of $(X_{t_1}, \ldots, X_{t_k}), (X_{t_1}, \ldots, X_{t_j})$ and $(X_{t_{j+1}}, \ldots, X_{t_k})$ respectively and let $\mu$ be the signed measure corresponding to the function

$$F(x_1, \ldots, x_k) - G_j(x_1, \ldots, x_j) H_j(x_{j+1}, \ldots, x_k)$$

51

which is of bounded variation. Then $|\mu| = \beta(t_{j+1} - t_j)$.

**Proof.** Let $\mathcal{B}(\mathbb{R}_j)$ denote the $\sigma$- field of $j$-dimensional Borel sets. Since sets of the form $\cup_{i=1}^{n} A_i \times B_i$ with $A_i \times B_i$ disjoint, $A_i \in \mathcal{B}(\mathbb{R}_j)$ and $B_i \in \mathcal{B}(\mathbb{R}_{k-j})$ generate $\mathcal{B}(\mathbb{R}_k)$, the variation of $\mu$ is given by

$$|\mu| = \sup \ \sum_{i=1}^{n} \mu(A_i \times B_i)$$

where the supremum is taken over the sets described above. Write $\mathbf{X}_1 = (X_{t_1}, \ldots, X_{t_j})$, $\mathbf{X}_2 = (X_{t_{j+1}}, \ldots, X_{t_k})$, $E_i = \mathbf{X}_1^{-1}(A_i)$, $F_i = \mathbf{X}_2^{-1}(B_i)$ so that $E_i \in \mathcal{M}(-\infty, t_j)$ and $F_i \in \mathcal{M}(t_{j+1}, \infty)$. We have

$$\sum_{i=1}^{n} |\mu(A_i \times B_i)| \leq \sum_{i=1}^{n} |P(\mathbf{X}_1^{-1}(A_i) \cap \mathbf{X}_2^{-1}(B_i)) - P\mathbf{X}_1^{-1}(A_i)P\mathbf{X}_2^{-1}(B_i)|$$

$$= \sum_{i=1}^{n} \left| \int_{E_i} \{P(F_i | \mathcal{M}(-\infty, t_j)) - P(F_i)\} \, dP \right|$$

$$\leq \sum_{i=1}^{n} \int_{E_i} \sup_{F_i \in \mathcal{M}(t_{j+1}, \infty)} |P(F_i | \mathcal{M}(-\infty, t_j) - P(F_i)| \, dP$$

$$\leq \beta(t_{j+1} - t_j)$$

since the sets $E_i$ are disjoint, and so $|\mu| \leq \beta(t_{j+1} - t_j)$.

The reverse inequality is not used in the sequel, so is not proved here. A proof may be found in Ibragimov and Rozanov (1978) p119. Note their definition of "variation" is a little different from the present one.

Our next lemma uses Lemma A, and is needed to establish (2).

**Lemma B.** *Let* $t_1 < \cdots < t_k, F, G_j$ *and* $H_j$ *be as in Lemma A. Let* $h$ *be a measurable function such that*

$$M = \max \left( \int |h|^{1+\delta} dF, \int \int |h|^{1+\delta} dG_j dH_j \right)$$

*is finite for some* $\delta > 0$. *Then*

$$\left| \int h \, dF - \int \int h \, dG_j dH_j \right| \leq 3M^{\frac{1}{1+\delta}} \beta^{\frac{\delta}{1+\delta}} (t_{j+1} - t_j).$$

52

**Proof.** Let $\nu = 1/(1+\delta)$ and set $B = \{\mathbf{x} : |h(x_1, \ldots, x_k)| < M^\nu \beta^{-\nu}\}$ where we write $\beta = \beta(t_{j+1} - t_j)$. If $\mu$ is the measure corresponding to $F - G_j H_j$, then by Lemma A we have

$$\left| \int_B h \, dF - \int \int_B h \, dG_j dH_j \right| = \left| \int_B h \, d\mu \right|$$

$$\leq \int_B |h| d|\mu|$$

$$\leq M^\nu \beta^{1-\nu}. \tag{5}$$

Also, on the complement $B'$ of $B$, we have $M^{-\nu} \beta^\nu |h| \geq 1$, so that $|h| \leq M^{-\nu\delta} \beta^{\nu\delta} |h|^{1+\delta}$ and hence

$$\left| \int_{B'} h \, dF - \int \int_{B'} h \, dG_j dH_j \right| \leq \int_{B'} |h| dF + \int \int_{B'} |h| dG_j dH_j$$

$$\leq 2M^{1-\nu\delta} \beta^{\nu\delta}$$

$$= 2M^\nu \beta^{1-\nu}. \tag{6}$$

Adding (5) and (6) gives the result.

We are now in a position to verify (2). For computational simplicity we assume the kernel of the $U$-statistic is of degree 2.

**Theorem 1.** *Let $X_t$ be a stationary process whose absolute regularity coefficient satisfies $\beta(n) = O(n^{-(2+\delta')/\delta'})$ for some $\delta' > 0$, and let $U_n$ be a $U$-statistic based on a kernel $\psi$ of degree 2 satisfying*

$$\sup_{i,j} E|\psi(X_i, X_j)|^{2+\delta} < \infty$$

*and*

$$\int |\psi(x_1, x_2)|^{2+\delta} dF(x_1) dF(x_2) < \infty$$

*where $F$ is the (common) d.f. if the $X$'s, and $\delta > \delta'$. Then*

$$Var \, U_n = 4\sigma^2 n^{-1} + O(n^{-2})$$

*where $\sigma^2 = \sigma_1^2 + 2\sum_{t=1}^\infty \sigma_{1t}^2$.*

**Proof.** Let $U_n = \theta + 2H_n^{(1)} + H_n^{(2)}$ be the $H$-decomposition of $U_n$. We first show that $Var \, H_n^{(2)} = O(n^{\lambda-4})$ where $\lambda = \max(2, 3 - \gamma)$ and $\gamma =$

$2(\delta - \delta')/\delta'(2 + \delta)$. To this end, let $\theta_t^{(2)} = Eh^{(2)}(X_1, X_{1+t})$. Unlike the independent case, $\theta_t^{(2)}$ is not zero, but applying the inequality $(E|Y|)^r \leq E(|Y|^r)$, valid for $r \geq 1$, we see that

$$
\begin{aligned}
E|\psi_1(X_1)|^{2+\delta} &= \int \left| \int \psi(x, y) dF(x) \right|^{2+\delta} dF(y) \\
&\leq \int \int |\psi(x, y)|^{2+\delta} dF(x) dF(y) \\
&< \infty
\end{aligned}
$$

and thus by (1) of Section 1.6, $E|h^{(2)}(X_i, X_j)|^{2+\delta}$ is finite. Thus we can apply Lemma B and use the fact that $\int h^{(2)}(x, y) dF(x) dF(y) = 0$ to conclude that $|\theta_t^{(2)}| \leq \beta^{\frac{1+\delta}{2+\delta}}(t) = O(t^{-\nu})$, where $\nu$ is now given by $\nu = (1 + \delta)(2 + \delta')/\delta'(2 + \delta)$. Hence

$$
\begin{aligned}
E(H_n^{(2)}) &= \binom{n}{2}^{-1} \sum_{1 \leq i < j \leq n} \theta_{j-i}^{(2)} \\
&= O(n^{-\nu}) \\
&= O(n^{-1-\gamma})
\end{aligned}
$$

since $1 + \gamma < \nu$.

Now we turn to bounding $E\{(H_n^{(2)})^2\}$. Consider the sum

$$
\sum_{(n,2)} \sum_{(n,2)} E\{h^{(2)}(X_i, X_j) h^{(2)}(X_k, X_l)\}; \tag{7}
$$

the terms of (7) can be split into two types: (a) those for which $i, j, k$ and $l$ are all distinct, and (b) those remaining.

For terms of type (a), there are six possible orderings of $i, j, k$ and $l$ to consider, but all are basically handled by the same technique. For example, if $i < k < j < l$, by Lemma B we can write, with $j = 1$ and $\delta/2$ in place of $\delta$

$$
\begin{aligned}
&\left| \int h^{(2)}(x_1 x_3) h^{(2)}(x_2, x_4) dF(x_1 x_2 x_3 x_4) \right. \\
&\left. - \int h^{(2)}(x_1, x_3) h^{(2)}(x_2, x_4) dG_1(x_1) dH_1(x_2, x_3, x_4) \right| \\
&\leq C \beta^{\frac{\delta}{2+\delta}}(k - i) \tag{8}
\end{aligned}
$$

for some generic constant $C$. Note that the application of Lemma B with $\delta/2$ in place of $\delta$ is legitimate since

$$E|h^{(2)}(X_1X_3)h^{(2)}(X_2, X_4)|^{1+\delta/2}$$
$$\leq \{E(h^{(2)}(X_1, X_3)^{2+\delta})E(h^{(2)}(X_2, X_4)^{2+\delta})\}^{\frac{1}{2}}$$

by the Cauchy-Schwartz inequality.

Since $\int h^{(2)}(x_1, x_3)dG_1(x_1) = E\,h^{(2)}(X_1, x_3) = 0$ by Theorem 2 of Section 1.6, it follows from (8) that

$$E\{h^{(2)}(X_i, X_j)h^{(2)}(X_k, X_l)\} \leq C\beta^{\delta/(2+\delta)}(k-i)$$

and the sum of the terms in (7) for which $i < k < j < l$ is in consequence less than

$$C \sum_{i<k<j<\ell} \beta^{\delta/(2+\delta)}(k-i)$$
$$\leq Cn^2 \sum_{i<k} \beta^{\delta/(2+\delta)}(k-i)$$
$$\leq Cn^2 \sum_{i=1}^{n-1}(n-i)i^{-\delta(2+\delta')/\delta'(2+\delta)}$$
$$= O(n^{3-\gamma}).$$

Similar arguments yield the same bound for the other orderings, so that (7) is $O(n^{3-\gamma})$.

Terms of type (b) are split up further into terms for which $i = k, i = l$ or $j = l$. Writing $\xi_{ijkl} = E|h^{(2)}(X_i, X_j)h^{(2)}(X_k, X_l)|$, the absolute value of the sum of terms of (7) for which $i = k$ is less than

$$\sum_{i<j\leq l} \xi_{ijil} + \sum_{i<l<j} \xi_{ijil}$$
$$= \sum_{i<j} \xi_{ijij} + \sum_{i<j<l} \xi_{ijil} + \sum_{i<l<j} \xi_{ijil}$$
$$= \sum_{i<j} \xi_{ijij} + 2 \sum_{i<j<l} \xi_{ijil}. \qquad (9)$$

Applying Lemma B, and using the fact that $\beta(n)$ is decreasing in $n$, when $i < j < l$ we get

$$\xi_{ijil} \leq C\beta^{\frac{\delta}{2+\delta}}(l-i)$$

55

and hence (9) is $O(n^2) + O(n^{2-\gamma}) = O(n^2)$. Similar arguments also serve for the cases $j = l$ and $i = l$.

Combining all these results shows that the expression (7) is $O(n^\lambda)$ where $\lambda = \max(2, 3 - \gamma)$, and hence

$$Var\, H_n^{(2)} = E(H_n^{(2)})^2 + (EH_n^{(2)})^2$$

$$\leq \binom{n}{2}^{-2} \sum_{i<j} \sum_{k<l} \xi_{ijkl} + O(n^{-2-2\gamma})$$

$$= O(n^{\lambda-4}) \tag{10}$$

as claimed.

Further,

$$Var\, H_n^{(1)} = n^{-2} \sum_{i=1}^n \sum_{j=1}^n Cov(h^{(1)}(X_i), h^{(1)}(X_j))$$

$$= n^{-1}\left(\sigma_1^2 + 2\sum_{t=1}^{n-1} \sigma_{1,t}^2\right) - 2n^{-2} \sum_{t=1}^{n-1} t\sigma_{1,t}^2. \tag{11}$$

But

$$Cov(h^{(1)}(X_1), h^{(1)}(X_{1+t})) = E(h^{(1)}(X_1)h^{(1)}(X_{1+t}))$$

$$\leq M\beta^{\frac{\delta}{2+\delta}}(t)$$

by Lemma B, so that the last term of (11) is less in absolute value than twice $Mn^{-2}\sum_{t=1}^{n-1} t\beta^{\frac{\delta}{2+\delta}}(t) = O(n^{-1-\gamma})$ and so converges to zero. The series $\sum_{t=1}^\infty \sigma_{1t}^2$ also converges by the Weierstrass $M$-test so that

$$Var\, H_n^{(1)} = \sigma^2 n^{-1} + o(n^{-1}) \tag{12}$$

where $\sigma^2 = \sigma_1^2 + 2\sum_{t=1}^\infty \sigma_{1,t}^2$.

Finally,

$$Var\, U_n = 4Var\, H_n^{(1)} + Var\, H_n^{(2)} + 4Cov(H_n^{(1)}, H_n^{(2)}) \tag{13}$$

where

$$|Cov(H_n^{(1)}, H_n^{(2)})| \leq (Var\, H_n^{(1)} Var\, H_n^{(2)})^{\frac{1}{2}}$$

$$= (O(n^{-1})O(n^{\lambda-4}))^{\frac{1}{2}}$$

$$= o(n^{-1})$$

56

and so the result follows from (13) using (10) and (12).

Note that Theorem 1 of Section 2.4.1 also follows from Theorem 1 of the present section since an $m$-dependent sequence has $\beta(n) = 0$ for $n > m$.

## 2.5 U-statistics based on sampling from finite populations

Consider a finite population of $N$ elements, labelled $x_1, x_2, \ldots, x_N$, and suppose that we draw a sample of $n$ at random without replacement. Let the random variables $X_1, \ldots, X_n$ denote the labels on the sampled units.

In this section we study $U$-statistics based on such sequences of random variables; the results presented are from Nandi and Sen (1963). Note that the random variables $X_1, \ldots, X_n$ are not independent, but they are exchangeable, in that the distribution of $(X_{i_1}, \ldots, X_{i_n})$ is invariant under all permutations $(i_1, \ldots, i_n)$ of $\{1, 2, \ldots, n\}$. Note also that

$$Pr(X_1 = x_{i_1}, \ldots, X_n = x_{i_n}) = 1/N_{(n)} \tag{1}$$

for all permutations $(i_1, \ldots, i_n)$ of $n$ indices chosen from $\{1, 2, \ldots, N\}$ (Here and in the sequel $N_{(n)} = N(N-1)\ldots(N-n+1)$, the number of such permutations. We will also use the notation $\sum_{[N,n]}$ to indicate summation overall $n$-permutations of $\{1, 2, \ldots, N\}$, so that the notation $(N)$ introduced in Section 1.2 is equivalent to $[N, N]$).

It follows directly from (1) and the exchangeability that the $U$-statistic based on a symmetric kernel $\psi$ and $X_1, \ldots, X_n$ is an unbiased estimator of

$$\theta = E(\psi(X_1, \ldots, X_k))$$
$$= \frac{1}{N_{(k)}} \sum_{[N,n]} \psi(x_{i_1}, \ldots, x_{i_k})$$
$$= \binom{N}{k}^{-1} \sum_{(N,k)} \psi(x_{i_1}, \ldots, x_{i_k}).$$

Thus $\theta$ is itself a $U$-statistic, computed from the labels. We write $\theta = U_N$ in the sequel.

The minimum variance properties of ordinary $U$-statistics described in Section 1.1 also hold good in the case of sampling from finite populations.

Specifically, let $f(X_1, \ldots, X_n)$ be an unbiased estimator of some parameter $\theta$:

$$Ef(X_1, \ldots, X_n) = \theta.$$

The $U$-statistic based on $f$ is $U_n = (n!)^{-1} \sum_{(n)} f(X_{i_1}, \ldots, X_{i_n})$. Note that $f$ could be a function of $k < n$ variables, in which case

$$U_n = \binom{n}{k}^{-1} \sum_{(n,k)} f^*(X_{i_1}, \ldots, X_{i_k})$$

where $f^*$ is a symmetrised version of $f$. In both cases $U_n$ has the same mean $\theta$ as $f$, due to the exchangeability of the $X$'s. Moreover, $U_n$ has a smaller variance than $f$ since

$$Ef^2 = N_{(n)}^{-1} \sum_{[N,n]} f^2(x_{i_1}, \ldots, x_{i_n})$$

$$= N_{(n)}^{-1} \sum_{[N,n]} (U_n(x_{i_1}, \ldots, x_{i_n}) + \{f(x_{i_1}, \ldots, x_{i_n}) - U_n(x_{i_1}, \ldots, x_{i_n})\})^2$$

$$= \binom{N}{n}^{-1} \sum_{(N,n)} U_n^2(x_{i_1}, \ldots, x_{i_n}) + N_{(n)}^{-1} \sum_{[N,n]} (U_n - f)^2$$

$$\geq EU_n^2.$$

The symmetric unbiased estimate $U_n$ is the only symmetric unbiased estimate, just as in the i.i.d. case. This follows from Lemma A below.

**Lemma A.** *Let $g(x_1, \ldots, x_n)$ be a symmetric function of $n$ variables, and suppose that $Eg(X_1, \ldots, X_n) = 0$ for all possible values of the label values $x_1, \ldots, x_n$ . Then $g$ is identically zero.*

**Proof.** Let $G(x_1, \ldots, x_N) = \sum_{(N,n)} g(x_{i_1}, \ldots, x_{i_n})$. We need to show that $G \equiv 0$ implies that $g \equiv 0$. In view of the symmetry of $g$ (and hence of $G$), it suffices to show that $G \equiv 0$ implies that $g(x_1, \ldots, x_n) = 0$ .

First set $x_1 = \cdots = x_N = x$, say. Then $G(x, \ldots, x) = N_{(n)} g(x, \ldots, x)$ so that $g(x, \ldots, x) = 0$ for all $x$. Now assume that $g(x_1, \ldots, x_j, x, \ldots, x) = 0$ for all $x_1, \ldots, x_j$ and $x$ for $j = 1, \ldots, k$. Consider the case where $x_{k+2} =$

$\cdots = x_N = x$ and let $S_{(j)}$ denote the sum over all terms in $G(x_1, \ldots, x_N)$ that have $j$ distinct indices less than $k + 2$. Suppose that

$$0 = G(x_1, \ldots, x_{k+1}, x, \ldots, x) = \sum_{j=1}^{k+1} S_{(j)}.$$

By the induction hypothesis, $S_{(j)} = 0$ for $j = 1, \ldots, k$ so that $S_{(k+1)} = 0$. But every term in $S_{(k+1)}$ is equal to $g(x_1, \ldots, x_{k+1}, x, \ldots, x)$, and so $g(x_1, \ldots, x_{k+1}, x, \ldots, x) = 0$. Hence by induction finally $g(x_1, \ldots, x_n) = 0$ for arbitrary $x$'s and the lemma is proved.

To compute the variance of $U_n$, define $\sigma_{c,N}^2$ to be the covariance between any two kernels having $c$ variables in common:

$$\sigma_{c,N}^2 = Cov\{\psi(X_1, \ldots, X_c, X_{c+1}, \ldots, X_k), \psi(X_1, \ldots, X_c, X_{k+1}, \ldots, X_{2k-c})\}$$

$$= \frac{1}{N_{(2k-c)}} \sum_{[N, 2k-c]} \psi(x_{i_1}, \ldots, x_{i_k}) \psi(x_{i_1}, \ldots, x_{i_c}, x_{i_{k+1}}, \ldots, x_{i_{2k-c}}) - U_N^2$$

$$= \binom{N}{2k-c}^{-1} \sum_{(N, 2k-c)} \psi^{[c]}(x_{i_1}, \ldots, x_{i_{2k-c}}) - U_N^2$$

where

$$\psi^{[c]}(x_1, \ldots, x_{2k-c})$$
$$= [(2k-c)!]^{-1} \sum_{[2k-c]} \psi(x_{j_1}, \ldots, x_{j_k}) \psi(x_{j_1}, \ldots, x_{j_c}, x_{j_{k+1}}, \ldots, x_{j_{2k-c}}),$$

the last sum being taken over all permutations of $1, 2, \ldots, 2k - c$.

Thus if $U_N^{(c)}$ is the $U$-statistic computed from the $N$ labels using kernel $\psi^{[c]}$, we have

$$\sigma_{c,N}^2 = U_N^{(c)} - U_N^2 \qquad c = 0, 1, 2, \ldots, k,$$

and so

$$Var\, U_n = \binom{n}{k}^{-1} \sum_{c=0}^{k} \binom{k}{c} \binom{n-k}{k-c} \sigma_{c,N}^2$$

$$= \binom{n}{k}^{-1} \sum_{c=0}^{k} \binom{k}{c} \binom{n-k}{k-c} U_N^{(c)} - U_N^2$$

since $\sum_{c=0}^{k} \binom{k}{c}\binom{n-k}{k-c} = \binom{n}{k}$.

The usual relationship between the covariances $\sigma_{c,N}^2$ of kernels with $c$ elements in common and the variances of the conditional expectations breaks down for finite population sampling. Define as usual

$$\psi_c(x_1,\ldots,x_c) = E\{\psi(X_1,\ldots,X_k)|X_1 = x_1,\ldots,X_c = x_c\}$$

then

$$\psi_c(x_1,\ldots,x_c) = \binom{N-c}{k-c}^{-1} \sum \psi(x_1,\ldots,x_c,x_{i_{c+1}},\ldots,x_{i_k}) \qquad (2)$$

the sum extending over all $(k-c)-$ subsets $\{i_{c+1},\ldots,i_k\}$ with $c < i_{c+1} < \ldots < i_k \leq N$. We have

$$E\psi_c(X_1,\ldots,X_c) = E[E\{\psi(X_1,\ldots,X_c,\ldots,X_k)|X_1,\ldots,X_c\}]$$
$$= E\{\psi(X_1,\ldots,X_k)\}$$
$$= U_N.$$

Define also for $c = 1,\ldots,k$

$$\bar{\sigma}_{c,N}^2 = Var(\psi_c(X_1,\ldots,X_c))$$

and

$$\bar{\sigma}_{0,N}^2 = 0.$$

**Theorem 1.** *The relationship between $\bar{\sigma}_{c,N}^2$ and $\sigma_{c,N}^2$ is given by*

$$\bar{\sigma}_{c,N}^2 = \binom{N-c}{k-c}^{-1} \sum_{d=0}^{k-c} \binom{k-c}{d}\binom{N-k}{k-c-d} \sigma_{c+d,N}^2. \qquad (3)$$

**Proof.** By the above considerations we can write

$$\bar{\sigma}_{c,N}^2 = Var\{\psi_c(X_1,\ldots,X_c)\}$$
$$= E\{\psi_c^2(X_1,\ldots,X_c)\} - U_N^2.$$

Now let $A = \{\alpha_1,\ldots,\alpha_c\}$ be a $c$-subset of $\{1,2,\ldots,N\}$ and let $\psi_c(A) = \psi_c(x_{\alpha_1},\ldots,x_{\alpha_c})$. Then

$$E\{\psi_c^2(X_1,\ldots,X_c)\} = N_{(c)}^{-1} \sum_{[N,c]} \psi_c^2(x_{\alpha_1},\ldots,x_{\alpha_c})$$

$$= \binom{N}{c}^{-1} \sum_{(N,c)} \psi_c^2(x_{\alpha_1},\ldots,x_{\alpha_c})$$

$$= \binom{N}{c}^{-1} \sum_{(N,c)} \psi_c^2(A). \tag{4}$$

From (2) we have

$$\psi_c(A) = \binom{N-c}{k-c}^{-1} \sum_{\{B:A\cap B=\emptyset\}} \psi(A,B) \tag{5}$$

where the sum ranges over all $(k-c)$-subsets $B = \{\beta_1,\ldots,\beta_{k-c}\}$ of the set $\{1,2,\ldots,N\}$ that are disjoint from $A$, and where

$$\psi(A,B) = \psi(x_{\alpha_1},\ldots,x_{\alpha_c},x_{\beta_1},\ldots,x_{\beta_{k-c}}).$$

Thus, from (4) and (5)

$$E\{\psi_c^2(X_1,\ldots,X_c)\} = \binom{N}{c}^{-1}\binom{N-c}{k-c}^{-2} {\sum_{(c)}}^{\dagger} \psi(A,B)\psi(A,C)$$

where the sum ${\sum_{(c)}}^{\dagger}$ ranges over all triples (A, B, C) of subsets of the set $\{1,2,\ldots,N\}$ having the properties that
(i) $A\cap B = \emptyset$, $A\cap C = \emptyset$;
(ii) $|A| = c$, $|B| = k-c$ and $|C| = k-c$.
Since there are $\binom{N}{c}\binom{N-c}{k-c}^2$ terms in this sum we have

$$Var\{\psi_c(X_1,\ldots,X_c)\}$$
$$= \binom{N}{c}^{-1}\binom{N-c}{k-c}^{-2} {\sum_{(c)}}^{\dagger} (\psi(A,B)\psi(A,C) - U_N^2) \tag{6}$$

We may split the terms of the sum into those having exactly $d$ elements in the intersection of $B$ and $C$ for $d = 0,1,\ldots,k-c$; and so

$${\sum_{(c)}}^{\dagger}\{\psi(A,B)\psi(A,C) - U_N^2\} = \sum_{d=0}^{k-c}{\sum_{\substack{(c)\\|B\cap C|=d}}}^{\dagger} \{\psi(A,B)\psi(A,C) - U_N^2\}$$

$$= \sum_{d=0}^{k-c}\binom{c+d}{d}{\sum_{(c+d)}}^{*}\{\psi(A^*,B^*)\psi(A^*,C^*) - U_N^2\} \tag{7}$$

61

where the summation $\sum_{(c+d)}^{*}$ is over all triples of sets $(A^*, B^*, C^*)$ satisfying

(i) $A^*, B^*$ and $C^*$ are pairwise disjoint;

(ii) $|A^*| = c + d, \quad |B^*| = |C^*| = k - c - d.$

Equation (7) follows by setting $A^* = A \cup (B \cap C)$, $B^* = B - (B \cap C)$ and $C^* = C - (B \cap C)$, and the factor $\binom{c+d}{d}$ arises from the fact that every triple $A^*, B^*, C^*$ in the right hand sum gives rise to $\binom{c+d}{d}$ terms in the left hand sum since the $d$ elements of $A^*$ that are transferred to $B$ and $C$ in the left hand sum can be chosen in $\binom{c+d}{d}$ ways.

Now consider

$$
\begin{aligned}
\sigma_{c,N}^2 &= Cov\{\psi(X_1, \ldots, X_k), \psi(X_1, \ldots, X_c, X_{k+1}, \ldots, X_{2k-c})\} \\
&= E\{\psi(X_1, \ldots, X_k)\psi(X_1, \ldots, X_c, X_{k+1}, \ldots, X_{2k-c})\} - U_N^2 \\
&= (N_{(2k-c)})^{-1} \sum_{[N,2k-c]} \{\psi(x_{i_1}, \ldots, x_{i_k}) \\
&\qquad \times \psi(x_{i_1}, \ldots, x_{i_c}, x_{i_{k+1}}, \ldots, x_{i_{2k-1}}) - U_N^2\} \\
&= \binom{N}{2k-c}^{-1} \frac{c!\{(k-c)!\}^2}{(2k-c)!} \sum_{(c)}^{*} \{\psi(A,B)\psi(A,C) - U_N^2\}
\end{aligned}
$$

so that

$$
\sum_{(c)}^{*} \{\psi(A,B)\psi(A,C) - U_N^2\} = \binom{N}{2k-c} \frac{(2k-c)!}{c!\{(k-c)!\}} \sigma_{c,N}^2 \qquad (8)
$$

Substituting (8) into (7) and using (6) we get

$$
\begin{aligned}
\bar{\sigma}_{c,N}^2 &= \binom{N}{c}^{-1} \binom{N-c}{k-c}^{-2} \sum_{d=0}^{k-c} \binom{c+d}{d} \binom{N}{2k-c-d} \\
&\qquad \times \frac{(2k-c-d)!}{(c+d)!\{(k-c-d)!\}^2} \sigma_{c+d,N}^2 \\
&= \binom{N-c}{k-c}^{-1} \sum_{d=0}^{k-c} \binom{k-c}{d} \binom{N-k}{k-c-d} \sigma_{c+d,N}^2
\end{aligned}
$$

upon simplifying the binomial coefficients.

We can invert the relationship between $\bar{\sigma}_{c,N}^2$ and the quantities $\sigma_{d,N}^2$:

**Theorem 2.** *The relationship between $\bar{\sigma}_{c,N}^2$ and $\sigma_{d,N}^2$ is given by*

$$\sigma_{d,N}^2 = \binom{N-k}{k-d}^{-1} \sum_{c=d}^{k} (-1)^{c-d} \binom{k-d}{c-d} \binom{N-c}{k-c} \bar{\sigma}_{c,N}^2.$$

**Proof.** From (3) we can write

$$\bar{\sigma}_{c,N}^2 \binom{N-c}{k-c} = \sum_{d=c}^{k} \binom{k-c}{d-c} \binom{N-k}{k-d} \sigma_{d,N}^2 \tag{9}$$

Multiplying (9) by $(-1)^{c-a} \binom{k-a}{c-a}$ and summing over $c$ from $a$ to $k$, we obtain

$$\sum_{c=a}^{k} (-1)^{c-a} \binom{k-a}{c-a} \binom{N-c}{k-c} \bar{\sigma}_{c,N}^2$$

$$= \sum_{c=a}^{k} \sum_{d=c}^{k} (-1)^{c-a} \binom{k-a}{c-a} \binom{k-c}{d-c} \binom{N-k}{k-d} \sigma_{d,N}^2$$

$$= \sum_{d=a}^{k} \left\{ \sum_{c=a}^{d} (-1)^{c-a} \binom{k-a}{c-a} \binom{k-c}{d-c} \right\} \binom{N-k}{k-d} \sigma_{d,N}^2. \tag{10}$$

The term in braces in (10) is zero unless $a = d$, in which case it equals unity, so (10) reduces to $\binom{N-k}{k-a} \sigma_{a,N}^2$, proving the theorem.

From the formula

$$Var\, U_n = \binom{n}{k}^{-1} \sum_{c=0}^{k} \binom{k}{c} \binom{n-k}{k-c} \sigma_{c,N}^2 \tag{11}$$

we deduce that, assuming that $\sigma_{c,N}^2$ is uniformly bounded in N,

$$Var\, U_n = \left(1 - \frac{k^2}{n}\right) \sigma_{0,N}^2 + \frac{k^2}{n} \sigma_{1,N}^2 + o(n^{-1}). \tag{12}$$

Using the fact that $\bar{\sigma}_{0,N}^2 = 0$ it follows from Theorem 2 that

$$\sigma_{0,N}^2 = \binom{N-k}{k}^{-1} \sum_{c=1}^{k} (-1)^c \binom{k}{c} \binom{N-c}{k-c} \bar{\sigma}_{c,N}^2$$

$$= -\frac{k^2}{N} \bar{\sigma}_{1,N}^2 + o(N^{-1}) \tag{13}$$

since the boundedness of $\sigma_{c,N}^2$ entails that of $\bar{\sigma}_{c,N}^2$ by Theorem 1. Similarly

$$\sigma_{1,N}^2 = \bar{\sigma}_{1,N}^2 - (k-1)^2 \bar{\sigma}_{2,N}^2 N^{-1} + o(N^{-1}) \tag{14}$$

and so from (12), (13) and (14) we get

$$Var\, U_n = \frac{k^2(N-n)}{nN} \bar{\sigma}_{1,N}^2 + o(n^{-1}) \tag{15}$$

assuming that $nN^{-1}$ converges to a finite positive number.

It is possible to derive an exact rather than asymptotic expression for $Var\, U_n$ in terms of the quantities $\bar{\sigma}_{c,N}^2$, by substituting the formula in Theorem 2 into (11). Nandi and Sen (1963) show by means of a heroic combinatorial calculation that

$$Var\, U_n = \sum_{c=1}^{k} \frac{\binom{k}{c}\binom{n-k}{k-c}\binom{N-n}{c}}{\binom{n}{k}\binom{N-2k+c}{c}} \bar{\sigma}_{c,N}^2.$$

## 2.6 Weighted $U$-statistics

Suppose with each $k$-subset $S = \{i_1, i_2, \dots, i_k\}$ of $\{1, 2, \dots, n\}$ we associate a weight $w(S)$. Consider the weighted $U$-statistic

$$W_n = \sum_{(n,k)} w(S)\psi(S)$$

where as before $\psi(S) = \psi(X_{i_1}, \dots, X_{i_k})$. If $\sum_{(n,k)} w(S) = 1$, the statistic $W_n$ is still unbiased for $\theta = E\psi(X_1, \dots, X_k)$, since

$$EW_n = \sum_{(n,k)} w(S)E\psi(S)$$
$$= \theta.$$

However, Equation (3) of Section 1.3 for $Var\, U_n$ must be modified. Define $w_c$ by the equation

$$w_c \binom{n}{k}^{-1} \binom{n-k}{k-c}\binom{k}{c} = \sum_{|S_1 \cap S_2|=c} w(S_1)w(S_2). \tag{1}$$

64

Then arguing as in Theorem 3 of Section 1.3, we obtain

$$Var\,W_n = \binom{n}{k}^{-1} \sum_{c=1}^{k} \binom{k}{c}\binom{n-k}{k-c} w_c \sigma_c^2.$$

## Example 1. Incomplete $U$-statistics.

The most important special case is that of incomplete $U$-statistics, treated in greater detail in Section 4.3 of Chapter 4. Let $\mathcal{D}$ be a class of $k$-subsets of $\{1,2,3,\ldots,n\}$ so that $\mathcal{D} \subseteq \mathcal{S}_{n,k}$. Define weights by

$$w(S) = \begin{cases} m^{-1} & S \in \mathcal{D}, \\ 0 & S \notin \mathcal{D}, \end{cases}$$

where $m$ is the number of sets in $\mathcal{D}$. Then the incomplete $U$-statistic based on $\mathcal{D}$ is

$$U_n^{(0)} = m^{-1} \sum_{s \in \mathcal{D}} \psi(S).$$

## Example 2. Expectation of order statistics.

Let $\theta = E(X_{1:k})$ where $X_{1:k}$ is the minimum value (smallest order statistic) in a sample of size $k$. Then an estimate of $\theta$ based on a sample of size $n$ is

$$U_n = \binom{n}{k}^{-1} \sum_{(n,k)} \min(X_{i_1},\ldots,X_{i_k}).$$

We can write this $U$-statistic as a more general form of a weighted $U$-statistic. Let $R_1,\ldots,R_n$ be the ranks of the $X$'s in the sample, so that $X_{R_1} < X_{R_2} \ldots < X_{R_n}$, with probability one, provided the $X$'s have an absolutely continuous distribution. Hence

$$U_n = \binom{n}{k}^{-1} \sum_{i=1}^{n} X_{R_i} \times \{\text{number of } k-\text{subsets}$$

$$\text{for which } X_i \text{ is the smallest element}\}$$

$$= \binom{n}{k}^{-1} \sum_{i=1}^{n-k} X_{R_i} \binom{n-i}{k-1}$$

$$= \sum_{i=1}^{n} w_{D_i} X_i$$

65

where $(D_1, \ldots, D_n)$ is the permutation inverse to $(R_1, \ldots, R_n)$ and

$$w_i = \begin{cases} \binom{n-i}{k-1}/\binom{n}{k}, & i = 1, 2, \ldots, n-k; \\ 0 & \text{otherwise.} \end{cases}$$

Note that, in contrast to Example 1, the weight depends on the $X's$ through their ordering.

This statistic is a special case of more general $U$-statistics constructed out of order statistics, which are discussed further in Section 2.7. It is also a member of the class of linear $U$-statistics, which have kernels that are linear combinations of order statistics of samples of size $k$. Such statistics have been considered by Blom (1980) and David and Rogers (1983). See also Example 4 of Section 1.2.

### Example 3. Random graphs.

A random graph $K_{n,p}$ is a set of $n$ points (called vertices) and a set of random edges, constructed by joining each of the $\binom{n}{2}$ possible pairs of points independently with probability $p$. The graph can be identified with its set of edges, which in this context can be thought of as a random subset of $N_p$ elements of $\mathcal{S}_{n,2}$, where $N_p$ has a binomial distribution, with parameters $N = \binom{n}{2}$ and $p$.

Let $G$ be a fixed graph on $\{1, 2, \ldots, n\}$, i.e. a fixed subset of $\mathcal{S}_{n,2}$. A subgraph of $K_{n,p}$ is just a subset of the set of edges, and two graphs are isomorphic (regarded as subsets of $\mathcal{S}_{n,2}$) if relabelling vertices changes one graph into another. The *subgraph count* $S_n(G)$ of $G$ is the number of subgraphs of $K_{n,p}$ isomorphic to $G$.

The distribution of $S_n(G)$ is of interest to graph theorists, and we can write it as a weighted $U$-statistic as follows.

Define a set of binary random variables $X_e, \ e \in \mathcal{S}_{n,2}$ by

$$X_e = \begin{cases} 1 & \text{if } e \text{ is an edge of } K_{n,p}; \\ 0 & \text{otherwise;} \end{cases}$$

and define indicator functions for any graph $A$ by

$$I(A) = \begin{cases} 1 & \text{if } A \text{ is a subgraph of } K_{n,p}; \\ 0 & \text{otherwise;} \end{cases}$$

and
$$I(A \sim G) = \begin{cases} 1 & \text{if } A \text{ is ismorphic to } G; \\ 0 & \text{otherwise.} \end{cases}$$

The relationship between these various quantities is

$$S_n(G) = \sum_A I(A \sim G)I(A) \qquad (2)$$

where the sum is taken over all graphs $A$ on $n$ points (i.e. all subsets of $\mathcal{S}_{n,2}$). Suppose now that $G$ has $k$ edges. Then we need only consider graphs $A$ with $k$ edges in (2), since no others can be isomorphic to $G$. If we identify $A$ with its edge set $\{e_1, \ldots, e_k\} \subseteq \mathcal{S}_{n,2}$, we can write $I(A) = \prod_{i=1}^{k} X_{e_i}$ and (2) becomes

$$S_n(G) = \sum_{e_1, \ldots, e_k} w(e_1, \ldots, e_k)\psi(X_{e_1}, \ldots, X_{e_k}) \qquad (3)$$

where

$$w(e_1, \ldots, e_k) = \begin{cases} 1 & \text{if } A = \{e_1, \ldots, e_k\} \text{ is isomorphic to } G, \\ 0 & \text{otherwise;} \end{cases}$$

and $\psi(x_1, \ldots, x_k) = \prod_{j=1}^{k} x_j$, the sum being taken over all choices of $k$ subsets of $\mathcal{S}_{n,k}$. If we identify the $N = \binom{n}{2}$ elements of $\mathcal{S}_{n,2}$ with the integers $\{1, 2, \ldots, N\}$, then we can write (3) as

$$S_n(G) = \sum_{(N,k)} w(e_1, \ldots, e_k) \; g(X_{e_1}, \ldots, X_{e_k})$$

and thus $S_n(G)$ is a weighted $U$-statistic.

The above example is due to Norwicki and Wierman (1987). Norwicki (1988), Baldi and Rinott (1989) and Norwicki (1989) discuss further connections between $U$-statistic theory and graph counting problems.

## 2.7 Generalised $L$-statistics

Consider a random sample $X_1, \ldots, X_n$ from a distribution $F$, and let $\psi$ be a symmetric kernel of order $k$. If the $N = \binom{n}{k}$ quantities $\psi(S), S \in \mathcal{S}_{n,k}$ are ordered, and denoted by $W_{1:n} \geq W_{2:n} \geq \cdots \geq W_{N:n}$, we may consider statistics of the form

$$\sum_{i=1}^{N} C_{n,i} \; W_{i:n} \qquad (1)$$

where the $C's$ are fixed constants. Statistics of the form (1) may be regarded either as a generalisation of $U$-statistics, or as a generalisation of the so-called $L$-statistics, and were introduced by Serfling (1984).

The form (1) is extremely general, as the following examples show.

**Example 1. $U$-statistics.**

Putting $C_{n,i} = N^{-1}$ for $i = 1, 2, \ldots, n$ yields the $U$-statistic with kernel $\psi$.

**Example 2. Trimmed $U$-statistics.**

Let $0 < \alpha < \frac{1}{2}$ and define the $C's$ by

$$C_{n,i} = \begin{cases} 0 & i = 1, 2, \ldots, [N\alpha]; \\ 1/(N - 2[N\alpha]) & i = [N\alpha] + 1, \ldots, N - [N\alpha]; \\ 0 & i = N - [N\alpha] + 1, \ldots, N. \end{cases}$$

The resulting statistic is a "trimmed" $U$-statistic where a proportion $2\alpha$ of the most extreme kernel values is discarded from the average. The result is a more robust estimator with some protection against outliers in the $X's$. Another approach to trimmed $U$-statstics is taken by Janssen, Serfling and Veraverbeke (1987), who discard extreme $x$'s before evaluating the kernel.

**Example 3. Order statistics revisited.**

In Example 2 of Section 2.6 we estimated the expectation of an order statistic by means of a weighted $U$-statistic whose weight depended on the subset of $X$'s chosen. Alternatively, we can write the relevant $U$-statistic as

$$U_n = \sum_{i=1}^{n} C_{n,i} \ W_{i:n}$$

when $W_{i:n} = X_{i:n}$ and $C_{n,i} = \binom{n-i}{k-1}/\binom{n}{k}$. Thus $U_n$ is a linear combination of order statistics since $X_{i:n}$ is the ith smallest of $X_1, \ldots, X_n$.

**Example 4. The Hodges-Lehmann estimator.**

Let $X_1, \ldots, X_n$ be a random sample from a symmetric distribution with median $\theta$. A well-known estimator of $\theta$ is the Hodges-Lehmann estimator $\xi_n$, which is the median of the Walsh averages introduced in Example 1 of Section 1.4:

$$\xi_n = \text{median} \left\{ \frac{1}{2}(X_i + X_j) : 1 \leq i < j \leq n \right\}.$$

The estimator $\xi_n$ is of the form (1) with

$$C_{n,i} = \begin{cases} 1 & i = [n/2] \text{ or } [n/2] + 1; \\ 0 & \text{otherwise}; \end{cases}$$

for $n$ even, and

$$C_{n,i} = \begin{cases} 1 & i = [\frac{n}{2}] + 1; \\ 0 & \text{otherwise}; \end{cases}$$

for $n$ odd, and kernel $\psi(x_1, x_2) = \frac{1}{2}(x_1 + x_2)$. More general versions with $\psi(x_1, \ldots, x_k) = (x_1 + \cdots + x_k)/k$ are also possible.

The theory of statistics of the form (1) draws on the theory of ordinary $U$-statistics, but more heavily on the theory of linear combinations of order statistics, or $L$-statistics, and for this reason, statistics of the form (1) are called generalised $L$-statistics. We provide a brief summary of the theory of (ordinary) $L$-statistics below, and the reader is referred to Serfling (1980), Chapter 8, for more details.

In the theory of (ordinary) $L$-statistics of the form

$$\sum_{i=1}^{n} C_{n,i} \, X_{i:n} \qquad (2)$$

where the $X_{i:n}$ are the order statistics derived from a random sample $X_1, \ldots, X_n$ with d.f. F, the weights $C_{n,i}$ are often derived from a smooth function $J$ integrable on [0,1] via the equation

$$C_{n,i} = \int_{(i-1)/n}^{i/n} J(t)dt \qquad (3)$$

where the function $J$ is normalised so that $\int_0^1 J(t)dt = 1$.

If $F_n$ denotes the empirical distribution function of the $X's$, i.e. if

$$F_n(x) = n^{-1} \sum_{i=1}^{n} I\{X_i \le x\},$$

then $X_{i:n} = F_n^{-1}(i/n)$ and (2) can be written

$$\sum_{i=1}^{n} \int_{(i-1)/n}^{i/n} J(t)F_n^{-1}(i/n)dt = \int_0^1 F_n^{-1}(t)J(t)dt.$$

69

The statistic (2) is thus seen to be a weighted average of the sample quantities $F_n^{-1}(i/n)$, with weighting function $J$. Since $F_n$ converges to $F$ in all reasonable senses, we might expect that (2) converges to $\int_0^1 F^{-1}(t)J(t)dt$. Under fairly weak conditions this turns out to be the case, and (2) is a consistent estimator of $\int_0^1 F^{-1}(t)J(t)dt$.

Now consider general functionals $T$ defined on some set $\mathcal{F}$ of distribution functions. In many situations a good estimator of $T(F)$ is $T(F_n)$, where $F_n$ is the empirical distribution function defined above. $L$-statistics are of this type, as are $V$-statistics, defined in Section 4.2.1. The asymptotic variance of $T(F_n)$ may be derived heuristically as follows.

If $G$ is a distribution function, the *Gateaux derivative of $T$ at $F$ in the direction $G$* is the quantity

$$\frac{d}{dt}T((1-t)F + tG)\,\Big|_{t=0}\,.$$

For many functionals $T$, the above derivative exists, and for some function $a$ is given by

$$\frac{d}{dt}T((1-t)F + tG)|_{t=0} = \int a(x)dG.$$

The function $a(x)$ depends on $F$ and $T$ but not on $G$, and is called the first kernel function of $T$ at $F$.

The first kernel function $a(x)$ is often called the *influence curve* and is also given by

$$IC(x; F, T) = \frac{d}{dt}T((1-t)F + tG_x)\,\Big|_{t=0}$$

where $G_x$ is the d.f. corresponding to a unit point mass at $x$. For functionals

$$T(F) = \int_0^1 F^{-1}(t)J(t)dt \tag{4}$$

the influence curve takes the form (see Serfling (1980) p265)

$$IC(x; F, T) = -\int_{-\infty}^{\infty} (I\{x \le y\} - F(y))J(F(y))dy.$$

These functionals may be expanded as a Taylor series about $t = 0$:

$$T((1-t)F + tG) = T(F) + t\int a(x)dG + R_G(t).$$

70

If we put $G = F_n$ and set $t = 1$ we obtain

$$T(F_n) = T(F) + n^{-1} \sum_{i=1}^{n} IC(X_i, F, T) + R_n \quad \text{say.}$$

If $R_n$ is asymptotically negligible, then the asymptotic behaviour of $T(F_n) - T(F)$ will be that of $n^{-1} \sum_{i=1}^{n} IC(X_i; F, T)$, which will be asymptotically normal with mean zero (since $E(IC(X_i; F, T)) = 0$ and asymptotic variance given by

$$n^{-1} Var(IC(X_i; F, T)) = n^{-1} \int_{-\infty}^{\infty} (IC(x; F, T))^2 dF(x).$$

For functionals of the form (4) the asymptotic variance is

$$n^{-1} \int_{-\infty}^{\infty} (F(\min(x, y)) - F(x)F(y))J(F(x))J(F(y))dy,$$

provided the integral is positive. Turning now to the case of generalised $L$-statistics, let $H(y)$ be the distribution function of $\psi(X_1, \dots, X_k)$ and let $H_n(y)$ be the "empirical d.f." of the realisations $\psi(S), S \in \mathcal{S}_{n,k}$:

$$H_n(y) = \binom{n}{k}^{-1} \sum_{(n,k)} I\{\psi(S) \le y\}$$

so that $H_n(y)$ is a $U$-statistic. If the weights in (1) are generated by a function $J$ as in (3) and $T$ is a functional of type (4), then the generalised $L$-statistic (1) may be written as

$$\sum_{i=1}^{N} C_{n,i} \; H_n^{-1}(i/N) = \int_0^1 H_n^{-1}(t)J(t)dt = T(H_n) \tag{5}$$

which estimates $\int_0^1 H^{-1}(t)J(t)dt = T(H)$. As in the case of (ordinary) $L$-statistics, the quantity $T(H_n) - T(H)$ is approximated by

$$\binom{n}{k}^{-1} \sum_{(n,k)} IC(\psi(X_{i_1}, \dots, X_{i_k}), H, T).$$

whose asymptotic variance can be found by the methods of Section 1.3 and is in fact $k^2 \sigma_1^2 n^{-1}$ where

$$\sigma_1^2 = Var \; E(IC(\psi(X_1, \dots, X_k), H, T)|X_1). \tag{6}$$

71

**Example 5.  Trimmed $U$-statistics (continued).**

Asymptotically equivalent to the statistic of Example 2 is the statistic

$$\frac{1}{1-\alpha}\int_{\alpha}^{1-\alpha}H_n^{-1}(t)dt$$

which is of the form $T(H_n)$ where $T$ is as in (4) and

$$J(t) = \begin{cases} 1/(1-2\alpha) & \alpha < t < 1-\alpha; \\ 0 & \text{otherwise}; \end{cases}$$

where $0 < \alpha < \frac{1}{2}$.

Another type of (ordinary) $L$-statistic in common use is that of the form

$$\sum_{i=1}^{m}c_iX_{\{np_i\}:n} \tag{7}$$

where $c_1,\ldots,c_m$ are fixed constants, $p_1,\ldots,p_m$ are probabilities, and $\{x\}$ denotes the smallest integer greater than or equal to $x$. Since $X_{\{np_i\}:n} = F_n^{-1}(p_i)$, (7) can be written

$$\sum_{i=1}^{m}c_iF_n^{-1}(p_i)$$

and thus estimates the functional

$$\sum_{i=1}^{m}c_iF^{-1}(p_i). \tag{8}$$

The influence curve of such functionals is (see e.g. Serfling (1980) p265)

$$IC(x;F,T) = \sum_{i=1}^{m}c_i(p_i - I\{x < F^{-1}(p_i)\})/f(F^{-1}(p_i)) \tag{9}$$

where $f$ is the density of $F$, assumed positive at $F^{-1}(p_1),\ldots,F^{-1}(p_m)$.

Generalised $L$-statistics of this type can be handled as in the previous case. We illustrate with an example.

**Example 6. The Hodges-Lehmann estimator (continued).**

The Hodges-Lehmann estimator of Example 2 can be written

$$\xi_n = H_n^{-1}(1/2)$$

where $H_n$ is the empirical d.f. of the kernel $\psi(x_1, x_2) = \frac{1}{2}(x_1 + x_2)$, so that

$$H_n(y) = \binom{n}{2}^{-1} \sum_{1 \le i < j \le n} I\{X_i + X_j \le 2y\},$$

and $H(y) = Pr(X_1 + X_2 \le 2y)$. If the underlying d.f. F is symmetric with median $\theta$, then $H(y)$ will also be symmetric with median $\theta$, so that $H^{-1}\left(\frac{1}{2}\right) = \theta$. Using the above heuristic arguments, we might expect the asymptotic variance of $\xi_n$ to be $4\sigma_1^2 n^{-1}$, where $\sigma_1^2$ is given by (6).

If $f$ denotes the underlying density of the $X's$, write $f(x) = g(x - \theta)$, so that $g$ is the symmetric density of the r.v. $Y = X - \theta$ which has median zero. If $h$ is the density of the r.v. $\frac{1}{2}(X_1 + X_2)$ we have

$$h(x) = 2 \int_{-\infty}^{\infty} f(y) f(2x - y) dy$$

and so

$$h(\theta) = 2 \int_{-\infty}^{\infty} g(y - \theta) g(\theta - y) dy$$

$$= 2 \int_{-\infty}^{\infty} g^2(y) dy. \tag{10}$$

Thus by (6) and (9)

$$E[IC(\psi(X_1, x), H, T)]$$
$$= E\left[\left(\frac{1}{2} - I\left\{\psi(X, x) < H^{-1}\left(\frac{1}{2}\right)\right\}\right)\right] / h\left(H^{-1}\left(\frac{1}{2}\right)\right)$$
$$= E\left[\left(\frac{1}{2} - I\left\{\frac{1}{2}(X_1 + x) < \theta\right\}\right)\right] / h(\theta)$$
$$= \left(\frac{1}{2} - F(2\theta - x)\right) / h(\theta)$$
$$= \left(\frac{1}{2} - G(\theta - x)\right) / h(\theta)$$

where $G$ is the d.f. of $Y = X - \theta$. Hence

$$\sigma_1^2 = Var\{\left(\frac{1}{2} - G(\theta - x)\right) / h(\theta)\}$$
$$= Var\{G(-Y)\}/h^2(\theta)$$
$$= Var\{1 - G(Y)\}/h^2(\theta)$$
$$= \left(\int_{-\infty}^{\infty} g^2 dx\right)^{-2} / 48$$

73

by (10), since $Y$ is symmetric and $G(Y)$ is uniformly distributed, and so the asymptotic variance of $\xi_n$ is $(\int_{-\infty}^{\infty} g^2 dx)^{-2}/12$.

## 2.8 Bibliographic details

Generalised $U$-statistics were introduced in Hoeffding (1948a) and their properties further developed in Lehmann (1951). The results on $U$-statistics based on independent but not identically distributed r.v.s are again taken from Hoeffding (1948a). Section 2.4 is based on Sen (1963) for the M-dependent case and on Yoshihara (1976) in the absolutely regular case. Nandi and Sen (1963) provided the basis for Section 2.5 as did Serfling (1984) for Section 2.7. The interesting application in Section 2.6 of weighted $U$-statistics to random graphs is due to Nowicki and Weirman (1987).

# CHAPTER THREE

## Asymptotics

### 3.1 Introduction

Many of the classical results of probability theory dealing with of sums of independent and identically distributed random variables have generalisations to $U$-statistics.

In this connection, we mention the Central Limit Theorem, together with associated results concerning rates of convergence; Poisson convergence of the binomial, the weak and strong laws of large numbers and the law of the iterated logarithm, together with related invariance principles, all of which have "$U$-statistic" versions.

The limit laws for sequences of $U$-statistics are in fact only partly analogous to those for sequences of sums of i.i.d. random variables. Many but not all $U$-statistics can be written as a sum of i.i.d. random variables plus a small perturbation, and for these the limit theory is more or less the same as for the i.i.d. case. However for others the limit theory is rather more complex. We begin the Chapter with a treatment of convergence in distribution, then proceed to $U$-statistic versions of the SLLN and the LIL, and then to invariance principles. The chapter continues with a discussion of asymptotic results for the $U$-statistic variations discussed in Chapter 2, and concludes with a brief treatment of $U$-statistics whose kernels contain estimated parameters. Our main tool throughout will be the $H$-decomposition.

### 3.2 Convergence in distribution of $U$-statistics

The basic tool used in this section is the $H$-decomposition, the properties of which determine the nature of the asymptotic distributions. We begin by looking at the class of $U$-statistics having the simplest asymptotic behaviour, those having asymptotically normal distributions.

#### 3.2.1 Asymptotic normality

The basic result, due to Hoeffding (1948a), is the asymptotic normality of $U$-statistics with $\sigma_1^2 > 0$, which thus have a non-zero first component in the $H$-decomposition.

**Theorem 1.** *Let $\sigma_1^2 > 0$. Then $n^{\frac{1}{2}}(U_n - \theta)$ is asymptotically normal with mean zero and asymptotic variance $k^2 \sigma_1^2$.*

**Proof.** From the $H$-decomposition, we may write

$$\sqrt{n}(U_n - \theta) = \sqrt{n}(kH_n^{(1)} + R_n)$$

$$= \frac{k}{\sqrt{n}} \sum_{i=1}^{n} h^{(1)}(X_i) + \sqrt{n} R_n$$

where $R_n = \sum_{j=2}^{k} \binom{k}{j} H_n^{(j)}$. The variance of $\sqrt{n} R_n$ is thus $n \sum_{j=2}^{k} \binom{k}{j}^2 \binom{n}{j}^{-1} \delta_j^2$ by Theorem 4 of Section 1.6, so $Var\, n^{1/2} R_n = O(n^{-1})$. Hence by Slutsky's theorem, the asymptotic behaviour of $n^{1/2}(U_n - \theta)$ is the same as that of $kn^{-1/2} \sum_{i=1}^{n} h^{(1)}(X_i)$. Since $Eh^{(1)}(X_1) = 0$ and $Var\, h^{(1)}(X_i) = \sigma_1^2 > 0$ by (7) and (8) of Section 1.6, the result follows from the central limit theorem for independent random variables.

**Example 1. Sample variance.**

From Example 2 of Section 1.3 we see that the sample variance will be asymptotically normally distributed with mean $\sigma^2$ and asymptotic variance $(\mu_4 - \sigma^4)/n$ provided $\mu_4 > \sigma^4$.

**Example 2. Kendall's Tau.**

From Example 5 of Section 1.3 the asymptotic distribution of $t_n$ under independence is normal with mean zero and variance $4/9n$.

Our next result is a multivariate version of Theorem 1.

**Theorem 2.** *Let $U_n^{(j)}$, $j = 1, \ldots, m$ be $U$-statistics having expectations $\theta_j$ and kernels $\psi^{(j)}$ of degrees $k_j$. Also let $\Sigma = (\sigma_{i,j})$ where*

$$\sigma_{i,j} = k_i k_j Cov(\psi^{(i)}(X_1, \ldots, X_{k_i}), \psi^{(j)}(X_{k_i}, \ldots, X_{k_i + k_j - 1})),$$

*and denote by $\mathbf{U}_n$ and $\boldsymbol{\theta}$ the $m$-vectors $(U_n^{(1)}, \ldots, U_n^{(m)})$ and $(\theta_1, \ldots, \theta_m)$ respectively. Then $n^{\frac{1}{2}}(\mathbf{U}_n - \boldsymbol{\theta})$ converges in distribution to a multivariate normal distribution with mean vector zero and covariance matrix $\Sigma$.*

**Proof.** Let $h_j^{(1)}$ be the first kernel function appearing in the $H$-decomposition of $U_n^{(j)}$, so that for each $j$ we have

$$n^{\frac{1}{2}}(U_n^{(j)} - \theta_j) = n^{-\frac{1}{2}} k_j \sum_{l-1}^{n} h_j^{(1)}(X_l) + o_p(1), \quad j = 1, \ldots, m. \qquad (1)$$

If $\mathbf{Y}_l = (k_1 h_1^{(1)}(X_l), \ldots, k_m h_m^{(1)}(X_l))$, then from (1) we get

$$n^{\frac{1}{2}}(\mathbf{U}_n - \theta) = n^{-\frac{1}{2}} \sum_{l=1}^{n} \mathbf{Y}_l + o_p(1).$$

Since the results of Section 2.2 imply that

$$\sigma_{i,j} = k_i k_j Cov(h^{(i)}(X_1), h^{(j)}(X_1))$$

the theorem now follows from the multivariate CLT for i.i.d. random vectors.

Normal limits are possible under more general conditions, where the kernels are allowed to depend on the indices of the random variables. Barbour and Eagleson (1985) discuss this situation, and consider applications to multiple comparison problems. See Chapter Six for more on this subject. Rao Jammalamadaka and Janson (1986) study triangular schemes of $U$-statistics and obtain infinitely divisible limit laws. See also Weber (1983). Frees (1989) allows the degree of the kernel to increase with the sample size. Nolan and Pollard (1986,1987) consider "$U$-processes" where the $U$-statistic sequence is regarded as a sequence of random processes indexed by its kernel.

In the classical case of sums $S_n$ of i.i.d. random variables, the ideas behind the central limit theorem can be extended in various ways. For example, we may consider rates of convergence, leading to the Berry-Esseen theorem and asymptotic expansions. $U$-statistic versions of these results are considered in Section 3.3. If the second moment of $X_1$ does not exist, the CLT is not directly applicable, but if the the the d.f. $F$ behaves correctly when $x \to \pm\infty$, then it is possible to find norming constants $A_n, B_n$ such that $A_n(S_n - B_n)$ converge in distribution to a so-called stable law. A $U$-statistic version of this theory is given in Malevich and Abdalimov (1977).

In another direction, considerable effort has been devoted to establishing the rate of convergence to zero of large deviation probabilities of the form $Pr(\sqrt{n}(\overline{X} - \mu)/\sigma > x_n)$ as both $n$ and $x_n$ increase. Such results have applications in several places in statistics, for example in the calculation of asymptotic relative efficiency. Serfling (1980) Section 5.6 has a discussion of similar results for $U$-statistics. Malevich and Abdalimov (1979) and Vandemaele (1982) discuss further refinements.

### 3.2.2 First-order degeneracy

When $\sigma_1^2 = 0$, but $\sigma_2^2 > 0$, the $U$-statistic is said to possess *first order degeneracy*. Under these circumstances, the first term in the $H$-decomposition vanishes almost surely, since $\sigma_1^2 = 0$ entails $h^{(1)}(x) = 0$ a.s. We may then write

$$n(U_n - \theta) = \binom{k}{2} n H_n^{(2)} + n R_n$$

where now $R_n = \sum_{j=3}^{n} \binom{k}{j} H_n^{(j)}$. It is readily seen that $Var\, n R_n = O(n^{-1})$ so that $n(U_n - \theta)$ and $\binom{k}{2} n H_n^{(2)}$ have the same asymptotic behaviour. Thus without loss of generality we may take $\theta = 0$ and $k = 2$ in studying the asymptotic behaviour of $U$-statistics with first order degeneracy.

We begin by studying several special cases, which indicate the method to be followed in general:

**Example 1.**

Let $\mu = EX_1$, $\sigma^2 = Var\, X_1$ and $U_n$ the $U$-statistic based on the kernel $\psi(X_i, X_j) = X_i X_j$. Then $\sigma_1^2 = Cov(X_1 X_2, X_2 X_3) = \mu^2 \sigma^2$ and $\sigma_2^2 = (\sigma^2 + \mu^2)^2 - \mu^4$, so if $\mu = 0$ the $U$-statistic has first-order degeneracy. Consider

$$nU_n = n \binom{n}{2}^{-1} \sum_{1 \le i < j \le n} X_i X_j$$

$$= \frac{n}{n-1} \left\{ \left( \sum_{i=1}^{n} X_i / \sqrt{n} \right)^2 - \sum_{i=1}^{n} X_i^2 / n \right\}.$$

Provided $\mu = 0$, the CLT ensures that the first term in the braces converges in distribution to a normal r.v. with mean 0 and variance $\sigma^2$; the second term converges in probability to $\sigma^2$ by the WLLN. Hence by Slutsky's theorem $U_n \xrightarrow{D} \sigma^2(Z^2 - 1)$, where $Z$ is a standard normal r.v., and where $\xrightarrow{D}$ denotes convergence in distribution. We will also use the notation $\xrightarrow{p}$ to denote convergence in probability.

**Example 2.**

Now let $\psi(x_1, x_2) = f(X_1) f(X_2)$. By the same arguments as those used above, if $Ef(X_1) = 0$ then $nU_n \xrightarrow{D} \sigma^2(Z^2 - 1)$ where $\sigma^2 = Var\, f(X_1)$.

78

**Example 3.**

Let $\psi(x_1, x_2) = af(x_1)f(x_2) + bg(x_1)g(x_2)$ and suppose that $Ef(X_1) = Eg(X_1) = Ef(X_1)g(X_1) = 0$ and that $Ef^2(X_1) = Eg^2(X_1) = 1$. Then $\sigma_1^2 = 0$ and

$$\sigma_2^2 = Var(af(X_1)f(X_2) + bg(X_1)g(X_2))$$
$$= a^2 Var f(X_1)f(X_2) + b^2 Var g(X_1)g(X_2)$$
$$= a^2 + b^2 > 0$$

provided $\psi(x_1, x_2)$ is not identically zero. Computing as in Example 1, we obtain

$$nU_n = \frac{n}{n-1}\left\{a\left(\sum_{i=1}^{n} \frac{f(X_i)}{\sqrt{n}}\right)^2 + b\left(\sum_{i=1}^{n} \frac{g(X_i)}{\sqrt{n}}\right)^2 \right.$$
$$\left. - a\sum_{i=1}^{n} \frac{f^2(X_i)}{n} - b\sum_{i=1}^{n} \frac{g^2(X_i)}{n}\right\}.$$

Now by the multivariate CLT, $(\sum f(X_i)/\sqrt{n}, \sum g(X_i)/\sqrt{n}) \xrightarrow{D} (Z_1, Z_2)$ where $Z_1$ and $Z_2$ are independently distributed as $N(0,1)$ and

$$a\sum f^2(X_i)/n + b\sum g^2(X_i)/n \xrightarrow{P} a + b$$

by the WLLN and Slutsky's theorem. Thus, since $az_1^2 + bz_2^2$ is a continuous function of $(z_1, z_2)$, $nU_n$ converges in distribution to $a(Z_1^2 - 1) + b(Z_2^2 - 1)$.

Now as described below, it follows from the Fredholm theory of integral equations that any symmetric function of two variables admits a series expansion of the form

$$\psi(x_1, x_2) = \sum_{\nu=1}^{\infty} \lambda_\nu f_\nu(x_1) f_\nu(x_2). \tag{1}$$

This suggests we may be able to apply the arguments of Example 3 to the terms of the series (1) and obtain a limit law of the form $\sum \lambda_\nu(Z_\nu^2 - 1)$. This is indeed the case, and the details are given in Theorem 1.

**Theorem 1.** Let $U_n$ be a $U$-statistic with mean zero based on a kernel $h(x_1, x_2)$ with $Eh(x_1, X_2) = 0$ (and hence $\sigma_1^2 = 0$), $Eh^2(X_1, X_2) < \infty$, and

79

$Eh(X_1, X_2) = 0$. Then the normalised statistic $nU_n$ converges in distribution to a r.v. of the form

$$\sum_{\nu=1}^{\infty} \lambda_\nu (Z_\nu^2 - 1) \qquad (2)$$

where $Z_1, Z_2 \ldots$ are independent standard normal random variables, and the $\lambda_\nu$ are the eigenvalues of the integral equation

$$\int h(x_1, x_2) f(x_2) dF(x_2) = \lambda f(x_1).$$

**Proof.** From the Fredholm theory of integral equations, it follows that there exist possibly finite sequences of eigenvalues and eigenfunctions which we denote repectively by $\lambda_\nu$ and $f_\nu$, such that $\int h(x_1, x_2) f_2(x_2) dF(x_2) = \lambda_\nu f_\nu(x_2)$ for $\nu = 0, 1, \ldots$ . The kernel $h$ admits the expansion

$$h(x_1, x_2) = \sum_{\nu=1}^{\infty} \lambda_\nu f_\nu(x_1) f_\nu(x_2)$$

which converges in mean square in the sense that

$$\int |h(x_1, x_2) - \sum_{\nu=1}^{K} \lambda_\nu f_\nu(x_1) f_\nu(x_2)|^2 dF(x_1) dF(x_2) \to 0$$

as $K \to \infty$. Moreover, the eigenfunctions form an orthonormal set, so that

$$\int f_\nu(x_1) f_\mu(x_1) dF(x_1) = \begin{cases} 0 & \mu \neq \nu, \\ 1 & \mu = \nu. \end{cases}$$

Note also since $\int h(x_1, x_2) dF(x_1) \equiv 0$, the constant function 1 is an eigenfunction corresponding to the eigenvalue zero. Further, the sequence of eigenvalues is square-summable.

Now for each K, define the kernel $h_K(x_1, x_2)$ by

$$h_K(x_1, x_2) = \sum_{\nu=1}^{K} \lambda_\nu f_\nu(x_1) f_\nu(x_2)$$

and let $U_{K,n}$ be the corresponding $U$-statistic:

$$U_{K,n} = \binom{n}{2}^{-1} \sum_{(n,2)} h_K(X_i, X_j) = \sum_{\nu=1}^{K} \lambda_\nu \binom{n}{2}^{-1} \sum_{(n,2)} f_\nu(X_i) f_\nu(X_j)$$

$$= \sum_{\nu=1}^{K} \lambda_\nu T_{\nu,n}, \qquad (3)$$

80

where $T_{\nu,n}$ is the $U$-statistic based on the kernel $f_\nu(x_1)f_\nu(x_2)$.

Note that $nT_{\nu,n}$ can be written

$$nT_{\nu,n} = \frac{1}{n-1}\left(\sum_{i=1}^{n}\sum_{j=1}^{n}f_\nu(X_i)f_\nu(X_j) - \sum_{i=1}^{n}f_\nu^2(X_i)\right)$$

$$= \frac{n}{n-1}\left\{\left(\sum_{i=1}^{n}f_\nu(X_i)/\sqrt{n}\right)^2 - \sum_{i=1}^{n}f_\nu^2(X_i)/n\right\}.$$

Now set $\mathbf{Z}_i^T = (Z_{1i},\ldots,Z_{Ki})$ where $Z_{\nu i} = f_\nu(X_i)$. The random vectors $\mathbf{Z}_i$ are independently and identically distributed with mean vector zero $(Ef_\nu(X_i) = 0$ since the $f_\nu$ are orthonormal and 1 is an eigenfunction) and variance-covariance matrix the identity, since

$$Cov(f_\nu(X_i), f_\mu(X_i)) = \int_{-\infty}^{\infty}f_\nu(x_1)f_\mu(x_1)dF(x_1)$$

and the eigenfunctions are orthonormal.

Thus by the multivariate central limit theorem, the vector $\overline{\mathbf{Z}}_n$ defined by

$$\overline{\mathbf{Z}}_n = n^{-\frac{1}{2}}\sum_{i=1}^{n}\mathbf{Z}_i^T = (\overline{Z}_{1n},\ldots,\overline{Z}_{Kn})$$

converges in distribution to a vector $\mathbf{Z} = (Z_1,\ldots,Z_K)$ having a multivariate normal distribution with mean vector 0 and dispersion matrix equal to an identity matrix.

Now we can write

$$nT_{\nu,n} = \frac{n}{n-1}(\overline{Z}_{\nu,n}^2 - S_{\nu,n}) \tag{4}$$

where $S_{\nu,n} = \sum_{i=1}^{n}f_\nu^2(X_i)/n$. The r.v. $S_{\nu,n}$ converges in probability to $Ef_\nu^2(X_i) = 1$ as $n \to \infty$ for each fixed $\nu$, by the weak law of large numbers. Using (3) and (4) we get

$$nU_{n,K} = \sum_{\nu=1}^{K}\lambda_\nu nT_{\nu,n} = \frac{n}{n-1}\sum_{\nu=1}^{K}\lambda_\nu\overline{Z}_{\nu,n}^2 - \frac{n}{n-1}\sum_{\nu=1}^{K}\lambda_\nu S_{\nu n}.$$

The first term converges to $\sum_{\nu=1}^{K} \lambda_\nu Z_i^2$ since the term is a continuous function of the $\overline{Z}_{\nu,n}$'s and the second to $\sum_{\nu=1}^{K} \lambda_\nu$, because the limit in probability of a finite sum is the sum of the limits in probability. Hence

$$nU_{n,K} \xrightarrow{D} \sum_{\nu=1}^{K} \lambda_\nu(Z_\nu^2 - 1).$$

Now we turn to an analysis of $nU_{n,K}$ and $nU_n$. Consider

$$E|nU_{n,K} - nU_n|^2 = E\left|n \sum_{\nu=K+1}^{\infty} \lambda_\nu T_{\nu,n}\right|^2$$

$$= \frac{2n}{n-1} \sum_{\nu=K+1}^{\infty} \lambda_\nu^2 \tag{5}$$

since $Cov(T_{\nu,n}, T_{\mu,n}) = \binom{n}{2}^{-1}$ if $\mu = \nu$, and zero otherwise, by Theorem 2 of Section 1.4.

Since the series in (5) converges, it follows that $nU_{n,K} \to nU_n$ uniformly in mean square as $K \to \infty$, and hence in distribution. Now let $\phi_n$ and $\phi_{n,K}$ be the characteristic functions of $nU_n$ and $nU_{n,K}$ respectively. Then

$$|\phi_n(t) - \phi_{n,K}(t)| = |E(e^{itnU_n} - e^{itnU_{n,K}})|$$

$$\leq E|e^{itn(U_n - U_{n,K})} - 1|$$

$$\leq |t|E|nU_n - nU_{n,K}|$$

$$\leq |t|\{E|nU_n - nU_{n,K}|^2\}^{\frac{1}{2}}$$

$$\leq |t|\left\{\frac{2n}{n-1} \sum_{\nu=K+1}^{\infty} \lambda_\nu^2\right\}^{\frac{1}{2}}$$

by (5). Hence, given $\varepsilon > 0$ we can choose $K_1$ so that $|\phi_n(t) - \phi_{n,K}(t)| < \varepsilon/3$ for all $n$ and all $K \geq K_1$. Similarly, we may choose $K_2$ so that

$$|\phi_K(t) - \phi(t)| < \varepsilon/3 \quad \text{for all} \quad K \geq K_2,$$

where $\phi$ is the c.f. of the r.v. (2), and $\phi_K$ is the c.f. of the r.v. (2) truncated to $K$ terms. Finally, since $nU_{n,K}$ converges in distribution to $\sum_{\nu=1}^{K} \lambda_\nu(Z_\nu^2 - 1)$ for all $K$, we can find $N$ such that for all $n > N$

$$|\phi_{n,K_0}(t) - \phi_{K_0}(t)| < \varepsilon/3$$

82

where $K_0 = \max(K_1, K_2)$. Then for all $n > N$,

$$|\phi_n(t) - \phi(t)| \leq |\phi_n(t) - \phi_{n,K_0}(t)| + |\phi_{n,K_0}(t) - \phi_{K_0}(t)| + |\phi_{K_0}(t) - \phi(t)| < \varepsilon$$

proving the theorem.

**Corollary 1.** Let $U_n$ be a $U$-statistic based on kernel $\psi(x_1, \ldots, x_k)$, and suppose that $\sigma_1^2 = 0, \sigma_2^2 > 0$. Then

$$n(U_n - \theta) \xrightarrow{D} \binom{k}{2} \sum_{\nu=1}^{\infty} \lambda_\nu (Z_\nu^2 - 1)$$

where the $Z_\nu$ are i.i.d. $N(0,1)$ and the $\lambda_\nu$ are the eigenvalues of the integral equation

$$\int_{-\infty}^{\infty} (\psi_2(x_1, x_2) - \psi_1(x_1) - \psi_1(x_2) + \theta) f(x_2) dF(x_2) = \lambda f(x_1).$$

**Proof.** The function $\psi_2(x_1, x_2) - \psi_1(x_1) - \psi_1(x_2) + \theta$ is just the function $h^{(2)}(x_1, x_2)$ which is the kernel of $H_n^{(2)}$ in the $H$-decomposition of $U_n$. Since $n(U_n - \theta)$ and $\binom{k}{2} n H_n^{(2)}$ have the same asymptotic distribution, the result follows from Theorem 1.

### 3.2.3 The general case.

Suppose that $U_n$ is a $U$-statistic based on a kernel $\psi$ of degree $k$, and suppose that $U_n$ has a degeneracy of order $d - 1$, i.e. that $\sigma_c^2 = 0$ for $c = 0, \ldots, d-1$ and $\sigma_d^2 > 0$. Then from Section 1.6, $n^{\frac{d}{2}}(U_n - \theta) = \binom{k}{d} n^{\frac{d}{2}} H_n^{(d)} + n^{\frac{d}{2}} R_n$ where $n^{\frac{d}{2}} R_n \xrightarrow{P} 0$. The $U$-statistic $H_n^{(d)}$ has kernel $h^{(d)}(x_1, \ldots, x_d)$ which satisfies

$$\int \cdots \int h^{(d)}(x_1, \ldots, x_d) \prod_{i=1}^{c} dF(x_i) \equiv 0$$

for $c = 1, 2, \ldots, d-1$. Thus in deriving the limit distribution of $U$-statistics with higher-order degeneracies, we can confine attention to the case when the order of the degeneracy is one less than the degree of the kernel.

As the order of the degeneracy increases, the asymptotic distributions become progressively more complicated, although they may all be derived by the same methods. Two possible derivations are known; one is a generalization of the methods used in the case $d = 2$, due to Rubin and Vitale (1980) , while the other, due to Dynkin and Mandelbaum (1983) uses the ingenious device of Poisson randomisation of the sample size and expresses the limit in the form of a Wiener integral.

In this section we present a discussion of the former method, while the latter is treated in Section 4.2 in the more general context of symmetric statistics. It is instructive to first consider a series of special cases.

**Example 1.**

Let $X_1, X_2, \ldots$ be i.i.d. with $EX_i = 0$ and $Var\, X_i = 1$, and define a kernel $\psi$ by $\psi(x_1, x_2, x_3) = x_1 x_2 x_3$. Then $\sigma_1^2 = \sigma_2^2 = 0$ but $\sigma_3^2 = 1 > 0$ so that the order of the degeneracy is 2. We may write

$$\sum_{(n,3)} X_{i_1} X_{i_2} X_{i_3} = \frac{1}{6}\left\{ (\sum_{i=1}^n X_i)^3 - 3(\sum_{i=1}^n X_i^2)(\sum_{i=1}^n X_i) + 2\sum_{i=1}^n X_i^3 \right\}$$

so that if $S_\nu = \sum_{i=1}^n X_i^\nu, \nu = 1, 2, 3$ we have

$$n^{\frac{3}{2}} \, U_n = n^3 \{n(n-1)(n-2)\}^{-1} \Big\{ (n^{-\frac{1}{2}} S_1)^3 - $$
$$3(n^{-1} S_2)(n^{-\frac{1}{2}} S_1) + 2(n^{-1} S_3)n^{-\frac{1}{2}} \Big\}.$$

Let $Z$ be a standard normal random variable. By the the central limit theorem and the weak law of large numbers, $n^{-\frac{1}{2}} S_1 \xrightarrow{D} Z$, $n^{-1} S_2 \xrightarrow{P} E(X_1^2)$ and $n^{-1} S_3 \xrightarrow{P} E(X_1^3)$ and so $n^{\frac{3}{2}} U_n \xrightarrow{D} Z^3 - 3Z$.

**Example 2.**

In a similar manner, if $\psi(x_1, x_2, x_3, x_4) = x_1 x_2 x_3 x_4$ then $0 = \sigma_1^2 = \sigma_2^2 = \sigma_3^2 < \sigma_4^2$ and $n^2 U_n \xrightarrow{D} Z^4 - 6Z^2 + 3$.

**Example 3.**

More generally, consider a kernel of the form

$$\psi(x_1, \ldots, x_k) = f_1(x_1) f_2(x_2) \ldots f_k(x_k)$$

where for the moment, the functions $f_1, \ldots, f_k$ are assumed to be distinct non-constant functions satisfying

$$\int |f_{i_1}(x) \ldots f_{i_j}(x)| \, dF(x) < \infty$$

84

for any subset $\{i_1, \ldots, i_j\}$ of $\{1, 2, \ldots, k\}$, and are further assumed to be orthonormal with respect to $F$ and in addition to satisfy $Ef_i(X_1) = 0$. If $U_n$ is the $U$-statistic based on a symmetrised version of $\psi$, then $U_n$ will have $\sigma_c^2 = 0$ for $c = 1, 2, \ldots, k-1$ and $\sigma_k^2 > 0$. We would thus expect $n^{\frac{k}{2}} U_n$ to have a non-degenerate limit and this turns out to be the case. Using a formula of Rubin and Vitale (1980), we can write (the sum $\sum_{[n,k]}$ denoting summatation of all permutations $(i_1, \ldots, i_k)$ of $\{1, 2, \ldots, n\}$)

$$\sum_{[n,k]} f_1(x_{i_1}) \ldots f_k(x_{i_k}) = \sum_{\mathcal{P}} \prod_{V \in \mathcal{P}} (-1)^{|V|-1} (|V| - 1)! \; S(V)$$

where the sum on the right is taken over all partitions $\mathcal{P}$ of $\{1, 2, \ldots, k\}$ into disjoint subsets, and $S(V) = \sum_{i=1}^{n} f_{\nu_1}(X_i) \ldots f_{\nu_p}(X_i)$ for $V = \{\nu_1, \ldots, \nu_p\}$. Consider a partition $\mathcal{P}$ having $j$ subsets, for which $j_1$ are of size $1, \ldots, j_k$ are of size $k$ so that $j_1 + \cdots + j_k = j$ and $k = j_1 + 2j_2 + \cdots + kj_k$. Then

$$n^{-k/2} \prod_{V \in \mathcal{P}} (-1)^{|V|-1} (|V| - 1)! S(V) = \prod_{l=1}^{k} \{(-1)^{l-1} (l-1)!\}^{j_l} \prod_{|V|=1} \frac{S(V)}{n^{\frac{1}{2}}}$$

$$\times \prod_{|V|>1} \frac{S(V)}{n} n^{-(k - j_1 - 2j_2 - \cdots - 2j_k)/2}. \quad (1)$$

If $|V| = 1, V = \{\nu\}$ say, then $n^{-\frac{1}{2}} S(V) = n^{-\frac{1}{2}} \sum_i f_\nu(X_i)$ converges in distribution to a $N(0,1)$ variate $Z_V$ say. Moreover, for any $V$ with $|V| > 1$, $n^{-1} S(V)$ converges in probability to $\mu(V)$ by the weak law of large numbers, where $\mu(V)$ is an abbreviation for $Ef_{\nu_1}(X_1) \cdots f_{\nu_p}(X_1)$, and where $V = \{\nu_1, \ldots, \nu_p\}$. Thus if $k = j_1 + 2j_2 + \cdots + 2j_k$, (1) converges in distribution to

$$\prod_{l=1}^{k} \{(-1)^{l-1} (l-1)!\}^{j_l} \prod_{|V|=1} Z_V \prod_{|V|>1} \mu(V)$$

and to zero otherwise. But the condition $k = j_1 + 2j_2 + \cdots + 2j_k$ only holds for partitions of size one or two, and moreover $\mu(V) = 0$ if $|V| = 2$ since the functions are orthonormal. Thus the only partitions leading to non-zero limits are those for which all the component sets are singletons. But there is only one such partition, and the limit of (1) reduces to $\prod_{i=1}^{k} Z_i$ where the $Z_i$ are independent $N(0,1)$ r.vs. (The independence derives from the fact that the random vector with elements $n^{-\frac{1}{2}} \sum_{i=1}^{n} f_\nu(X_i), \nu = 1, 2, \ldots, k$ converges in distribution to a vector of independent $N(0,1)$ r.v.s.)

What happens if the functions $f_1, \ldots, f_k$ are not distinct? Suppose now that

$$\psi(x_1, \ldots, x_k) = f_1(x_1) \cdots f_k(x_k)$$

where now

$$f_1 = \cdots = f_{r_1}, f_{r_1+1} = \cdots = f_{r_1+r_2}, \cdots,$$
$$f_{r_1+\cdots+r_{m-1}+1} = \cdots = f_{r_1+\cdots+r_m}$$

so there are exactly $m$ distinct functions, assumed orthonormal, having the property $\int f_j(x) dF(x) = 0$. We can still apply the above theory, with the exception that now not all the $\mu(V)'s$ for two-point sets will be zero. We need to consider partitions consisting of singletons and two-point sets $\{\nu_1, \nu_2\}$ for which the corresponding $f's$ are identical. Every such partition is the union of $m$ subpartitions $\mathcal{P}_1, \ldots, \mathcal{P}_m$ each consisting of sets containing only indices $\nu$ for which the corresponding functions $f_\nu$ are identical. For such a sub-partition, having $r_l$ elements say, suppose there are $p_l$ two-point sets and $r_l - 2p_l$ one-point sets, and so for the whole partition

$$n^{-\frac{k}{2}} \prod_{V \varepsilon \mathcal{P}} (-1)^{|V|} (|V| - 1)! \; S(V) = \prod_{l=1}^{m} (-1)^{p_l} T_l^{r_l - 2p_l} + o_p(1) \qquad (2)$$

where $T_l = n^{-\frac{1}{2}} \sum_{i=1}^{n} f_\nu(X_i), \nu = r_1 + \cdots + r_l$.

The number of partitions satisfying the requirements for a non-zero limit is

$$\prod_{l=1}^{m} \frac{r_l!}{2^{p_l}(r_l - 2p_l)! p_l!}$$

since any such partition can be constructed by choosing the subpartitions independently, and there are $r_l!/2^{p_l}(r_l - 2p_l)! p_l!$ ways of partitioning $r_l$ objects into $r_l - 2p_l$ one-point sets and $p_l$ two-point sets.

Hence

$$n^{\frac{k}{2}} U_n = \sum_{\mathcal{P}} n^k \{n(n-1)\ldots(n-k+1)\}^{-1} n^{-\frac{k}{2}} \prod_{V \varepsilon P} (-1)^{|V|} (|V| - 1)! S(V)$$

$$= \sum_{p_1=0}^{[r_1/2]} \cdots \sum_{p_m=0}^{[r_m/2]} \prod_{l=1}^{m} r_l!/[2^{p_l}(r_l - 2p_l)!p_l!] \prod_{l=1}^{m}(-1)^{p_l} T_l^{r_l-2p_l} + o_p(1)$$

$$= \prod_{l=1}^{m} \sum_{p_l=0}^{[r_l/2]} r_l!/[2^{p_l}(r_l - 2p_l)p_l!](-1)^{p_l} T_l^{r_l-2p_l} + o_p(1)$$

$$= \prod_{l=1}^{m} H_{r_l}(T_l) + o_p(1)$$

where $H_r$ is the $r$th Hermite polynomial, and since $T_1, \ldots, T_m$ converge in distribution to independent standard normal r.v.s $Z_1, \ldots, Z_m$, it follows that

$$n^{\frac{k}{2}} U_n \xrightarrow{D} \prod_{l=1}^{m} H_{r_l}(Z_l).$$

As an illustration, consider the kernel $\psi(x_1, \ldots, x_k) = x_1 x_2 \ldots x_k$. Assuming that $E(X) = 0$ and $E(X^2) = 1$ we see that

$$n^{\frac{k}{2}} U_n \xrightarrow{D} H_k(Z)$$

where $Z$ is $N(0,1)$. Examples 1 and 2 are special cases of this result.

Note that the assumption that none of the functions $f_1, \ldots, f_k$ is constant is crucial, for in equation (2) a constant function, $f_1 \equiv 1$ say, will introduce a factor $n^{-\frac{1}{2}} \sum_{i=1}^{n} f_1(x_i) = n^{\frac{1}{2}}$ which does not converge.

The examples above suggest that if we could express a general $k$-degree kernel in terms of a series expansion similar to that used in the case $k = 2$, we could obtain an albeit complicated expression for the limit distribution of $n^{\frac{k}{2}}(U_n - \theta)$.

We can in fact do this by using some ideas from the theory of Hilbert space. Consider $L_2(F)$, the space of all functions square integrable with respect to $F$, i.e. the space of all functions $g$ satisfying

$$\int_{-\infty}^{\infty} |g(x)|^2 dF(x) < \infty.$$

Equipped with the inner product $(g, h) = \int_{-\infty}^{\infty} g(x)h(x)dF(x)$, $L_2(F)$ is a separable Hilbert space, in that there exists an orthonormal set of functions $\{e_\nu(x)\}_{\nu=0}^{\infty}$ such that

$$\lim_{n \to \infty} \int_{-\infty}^{\infty} |h(x) - \sum_{\nu=0}^{n}(h, e_\nu)e_\nu(x)|^2 dF(x) = 0.$$

for every $h \in L_2(F)$; the set $\{e_\nu\}$ is called an orthonormal basis for $L_2(F)$. We may assume that the constant function 1 is in the basis. The coefficients $(h, e_\nu)$ satisfy $\sum_{\nu=1}^{\infty} |(h, e_\nu)|^2 = \int |h(x)|^2 dF(x)$. Consider also the space $L_2(F^k)$ of all functions $h(x_1, \ldots, x_k)$ satisfying

$$\int_{\mathbb{R}_k} |h(x_1, \ldots, x_k)|^2 \prod_{i=1}^{k} dF(x_i) < \infty.$$

With the inner product

$$(g, h) = \int g(x_1, \ldots, x_k) h(x_1, \ldots, x_k) \prod_{i=1}^{k} dF(x_i),$$

the space $L_2(F^k)$ is also a separable Hilbert space, and the set of functions of the form

$$e_{i_1}(x_1) \ldots e_{i_k}(x_k) \tag{3}$$

is an orthonormal basis for $L_2(F^k)$. Thus, for every $h$ in $L_2(F^k)$,

$$h(x_1, \ldots, x_k) = \sum_{i_1=1}^{\infty} \ldots \sum_{i_k=1}^{\infty} (h, e_{i_1} \ldots e_{i_k}) e_{i_1}(x_1) \ldots e_{i_k}(x_k)$$

the series converging in mean square.

We now turn to the problem of determining the limit distribution of a $U$-statistic with kernel $h(x_1, \ldots, x_k)$ satisfying $Eh(X_1, \ldots, X_k) = 0$ and $0 = \sigma_1^2 = \ldots = \sigma_{k-1}^2 < \sigma_k^2$. This suggests considering the kernel

$$h_K(x_1, \ldots, x_k) = \sum_{i_1=1}^{K} \ldots \sum_{i_k=1}^{K} (h, e_{i_1} \ldots e_{i_k}) e_{i_1}(x_1) \ldots e_{i_k}(x_k)$$

and using the asymptotic distribution of the $U$-statistics $U_{n,K}$ based on $h_K$ to approximate that of $U_n$ based on $h$. To carry out this program, we need only

  (i) Prove that $U_{n,K}$ converges to $U_n$ in mean square as $K \to \infty$, uniformly in $n$

 (ii) Compute the asymptotic distribution of $U_{n,K}$.

The desired result will then follow as for the case $k = 2$. Result (i) is proved in exactly the same manner as the case $k = 2$; we require only the property

$$\sum_{i_1=1}^{\infty} \cdots \sum_{i_k=1}^{\infty} |(h, e_{i_1} \ldots e_{i_k})|^2 < \infty$$

which follows from the fact that the functions (3) are an orthonormal basis. The calculation (ii) can be done using the methods of the examples. Let $\mathbf{i} = (i_1, \ldots, i_k)$ and let $W_{n,\mathbf{i}}$ be the $U$-statistic based on $e_{i_1}(x_1) \ldots e_{i_k}(x_k)$. Then

$$n^{\frac{k}{2}} U_{n,K} = \sum_{i_1=1}^{K} \cdots \sum_{i_k=1}^{K} (h, e_{i_1} \ldots e_{i_k}) n^{\frac{k}{2}} W_{n,\mathbf{i}} \tag{4}$$

In order for the method used in the examples to be applicable, it is essential that none of the functions $e_1(x), \ldots, e_k(x)$ is constant. However, one of the basis functions is unity. In this case, in order to apply the method we require $(h, e_{i_1} \ldots e_{i_k}) = 0$ whenever one or more of the $e_{i_j}$ is unity. Since the $U$-statistic based on $h$ has a degeneracy of order $k - 1$, we have $h_c(x_1, \ldots, x_c) = Eh(x_1, \ldots, x_c, X_{c+1}, \ldots X_k) = 0$ for $c = 1, 2, \ldots, k-1$ and so if $e_{i_1}, \ldots, e_{i_c}$ are unity, then

$$(h, e_{i_1}, \ldots, e_{i_k}) = \int_{\mathbb{R}^k} h(x_1, \ldots, x_k) e_{i_{c+1}}(x_{c+1}) \ldots e_{i_k}(x_{i_k}) \prod_{i=1}^{k} dF(x_i)$$

$$= \int_{\mathbb{R}^{k-c}} h_{k-c}(x_1, \ldots, x_{k-c}) e_{i_{c+1}}(x_{c+1}) \ldots e_{i_k}(x_k) \prod_{i=c+1}^{k} dF(x_i)$$

$$= 0.$$

Thus in (4), only terms with non-constant $e_j$ need be considered. For this type of term, $n^{\frac{k}{2}} W_{n,\mathbf{i}} = \prod_{l=1}^{\infty} H_{r_l(\mathbf{i})}(T_l) + o_p(1)$ by the arguments employed in Example 3, where $r_l(\mathbf{i})$ denotes the number of indices in $\mathbf{i}$ that are equal to $l$, and $T_l = n^{-\frac{1}{2}} \sum_{i=1}^{n} e_l(X_i)$. Thus

$$n^{\frac{k}{2}} U_n = \sum_{i_1=1}^{K} \cdots \sum_{i_k=1}^{K} (h, e_{i_1} \ldots e_{i_k}) W_{n,\mathbf{i}}$$

$$= \sum_{i_1=1}^{K} \cdots \sum_{i_k=1}^{K} (h, e_{1_i} \ldots e_{i_k}) \prod_{l=1}^{K} H_{r_l(\mathbf{i})}(T_l) + o_p(1) \tag{5}$$

89

which converges in distribution to

$$\sum_{i_1=1}^{K} \cdots \sum_{i_k=1}^{K} (h, e_{i_1} \ldots e_{i_k}) \prod_{l=1}^{K} H_{r_l(\mathbf{i})}(Z_l) \qquad (6)$$

since $T_1, \ldots, T_K$ converge in distribution to independent standard normal random variables $Z_1, \ldots, Z_K$.

In any particular application, we would try to choose a basis to make as many of the coefficients $(h, e_{i_1} \ldots e_{i_k})$ as possible equal zero. For example, when $k = 2$, we can choose a basis such that $(h, e_{i_1} e_{i_2}) = 0$ unless $i_1 = i_2$, and so the limit is of the form $\binom{k}{2} \sum_{i=1}^{\infty} \lambda_i H_2(Z_i)$. This agrees with Theorem 1 of Section 3.2.2 since $H_2(Z) = Z^2 - 1$.

Based on the forgoing considerations, we can now state the general result:

**Theorem 1.** Let $U_n$ be a $U$-statistic based on the kernel $\psi(x_1, \ldots, x_k)$, and a random sample $X_1, \ldots, X_n$ with distribution function $F$. Suppose that $0 = \sigma_1^2 = \ldots = \sigma_{d-1}^2 < \sigma_d^2$. Then the asymptotic distribution of $n^{d/2}(U_n - \theta)$ is that of

$$\binom{k}{d} \sum_{i_1=1}^{\infty} \cdots \sum_{i_d=1}^{\infty} (h^{(d)}, e_{i_1} \ldots e_{i_d}) \prod_{l=1}^{\infty} H_{r_l(\mathbf{i})}(Z_l)$$

where $e_1, e_2, \ldots$ is an orthonormal basis for $L_2(F)$, $h^{(d)}$ is the kernel for the $U$-statistic $H_n^{(d)}$ in the $H$-decomposition of $U_n, Z_1, Z_2, \ldots$ is a sequence of independent standard normal random variables, and $r_l(\mathbf{i})$ the number of indices among the $\mathbf{i} = (i_1, \ldots, i_d)$ equal to $l$.

**Proof.** In the $H$-decomposition the $U$-statistic $H_n^{(d)}$ has a kernel $h^{(d)}$ of degree $d$, a degeneracy of order $d - 1$ and expectation zero. Applying the truncation argument to $h^{(d)}$, and using (5) and (6), we obtain the result.

### 3.2.4   Poisson Convergence

The subject of this section is an extension to the $U$-statistic context of the Poisson convergence of sums of independent zero-one random variables, and is the Poisson counterpart of Theorem 1 of Section 3.2.1 which generalises the Central Limit Theorem. Specifically, if $X_{1n}, \ldots, X_{nn}$ are independent zero-one random variables, with

$$Pr(X_{in} = 1) = p_n, \ \ Pr(X_{in} = 0) = 1 - p_n$$

then the convergence of $\sum_{i=1}^{n} X_{in}$ to a Poisson law with parameter $\lambda = \lim np_n$ is just the familiar Poisson convergence of the binomial distribution discussed in elementary textbooks. For a $U$-statistic generalisation of this result, suppose that $X_1, X_2, \ldots$ is a sequence of i.i.d. random variables, and $\psi_n$ is a sequence of kernels, each of degree $k$, having value zero or one, with

$$Pr\big(\psi_n(X_1, \ldots, X_k) = 1\big) = p_n, \ \ Pr\big(\psi_n(X_1, \ldots, X_k) = 0\big) = 1 - p_n.$$

The main theorem of this section gives conditions under which the random variable

$$T_n = \sum_{(n,k)} \psi_n(X_{i_1}, \ldots, X_{i_k}) \tag{1}$$

converges to a Poisson law. We will need some preliminary results.

The r.v. $T_n$ in (1) takes values in the set of non-negative integers, denoted hereafter by $\mathbb{Z}_{(-)}$, and the weak convergence of such random variables is conveniently expressed in terms of the concept of *total variation distance*. If $T$ takes values in $\mathbb{Z}_{(-)}$, and $\mu$ is a probability measure on $\mathbb{Z}_{(-)}$, then the total variation distance between $T$ and $\mu$ is defined by

$$TVD(T, \mu) = \sup_{A} \big| Pr(T \epsilon A) - \mu(A) \big|$$

where the supremum is taken over all subsets $A$ of $\mathbb{Z}_{(-)}$. The connection with weak convergence is given in Theorem 1:

**Theorem 1.** *Let $\{T_n\}$ be a sequence of r.v.s taking values in $\mathbb{Z}_{(-)}$, and $\mu$ a probability measure on $\mathbb{Z}_{(-)}$. Then $T_n$ converges in distribution to $\mu$ if and only if $TVD(T_n, \mu) \to 0$.*

**Proof.** Let $\mu(k)$ be the probability that $\mu$ assigns to $\{k\}$. Weak convergence in the present context is equivalent to

$$\lim_{n \to \infty} Pr(T_n = k) = \mu(k) \tag{2}$$

91

for all $k$ in $\mathbb{Z}_{(-)}$ and (2) is obviously implied by $TVD(T_n, \mu) \to 0$ since for all $k$,

$$|Pr(T_n = k) - \mu(k)| \leq TVD(T_n, \mu).$$

Conversely, using Lemma A below, we obtain

$$\lim_{n \to \infty} \sum_{k=0}^{\infty} |Pr(T_n = k) - \mu(k)| = 0. \tag{3}$$

In view of the inequality

$$|Pr(T_n \epsilon A) - \mu(A)| \leq \sum_{k=0}^{\infty} |Pr(T_n = k) - \mu(k)|$$

which is valid for all subsets $A$ of $\mathbb{Z}_{(-)}$, it follows that

$$TVD(T_n, \mu) \leq \sum_{k=0}^{\infty} |Pr(T_n = k) - \mu(k)|$$

and so the theorem follows from (3).

**Lemma A.** *Suppose that*

(i) *For all* $n$, $\sum_{k=0}^{\infty} A_{nk} = 1$ *and* $A_{nk} \geq 0$,

(ii) $\sum_{k=0}^{\infty} b_k = 1$;

(iii) $\lim_{n \to \infty} A_{nk} = b_k$ *for all* $k$.

Then $\lim_{n \to \infty} \sum_{k=0}^{\infty} |A_{nk} - b_k| = 0$.

**Proof.** Given $\varepsilon > 0$, by (ii) and (iii) we can find integers $N$ and $K$ satisfying the inequalities $|\sum_{k>K} b_k| < \varepsilon/4$ and $|A_{nk} - b_k| < \varepsilon/4K$ for $n > N$ and $k = 1, 2, \ldots K$. Further,

$$\sum_{k>K} A_{nk} = 1 - \sum_{k=1}^{K} A_{nk} = 1 - \sum_{k=1}^{K}(A_{nk} - b_k) - \sum_{k=1}^{K} b_k$$

$$= \sum_{k>K} b_k - \sum_{k=1}^{K}(A_{nk} - b_k)$$

so that for all $n > N$,

$$\sum_{k=1}^{\infty} |A_{nk} - b_k| \leq \sum_{k=1}^{K} |A_{nk} - b_k| + |\sum_{k>K} A_{nk}| + |\sum_{k>K} b_k|$$

$$\leq 2(\sum_{k=1}^{K} |A_{nk} - b_k| + |\sum_{k>K} b_k|)$$

$$< \varepsilon,$$

92

proving the lemma.

We now state the main theorem of this section, which is due to Silverman and Brown (1978). The method of proof is taken from Barbour and Eagleson (1984).

**Theorem 2.** *Let* $X_1, X_2, \ldots$ *be a sequence of i.i.d. random variables, and let* $\psi_n(x_1, \ldots, x_k)$ *be a sequence of symmetric functions which take only values zero and one, having expectations* $E\psi_n(X_1, \ldots, X_k) = p_n$. *Define*

$$T_n = \sum_{(n,k)} \psi_n(X_{i_1}, \ldots, X_{i_k}) \tag{4}$$

*and suppose that*

$$(i) \quad \lim_{n \to \infty} \binom{n}{k} p_n = \lambda$$

*and*

$$(ii) \quad \lim_{n \to \infty} n^{2k-1} E\{\psi_n(X_1, \ldots, X_k)\psi_n(X_2, \ldots, X_{k+1})\} = 0.$$

*Then* $T_n$ *converges to a Poisson distribution with parameter* $\lambda$.

**Proof.** A key role in the proof is played by the function $x$ defined on $\mathbb{Z}_{(-)}$ by

$$x(0) = 0,$$

and for $m \geq 0$ by

$$x(m+1) = \lambda^{-1}\{p_\lambda(A \cap \mathbb{Z}_m) - p_\lambda(A)p_\lambda(\mathbb{Z}_m)\}/p_\lambda(\{m\})$$

where $A$ is a fixed subset of $\mathbb{Z}_{(-)}$, $\mathbb{Z}_m = \{0, 1, 2, \ldots, m\}$ and $p_\lambda(S)$ is the probability assigned to the set $S \subseteq \mathbb{Z}_{(-)}$ by the Poisson distribution with parameter $\lambda$. The function $x$ has the properties

(i) $\lambda x(m+1) - mx(m) = \begin{cases} 1 - p_\lambda(A) & \text{if } m \in A, \\ -p_\lambda(A) & \text{otherwise} \end{cases}$ ;

(ii) $x(m)$ is bounded, and

(iii) $\sup_m |x(m+1) - x(m)| < \min(1, \lambda^{-1})$.

Details may be found in the appendix of Barbour and Eagleson (1983).

The function $x$ is used to prove the following inequality, which is the basis of the proof of Theorem 2 : we show below that

$$TVD(T_n, p_{\lambda_n}) \le \min(1, \lambda_n^{-1}) \binom{n}{k}$$

$$\left[ p_n^2 \left\{ \binom{n}{k} - \binom{n-k}{k} \right\} + \sum_{c=1}^{k-1} \binom{k}{c} \binom{n-k}{k-c} \eta_{n,c}^2 \right] \quad (5)$$

where $\eta_{n,c}$ is defined by

$$\eta_{n,c} = E\psi_n(X_1, \ldots, X_k)\psi(X_1, \ldots, X_c, X_{k+1}, \ldots, X_{2k-c+1})$$

and $\lambda_n = \binom{n}{k} p_n$.

To prove (5), let $\psi_n(S)$ de$\{i_1, \ldots, i_k\}$, and let $J$ be an arbitrary $k$-subset of $\{1, 2, \ldots, n\}$. Then for any $J$, we can write (4) as

$$T_n = \sum_{c=0}^{k} \left\{ \sum_{|S \cap J| = c} \psi_n(S) \right\} = \sum_{c=0}^{k-1} T_J^{(c)} + \psi_n(J),$$

where $T_J^{(c)}$ consists of the sum of the kernels evaluated for all sets $S$ that have $c$ elements in common with $J$. The r.v. $T_J^{(0)}$ is independent of $\psi_n(J)$, and $T_J^{(c)}$ is the sum of $\binom{k}{c}\binom{n-k}{k-c}$ terms $\psi_n(S)$. Now consider

$$\lambda_n x(T_n + 1) - T_n x(T_n)$$

$$= \binom{n}{k} p_n x(T_n + 1) - \sum_{(n,k)} \psi_n(J) \{ x(T_J^{(0)} + 1) + (x(T_n) - x(T_J^{(0)} + 1)) \}$$

$$= \sum_{(n,k)} [ p_n x(T_n + 1) - \psi_n(J) \{ x(T_J^{(0)} + 1) + (x(T_n) - x(T_J^{(0)} + 1)) \} ].$$

and so, due to the independence of $\psi_n(J)$ and $T_J^{(0)}$, we can write

$$E\{ \lambda_n x(T_n + 1) - T_n x(T_n) \}$$

$$= \sum_{(n,k)} [ p_n E\{ x(T_n + 1) - x(T_J^{(0)} + 1) \} - E\{ \psi_n(J)(x(T_n) - x(T_J^{(0)} + 1)) \} ]. \quad (6)$$

Denote $T_n - T_J^{(0)}$ by $Z_J$ and $\sup_m |x(m+1) - x(m)|$ by $\Delta x$. The r.v. $Z_J$ takes values in $\mathbb{Z}_{(-)}$ by its definition, so we can write

$$|E\{ x(T_n + 1) - x(T_J^{(0)} + 1) \}|$$

94

$$\leq \sum_{k=0}^{\infty} \left| E\{x(T_J^{(0)} + Z_J + 1) - x(T_J^{(0)} + 1)|Z_J = k\} \right| Pr(Z_J = k)$$

$$\leq \sum_{k=0}^{\infty} k \, \triangle \, x Pr(Z_J = k)$$

$$= E(Z_J) \, \triangle \, x$$

$$= p_n \left\{ \binom{n}{k} - \binom{n-k}{k} \right\} \triangle \, x. \qquad (7)$$

Also, since $Z_J = 0$ implies that $T_J^{(k)} = \psi_n(J)$ is zero, we have

$$\left| E\{\psi_n(J)(x(T_n) - x(T_J^{(0)} + 1))\} \right|$$

$$= \left| E\{\psi_n(J)(x(T_J^{(0)} + Z_J) - x(T_J^{(0)} + 1))\} \right|$$

$$= \sum_{k=2}^{\infty} \left| E\{\psi_n(J)(x(T_J^{(0)} + Z_J) - x(T_J^{(0)} + 1))|Z_J = k\} \right| Pr(Z_J = k)$$

$$\leq \sum_{k=2}^{\infty} E\{\psi_n(J)(k-1) \triangle x|Z_J = k\} Pr(Z_J = k)$$

$$= E\{\psi_n(J)(Z_J - 1)\} \triangle x$$

$$\leq \sum_{c=1}^{k-1} E\{\psi_n(J) \sum_{|S \cap J|=c} \psi_n(S)\} \triangle x$$

$$= \sum_{c=1}^{k-1} \binom{k}{c} \binom{n-k}{k-c} \eta_{n,c} \triangle x. \qquad (8)$$

Combining (7) and (8) and using (6) we get

$$\left| E\{\lambda_n x(T_n + 1) - T_n x(T_n)\} \right|$$

$$\leq \triangle x \binom{n}{k} \left[ p_n^2 \left\{ \binom{n}{k} - \binom{n-k}{k} \right\} + \sum_{c=1}^{k-1} \binom{k}{c} \binom{n-k}{k-c} \eta_{n,c} \right]. \qquad (9)$$

The proof of (5) is completed by noting that by property (i) of $x$,

$$\left| E\{\lambda_n x(T_n + 1) - T_n x(T_n)\} \right| = \left| Pr(T_n \in A) - P_{\lambda_n}(A) \right|$$

so that (5) follows from (9).

To complete the proof of the theorem, note that by Theorem 4 of Section 1.3

$$\eta_{n,c} = \sigma_{n,c}^2 + p_n^2 \leq c(k-1)^{-1} \sigma_{n,k-1}^2 + p_n^2$$

$$\leq \sigma_{n,k-1}^2 + p_n^2$$

$$\leq \eta_{n,k-1}$$

for $c = 1, 2, \ldots, k - 1$, so that

$$\sum_{c=1}^{k-1} \binom{n}{k}\binom{k}{c}\binom{n-k}{k-c}\eta_{n,c} \leq \sum_{c=1}^{k-1} \binom{n}{k}\binom{k}{c}\binom{n-k}{k-c}\eta_{n,k-1}$$
$$= O(n^{2k-1})\eta_{n,k-1}$$

which converges to zero by assumption (ii). Also

$$\binom{n}{k}p_n^2\left\{\binom{n}{k} - \binom{n-k}{k}\right\} = \binom{n}{k}^2 p_n^2 O(n^{-1})$$

and hence the right hand side of (9) converges to zero since $\Delta x$ is less than $\min(1, \lambda_n^{-1})$. Since $p_{\lambda_n}(\{k\}) \to p_\lambda(\{k\})$ for each $k$, the theorem is proved.

Brown and Silverman (1979) prove a rate of convergence result which complements Theorem 2. Using the notation of that theorem, they show that if

$$\rho_n = n^{2k} Cov\left(\psi(X_1, \ldots, X_k), \psi(X_2, \ldots, X_{k+1})\right)$$

then there exist constants $c_1$ and $c_2$ such that

$$TVD(T_n, p_{\lambda_n}) \leq c_1\lambda_n^2 n^{-1} + c_2\rho_n^{1/2}n^{-1/2}.$$

Silverman and Brown (1978) and Grusho (1986) also study the convergence to a Poisson process associated with $U$-statistics. For more on the basic convergence result, see Barbour and Eagleson (1987) and Grusho (1988).

## 3.3. Rates of convergence in the $U$-statistic central limit theorem

### 3.3.1 Introduction

Suppose, as usual, that $X_1, \ldots, X_n$ is a sequence of independent and identically distributed random variables having mean $\mu$ and variance $\sigma^2$. The classic result describing the rate of convergence of $S_n = \sum_{i=1}^n (X_i - \mu)/\sigma\sqrt{n}$ to the normal distribution is the Berry-Esseen theorem, which states that provided the third moment $\nu_3 = E|X_1 - \mu|^3$ is finite, then

$$\sup_x |F_n(x) - \Phi(x)| \leq \frac{C\nu_3}{\sigma^3\sqrt{n}}. \tag{1}$$

96

Here $F_n$ is the distribution function of $S_n$, $\Phi$ is the distribution function of the standard normal distribution, and $C$ is a constant independent of $n$ and the distribution of the $X$'s.

A sharper result is as follows: provided the common distribution of the $X$'s is non-lattice (i.e. not concentrated on points $0, \pm h, \pm 2h, \ldots$ for some number $h$) then $F_n$ admits the asymptotic expansion

$$F_n(x) = \Phi(x) + \frac{\lambda_3(1 - x^2)e^{-x^2/2}}{\sqrt{2\pi}n^{\frac{1}{2}}} + o(n^{-\frac{1}{2}}) \qquad (2)$$

uniformly in $x$, where the constant $\lambda_3$ in (2) is the third cumulant of the distribution of $(X_1 - \mu)/\sigma$. Expansions of higher order are possible for distributions possessing moments higher than the third. For a full discussion, see Feller (1971), Chapter XVI.

The reader might suspect that since non-degenerate $U$-statistics are "almost" the sum of i.i.d. random variables, these results should also apply to non-degenerate $U$-statistic sequences and this is indeed the case. We treat the Berry-Esseen theorem for $U$-statistics in some detail, and present a brief discussion of $U$-statistic versions of (2).

Similar results are true for degenerate $U$-statistics, and these also receive a brief treatment.

### 3.3.2 The Berry-Esseen theorem for $U$-statistics

Several authors have contributed to the problem of establishing a Berry-Esseen theorem for $U$-statistics. In this connection we mention Grams and Serfling (1973), Bickel (1974), Chan and Wierman (1977), Callaert and Janssen (1978), Ahmad (1981), Helmers and van Zwet (1982), Boroskikh (1984), Korolyuk and Boroskikh (1986) and Friedrich (1989). These authors prove a series of results with progressive weakening of moment conditions and sharper error rates. We will present an adaption of Friedrich's proof; in the interests of simplicity we consider only the case $k = 2$.

**Theorem 1.** *(The Berry-Esseen theorem for $U$-statistics.) Let $U_n$ be a non-degenerate $U$-statistic of degree 2, based on a sequence of i.i.d. random variables $\{X_n\}$. Suppose that the kernel $\psi$ has an H-decomposition*

$$\psi(x_1, x_2) = \theta + h^{(1)}(x_1) + h^{(1)}(x_2) + h^{(2)}(x_1, x_2)$$

where $E|h^{(1)}(X_1)|^3 < \infty$ and $E|h^{(2)}(X_1, X_2)|^{5/3} < \infty$. Let $\rho$ denote the quantity $E|h^{(1)}(X_1)|^3/\sigma_1^3$ and $\lambda_p = E|h^{(2)}(X_1, X_2)|^p/\sigma_1^p$. Then there exist constants $C_1$, $C_2$ and $C_3$ depending neither on $n$, $\psi$ nor the distribution of the $X's$ such that

$$\sup_x \left| Pr(\sqrt{n}(U_n - \theta)/2\sigma_1 \leq x) - \Phi(x) \right| \leq \{C_1\rho + C_2\lambda_{5/3} + C_3(\rho\lambda_{3/2})^{2/3}\}n^{-\frac{1}{2}}$$

for all $n \geq 2$.

The proof of this theorem is rather lengthy, but is included to give some insight into the techniques required to establish such results. The proof itself is not difficult, but a brief summary of the strategy used may help the reader to find a way through the details. The basic idea is to split the $U$-statistic into its projection and a remainder, which are treated using characteristic functions and the Esseen smoothing lemma (see Feller (1971), p538). The projection term on the right of (9) below is dealt with as in the standard proof for the Berry-Esseen theorem for i.i.d. summands, while the other term is estimated by exploiting the martingale structure of $U$-statistics, which is described further in Section 3.4.1. The estimates of the various terms of (9) are then combined to yield the result.

In the interests of simplicity, we make no attempt to calculate the values of the constants $C_1$, $C_2$ and $C_3$.

**Proof.** Let

$$\psi(x_1, x_2) = \theta + h^{(1)}(x_1) + h^{(1)}(x_2) + h^{(2)}(x_1, x_2)$$

be the usual $H$-decomposition of the kernel, with corresponding decomposition of the normalised $U$-statistic

$$\sqrt{n}(U_n - \theta)/2\sigma_1 = \sqrt{n}H_n^{(1)}/\sigma_1 + \sqrt{n}H_n^{(2)}/2\sigma_1. \qquad (1)$$

Define the quantities

$$\Delta_k = \frac{\sqrt{n}}{2\sigma_1}\binom{n}{2}^{-1} \sum_{j=k+1}^{n} h^{(2)}(X_k, X_j), \qquad k = 1, \dots, n-1, \qquad (2)$$

$$\Delta_{k,j} = \frac{\sqrt{n}}{2\sigma_1}\binom{n}{2}^{-1} h^{(2)}(X_k, X_j), \qquad 1 \leq k < j \leq n$$

and

98

$$S_k = \frac{\sqrt{n}(U_n - \theta)}{2\sigma_1} - (\Delta_1 + \cdots + \Delta_k), \qquad k = 1, \ldots, n-1. \tag{3}$$

Also define

$$S_0 = \frac{\sqrt{n}(U_n - \theta)}{2\sigma_1} \tag{4}$$

so that

$$\Delta_k = S_{k-1} - S_k, \qquad k = 1, \ldots, n-1. \tag{5}$$

Let $T_j = h^{(1)}(X_j)/\sigma_1\sqrt{n}$. Then $\sqrt{n}H_n^{(1)}/\sigma_1 = \sum_{j=1}^n T_j$, and from (2)–(5) we get

$$S_k = \sum_{j=1}^n T_j + \Delta_{k+1} + \cdots + \Delta_{n-1} \tag{6}$$

and in particular

$$S_{n-1} = \sqrt{n}H_n^{(1)}/\sigma_1. \tag{7}$$

Further define $\mathbf{Y}_k = (X_{k+1}, \ldots, X_n)$; then from (6) we get

$$E(S_k|\mathbf{Y}_k) = \sum_{j=k+1}^n T_j + \Delta_{k+1} + \cdots + \Delta_{n-1}$$

so that

$$S_k = \sum_{j=1}^k T_j + E(S_k|\mathbf{Y}_k). \tag{8}$$

With these preliminaries out of the way, let $\eta(t)$ be the characteristic function (c.f.) of the r.v. $h^{(1)}(X_1)$, and let $\phi_n(t)$ be the c.f. of $\sqrt{n}(U_n - \theta)/2\sigma_1$. Then $\eta^n(t/\sqrt{n}\sigma_1)$ is the c.f. of $\sqrt{n}H_n^{(1)}/\sigma_1$, and by the Esseen smoothing lemma, (see e.g. Feller (1971) p538)

$$\sup_x \left| Pr\left( \sqrt{n}(U_n - \theta)/2\sigma_1 \leq x \right) - \Phi(x) \right|$$

$$\leq \frac{1}{\pi} \int_{-\sqrt{n}/\rho}^{\sqrt{n}/\rho} \left| \phi_n(t) - e^{-t^2/2} \right| |t|^{-1} dt + \frac{24\rho}{\pi\sqrt{2\pi}} n^{-\frac{1}{2}}$$

$$\leq \frac{1}{\pi} \int_{-\sqrt{n}/\rho}^{\sqrt{n}/\rho} \left| \eta^n(t/n^{\frac{1}{2}}\sigma_1) - e^{-t^2/2} \right| |t|^{-1} dt + \frac{24\rho}{\pi\sqrt{2\pi}} n^{-\frac{1}{2}}$$

$$+ \frac{1}{\pi} \int_{-\sqrt{n}/\rho}^{\sqrt{n}/\rho} \left| \eta^n(t/n^{\frac{1}{2}}\sigma_1) - \phi_n(t) \right| |t|^{-1} dt. \tag{9}$$

99

The standard Berry-Esseen argument applied to the i.i.d. random variables $h^{(1)}(X_j)$ shows that there is a constant $C_1$ such that the first two terms of (9) are less than $C_1 \rho n^{-\frac{1}{2}}$. To complete the proof, we need to estimate the third term of (9). Consider the integrand of this term: using (4) and (7) we can write

$$
\begin{aligned}
\left| \phi_n(t) - \eta^n(t/n^{\frac{1}{2}}\sigma_1) \right| \\
= \left| E(e^{itS_0} - e^{itS_{n-1}}) \right| \\
\leq \sum_{k=1}^{n-1} \left| E(e^{itS_{k-1}} - e^{itS_k}) \right| \\
= \sum_{k=1}^{n-1} \left| E\{e^{itS_k}(e^{it\Delta_k} - 1)\} \right| \\
\leq \sum_{k=1}^{n-1} \left| E\{e^{itS_k}(e^{it\Delta_k} - 1 - it\Delta_k)\} \right| + |t| \sum_{k=1}^{n-1} \left| E(e^{itS_k}\Delta_k) \right| \\
= Z_1(t) + |t|Z_2(t), \quad \text{say}.
\end{aligned} \tag{10}
$$

Using (8), we see that a typical term of $Z_1(t)$ equals

$$
E\left[ \exp\{it(\sum_{j=1}^{k} T_j + E(S_k|\mathbf{Y}_k))\}(e^{it\Delta_k} - 1 - it\Delta_k) \right]
$$
$$
= \prod_{j=1}^{k-1} E(e^{itT_j})E\{e^{it(T_k + E(S_k|\mathbf{Y}_k))}(e^{it\Delta_k} - 1 - it\Delta_k)\} \tag{11}
$$

since $T_1, \ldots, T_{k-1}$ are independent of the other terms. (If $k = 1$ the product is taken to be unity.) From the standard treatment of the Berry-Esseen theorem in the i.i.d. case (see e.g. Feller (1971) p453) it follows that

$$
\left| \eta(t/\sigma_1 n^{\frac{1}{2}}) \right| \leq e^{-\frac{1}{3n}t^2}
$$

for $|t| < \sqrt{n}/\rho$. Also, using the inequality $|e^{ix} - 1 - ix| < 2|x|^p$ for $1 \leq p \leq 2$, we obtain

$$
E|e^{it\Delta_k} - 1 - it\Delta_k| \leq 2|t|^p E|\Delta_k|^p.
$$

100

For fixed $k$ and $n$, the sequence $\{\sum_{j=k+l}\Delta_{k,j}\}_{l=1}^{n-k}$ is a reverse martingale adapted to the $\sigma$-fields $\mathcal{F}_l = \sigma(X_k, X_{k+l}, X_{k+l+1}, \ldots)$, and so by Theorem 6 of Section 4.3.1 we get

$$E|\Delta_k|^p \leq 2 \sum_{j=k+1}^{n} |\Delta_{k,j}|^p$$

$$\leq Cn^{-\frac{3}{2}p} \sum_{j=k+1}^{n} E|h^{(2)}(X_k, X_j)|^p$$

$$= C\lambda_p n^{1-\frac{3}{2}p}$$

where here and subsequently $C$ is a generic constant varying according to context but not depending on $n$, $\psi$ or the distribution of the $X$'s. Hence, combining these estimates, from (11) we get

$$Z_1(t) \leq \sum_{k=1}^{n-1} C\lambda_p n^{1-\frac{3}{2}p} |t|^p e^{-\frac{(k-1)}{3n}t^2}. \tag{12}$$

The estimation of $Z_2$ requires more delicacy. Define for $1 \leq k \leq l < n$

$$W_{k,l} = \frac{\sqrt{n}}{2\sigma_1}\binom{n}{2}^{-1} \sum_{j=l+1}^{n} h^{(2)}(X_k, X_j)$$

so that $W_{k,l}$ is independent of $X_1, \ldots, X_{k-1}, X_{k+1}, \ldots X_l$. Note that $E(W_{k,l}|X_k) = 0$, $W_{k,k} = \Delta_k$ and $W_{k,n-1} = \Delta_{k,n}$. Also define $W_{k,j} = 0$ for $j \geq n$, and let $m(k)$ be the largest integer such that $km(k) < n$. In terms of these quantities, we can write

$$Z_2(t) = \sum_{k=1}^{n-1} \left| E(e^{itS_k}\Delta_k) \right|$$

$$= \sum_{k=1}^{n-1} \left| E(e^{itS_k}W_{k,k}) \right|$$

$$= \sum_{k=1}^{n-1} \left| \sum_{j}^{m(k)} E\left(e^{itS_{jk}}W_{k,jk}\right) - E\left(e^{itS_{(j+1)k}}W_{k,(j+1)k}\right) \right|$$

$$\leq \sum_{k=1}^{n-1} \sum_{j=1}^{m(k)} \left| E\{(e^{itS_{jk}} - e^{itS_{(j+1)k}})W_{k,jk}\} \right|$$

101

$$+ \sum_{k=1}^{n-1} \sum_{j=1}^{m(k)} \left| E\{e^{itS_{(j+1)k}}(W_{k,jk} - W_{k,(j+1)k})\} \right|$$

$$= Z_{21}(t) + Z_{22}(t), \quad \text{say.}$$

A typical term in $Z_{21}(t)$ can be written

$$E\{e^{itS_{jk}}(1 - e^{it(S_{(j+1)k} - S_{jk})})W_{k,jk}\} = E\left(\exp\left(it \sum_{\substack{l=1 \\ l \neq k}}^{jk} T_l\right)\right)$$

$$\times E\{e^{it(T_k + E(S_{jk}|Y_{jk}))}(1 - e^{it(S_{(j+1)k} - S_{jk})})W_{k,jk}\} \quad (13)$$

since $T_1, \ldots, T_{k-1}, T_{k+1}, \ldots, T_{jk}$ are independent of the other quantities in (13). Using the fact that $S_{(j+1)k} - S_{jk}$ is independent of $X_k$, we obtain

$$E\{(1 - e^{it(S_{(j+1)k} - S_{jk})})W_{k,jk}\} = E\{(1 - e^{it(S_{j+1)k} - S_{jk})})E(W_{k,jk}|X_k)\}$$

$$= 0,$$

and so by using the Holder inequality and the inequalities $|e^{ix} - 1| \leq |x|$ and $|e^{ix} - 1| \leq 2|x|^{p-1}$, it follows that the second factor of (13) is bounded by

$$\left| E\{(e^{itT_k} - 1)(1 - e^{it(S_{(j+1)k} - S_{jk})})W_{k,jk}\} \right|$$

$$\leq \left[ E\{|e^{itT_k} - 1|^q |1 - e^{it(S_{(j+1)k} - S_{jk})}|^q\} \right]^{1/q} \{E|W_{k,jk}|^p\}^{1/p}$$

$$\leq |t|^p \{E|T_k|^q\}^{1/q} \{E|S_{(j+1)k} - S_{jk}|^p\}^{1/q} \{E|W_{k,jk}|^p\}^{1/p} \quad (14)$$

where $1 \leq p \leq 2$ and $q = p/(1-p)$ and thus $q \leq 3$ if $p \geq 3/2$. Estimating the factors of (14), we get

$$\{E|T_k|^q\}^{1/q} \leq \{E|T_k|^3\}^{\frac{1}{3}} = n^{-\frac{1}{2}}\rho^{\frac{1}{3}},$$

and

$$\{E|S_{(j+1)k} - S_{jk}|^p\}^{1/q} = \{E|\Delta_{jk+1} + \cdots + \Delta_{(j+1)k}|^p\}^{1/q}$$

$$\leq \{2 \sum_{l=jk+1}^{(j+1)k} E|\Delta_l|^p\}^{1/q}$$

$$\leq Ck^{1-1/p}\lambda_p^{1-1/p}n^{(1-\frac{3}{2}p)(1-1/p)}$$

102

by applying Theorem 6 of Section 3.4.1 to the reverse martingale $\{\Delta_{jk+l}\}_{l=1}^{k}$. Finally,

$$
\begin{aligned}
\{E|W_{k,jk}|^p\}^{1/p} &= \{E|\Delta_{k,jk+1} + \cdots + \Delta_{k,n}|^p\}^{1/p} \\
&\leq \{CnE|\Delta_{k,l}|^p\}^{1/p} \\
&\leq Cn^{(p-3/2)}\lambda_p^{1/p}
\end{aligned}
$$

again using Theorem 6 of Section 4.3.1. Combining all these, we see that (14) is less than $C|t|^p \rho^{\frac{1}{3}} k^{1-1/p} \lambda_p n^{(1-\frac{3}{2}p)}$, and thus from (13) we get

$$
Z_{2,1}(t) \leq \sum_{k=1}^{n-1}\sum_{j=1}^{m(k)} C\rho^{\frac{1}{3}} k^{1-1/p}\lambda_p n^{(\frac{1}{2}-\frac{3}{2}p)}|t|^p e^{-\frac{(jk-1)}{3n}t^2}. \tag{15}
$$

Next we turn to the bounding of $Z_{22}(t)$. A typical term of $Z_{22}(t)$ is less than

$$
\left|Ee^{itS_{(j+1)k}}\sum_{l=jk+1}^{(j+1)k}\Delta_{k,l}\right|
$$

$$
\leq \sum_{l=jk+1}^{(j+1)k}\left|E\left[\exp(it\sum_{\substack{m=1\\m\neq k}}^{l-1} T_m)\right.\right.
$$

$$
\left.\left.\times \exp\{it(T_k + T_l + \sum_{m=l+1}^{(j+1)k} T_m + E(S_{j+1)k}|\mathbf{Y}_{(j+1)k})\}\Delta_{k,l}\right]\right|
$$

$$
\leq \sum_{l=jk+1}^{(j+1)k}\left|\eta^{l-2}(t/\sigma_1 n^{\frac{1}{2}})\right| E\left|(e^{itT_k} - 1)(e^{itT_l} - 1)\Delta_{k,l}\right|
$$

using the type of conditioning argument employed to estimate $Z_{21}$. By Holder's inequality, the expectation above is less than

$$
\left\{E|(e^{itT_k} - 1)(e^{itT_l} - 1)|^3\right\}^{\frac{1}{3}}\{E|\Delta_{k,l}|^{\frac{3}{2}}\}^{\frac{2}{3}}
$$

$$
\leq |t|^2\{|T_k|^3\}^{\frac{2}{3}}\{E|\Delta_{k,l}|^{\frac{3}{2}}\}^{\frac{2}{3}}
$$

$$
\leq C|t|^2 \rho^{\frac{2}{3}}\lambda_{\frac{3}{2}}^{\frac{2}{3}} n^{-\frac{5}{2}}
$$

and so

$$Z_{2,2}(t) \leq \sum_{k=1}^{n-1} \sum_{j=1}^{m(k)} C\rho^{\frac{2}{3}} \lambda_{\frac{3}{2}}^{\frac{2}{3}} n^{-\frac{5}{2}} |t|^2 \sum_{l=jk+1}^{\min(n,(j+1)k)} e^{-\frac{(l-2)}{3n} t^2}$$

$$\leq C\rho^{\frac{2}{3}} \lambda_{\frac{3}{2}}^{\frac{2}{3}} \left( n^{-\frac{5}{2}} |t|^2 + n^{-\frac{3}{2}} \sum_{l=1}^{n} e^{-\frac{l}{3n} t^2} \right). \tag{16}$$

To complete the proof, we use (12), (15) and (16) to bound the second integral in (9). From (10) we obtain

$$\frac{1}{\pi} \int_{-\sqrt{n}/\rho}^{\sqrt{n}/\rho} |\eta^n(t/\sigma_1 \sqrt{n}) - \phi_n(t)| |t|^{-1} \, dt \leq \frac{1}{\pi} \int_{-\sqrt{n}/\rho}^{\sqrt{n}/\rho} |Z_1(t)| |t|^{-1} dt$$

$$+ \frac{1}{\pi} \int_{-\sqrt{n}/\rho}^{\sqrt{n}/\rho} |Z_{21}(t)| \, dt + \frac{1}{\pi} \int_{-\sqrt{n}/\rho}^{\sqrt{n}/\rho} |Z_{2,2}(t)| \, dt. \tag{17}$$

Using (12), the first term on the right of (17) is less than

$$C\lambda_p n^{(1-\frac{3}{2}p)} \left\{ \int_{-\sqrt{n}/\rho}^{\sqrt{n}/\rho} |t|^{p-1} dt + \sum_{k=1}^{n} \int_{-\infty}^{\infty} |t|^{p-1} e^{-\frac{k}{3n} t^2} \right\}. \tag{18}$$

Using the integral formula

$$\int_{-\infty}^{\infty} |t|^{\nu} e^{-kt^2} \, dt \leq C k^{-\frac{\nu+1}{2}}$$

where C depends neither on $\nu$ nor $k$, we see that (18) is less than

$$C\lambda_p n^{1-\frac{3}{2}p} \left\{ C_1 n^{\frac{p}{2}} + C_2 \sum_{k=1}^{n} \left( \frac{k}{n} \right)^{-p/2} \right\} = C\lambda_p n^{1-p} \left\{ C_1 + C_2 n^{1-p/2} \right\}$$

which is less than $C\lambda_p n^{-\frac{1}{2}}$ when $p = 5/3$.

By (15) the second term of (17) is less than

$$C\rho^{\frac{1}{3}} \lambda_p n^{\frac{1}{2}-\frac{3}{2}p} \sum_{k=1}^{n-1} k^{1-1/p} \sum_{j=1}^{m(k)} \int_{-\sqrt{n}/\rho}^{\sqrt{n}/\rho} |t|^p e^{-\frac{(jk-1)}{3n} t^2} \, dt$$

$$\leq C\rho^{\frac{1}{3}} \lambda_p n^{\frac{1}{2}-\frac{3}{2}p} \left\{ \sum_{j=1}^{n-1} \int_{-\sqrt{n}/\rho}^{\sqrt{n}/\rho} |t|^p e^{-\frac{(j-1)}{3n} t^2} \, dt \right.$$

$$+ \sum_{k=2}^{n-1} k^{1-1/p} \sum_{j=1}^{m(k)} \int_{-\infty}^{\infty} |t|^p e^{-\frac{kj}{6n}t^2} \, dt \Bigg\}$$

$$\leq C\rho^{\frac{1}{3}}\lambda_p n^{\frac{1}{2}-\frac{3}{2}p} \Bigg\{ C_1 n^{\frac{p+1}{2}} \rho^{-(p+1)} + C_2 n^{\frac{p+1}{2}} \sum_{j=1}^{n} j^{-\frac{(p+1)}{2}}$$

$$+ C_3 n^{\frac{p+1}{2}} \sum_{k=2}^{n-1} k^{1-1/p} \sum_{j=1}^{m(k)} (kj)^{-\frac{(p+1)}{2}} \Bigg\}$$

$$\leq C\rho^{\frac{1}{3}}\lambda_p n^{1-p} \{ C_1 \rho^{-(p+1)} + C_2 n^{1-(\frac{p+1}{2})} + C_3 n^{2-\frac{p+1}{2}-1/p} \}. \quad (19)$$

Using the fact that $\rho > 1$, we see that for $p = 5/3$, (19) is less than $C\rho^{\frac{1}{3}}\lambda_p n^{-\frac{1}{2}}$.

Finally, to bound the third term in (17), we use (16), and get

$$\frac{1}{\pi} \int_{-\sqrt{n}/\rho}^{\sqrt{n}/\rho} Z_{2,2}(t) \, dt \leq C\rho^{\frac{2}{3}}\lambda_{\frac{3}{2}}^{\frac{2}{3}} \left( n^{-\frac{5}{2}} \int_{-\sqrt{n}/\rho}^{\sqrt{n}/\rho} |t|^2 \, dt \right.$$

$$\left. + n^{-\frac{3}{2}} \sum_{l=1}^{n} \int_{-\infty}^{\infty} |t|^2 e^{-\frac{l}{3n}t^2} \, dt \right)$$

$$\leq C\rho^{\frac{2}{3}}\lambda_{\frac{3}{2}}^{\frac{2}{3}} (C_1 n^{-1} + C_2 n^{-1/2})$$

$$\leq C\rho^{\frac{2}{3}}\lambda_{\frac{3}{2}}^{\frac{2}{3}} n^{-1/2}.$$

The proof is complete.

An interesting feature of this theorem is that the third moment assumption, which would seem natural for a Berry-Esseen theorem, is required only for the projection and not for the remainder. Calleart and Janssen (1978) proved the theorem assuming a finite third moment for the kernel. Helmers and van Zwet (1982) prove the result for $U$-statistics having $E|h^{(2)}(X_1, X_2)|^p < \infty$ for $p > 5/3$, and Korolyuk and Boroskikh (1985) prove the current theorem. The proof we have given is an adaption to the $U$-statistic case of a very general theorem by Friedrich (1989), who generalises a Berry-Esseen theorem for symmetric statistics by van Zwet (1984). Borovskikh (1984) and Ahmad (1981) consider even weaker assumptions leading to slower convergence rates. Korolyuk and Boroskikh (1988) consider a Berry-Esseen bound for degenerate $U$-statistics and obtain a bound of $o(n^{-\frac{1}{2}})$ in this case. See also de Wet (1987).

105

A Berry-Esseen theorem has been established by Ghosh (1985) for functions of non-degenerate $U$-statistics of degree two. He proves that, if $g$ is a function possessing a bounded second derivative, and

$$T_n = \sqrt{n}\{g(U_n) - g(\theta)\}/2\sigma_1 g'(\theta)$$

then $T_n$ obeys the Berry-Esseen theorem i.e.

$$\sup_x |Pr(T_n \leq x) - \Phi(x)| = O(n^{-\frac{1}{2}}).$$

Further, the results remain true if $g(U_n)$ is replaced by a jackknifed version of $g(U_n)$ (i.e. bias corrected) and if $2\sigma_1(g'(\theta))$ is replaced by a jackknife estimate of the standard error. For material on jackknifing $U$-statistics see Chapter 5.

### 3.3.3 Asymptotic expansions

We begin by reviewing the basic material on asymptotic expansions of Edgeworth type. We follow the discussion in Bickel (1974), using his notation.

If $\{T_n\}$ is a sequence of statistics, with distribution functions $F_n$, an *asymptotic expansion of Edgeworth type* for $F_n$ with $r + 1$ terms is one of the form

$$F_n(x) \sim A_0(x) + \sum_{j=1}^{r} A_j(x)n^{-\frac{j}{2}}.$$

The expansion is *valid to $r + 1$ terms* if

$$|F_n(x) - A_0(x) - \sum_{j=1}^{r} A_j(x)n^{-\frac{j}{2}}| = o(n^{-\frac{r}{2}}) \tag{1}$$

and *uniformly valid to $r + 1$ terms* if the supremum over $x$ of the left hand side of (1) is $o(n^{-\frac{r}{2}})$.

In the case where $T_n$ is asymptotically normal, the function $A_0(x)$ is $\Phi(x)$ and the function $A_j(x)$ can be defined in terms of expansions of the cumulants of $T_n$.

106

Specifically, suppose that the statistic $T_n$ has moments up to $r$th order, with $E(T_n) = 0$, $E(T_n^2) = 1$ and cumulants $\kappa_{j,n}$ admitting expansions of the form

$$\kappa_{j,n} = \sum_{l=0}^{r-j+2} K_j^{(l)} n^{-(j+l-2)/2} + o(n^{-\frac{r}{2}}), j \geq 3. \tag{2}$$

If $\phi_n(t)$ is the characteristic function of $T_n$, then

$$\log \phi_n(t) = \sum_{j=1}^{r} \frac{\kappa_{j,n}}{j!}(it)^j + o(t^r) \tag{3}$$

as $t \to 0$. Substituting (2) into (3) and rearranging terms gives

$$\log \phi_n(t) = -t^2/2 + P_1(it)n^{-\frac{1}{2}} + \cdots + P_r(it)n^{-\frac{r}{2}} + o(n^{-\frac{r}{2}})$$

where the $P's$ are polynomials

$$P_k(x) = \sum_{j=3}^{k+2} \frac{K_j^{(k+2-j)}}{j!} x^j.$$

An equivalent expression for $\phi_n(t)$ is

$$\phi_n(t) = e^{-t/2}\left(1 + \sum_{j=1}^{r} Q_j(it)n^{-\frac{i}{2}}\right) + o(n^{-\frac{r}{2}}) \tag{4}$$

where the $Q's$ are polynomials derived from the $P's$. For example,

$$Q_1(x) = P_1(x) \quad \text{and} \quad Q_2(x) = P_2(x) + \tfrac{1}{2}P_1^2(x).$$

Formal Fourier inversion of (4) gives

$$F_n(x) = \Phi(x) - \phi(x)\left\{\sum_{j=1}^{r}\left\{\sum_{k\geq 1} A_{jk}H_{k-1}(x)\right\}n^{-\frac{i}{2}}\right\} + o(n^{-\frac{r}{2}}) \tag{5}$$

where $Q_j(x) = \sum_{k\geq 1} A_{jk}x^k$, $\phi$ is the standard normal density, and $H_k$ is the $k$th Hermite polynomial defined by $\phi^{(k)}(x) = (-1)^k H_k(x)\phi(x)$.

For the case $r = 2$ we have

$$\kappa_{1,n} = 0,$$

$$\kappa_{2,n} = 1,$$

$$\kappa_{3,n} = K_3^{(0)} n^{-\frac{1}{2}} + K_3^{(1)} n^{-1} + o(n^{-1}), \tag{6}$$

$$\kappa_{4,n} = K_4^{(0)} n^{-1} + o(n^{-1}),$$

so that

$$Q_1(x) = \frac{1}{6} K_3^{(0)} x^3,$$

$$Q_2(x) = \frac{1}{6} K_3^{(1)} x^3 + \frac{1}{24} K_4^{(0)} x^4 + \frac{1}{72} (K_3^{(0)})^2 x^6.$$

Also

$$H_2(x) = x^2 - 1,$$

$$H_3(x) = x^3 - 3x,$$

$$H_5(x) = x^5 - 10x^3 + 15x,$$

so that the formal expansion (5) becomes

$$F_n(x) = \Phi(x) - \phi(x) \left[ \tfrac{1}{6} K_3^{(0)} (x^2 - 1) n^{-\frac{1}{2}} + \left\{ \tfrac{1}{6} K_3^{(1)} (x^2 - 1) \right. \right.$$
$$\left. \left. + \tfrac{1}{24} K_4^{(0)} (x^3 - 3x) + \tfrac{1}{72} (K_3^{(0)})^2 (x^5 - 10x^3 + 15) \right\} n^{-1} \right] + o(n^{-1}). (7)$$

Applying (7) to the classical case where $T_n = \sum(X_i - \mu)/\sqrt{n}\sigma$ for i.i.d. summands gives (2) in Section 3.3.1. To extend this to the $U$-statistic case, where $T_n = \sqrt{n}(U_n - \theta)/s.d.(U_n)$ we need to establish the formula (2) for the asymptotic expansion of the cumulants of $T_n$. According to Bhattacharya and Puri (1983), the expansion (2) is valid for $r = 2$ under fourth moment assumptions, so it remains to compute the $K_j^{(l)}$. For the third cumulant, we have, for kernels of degree 2,

$$\kappa_{3,n} = E\{(U_n - \theta)/s.d.(U_n)\}^3$$
$$= \sigma_1^{-3} n^{-\frac{9}{2}} \left\{ \sum_{(n,2)} (\psi(X_i, X_j) - \theta) \right\}^3 + o(n^{-1}) \qquad (8)$$

since $Var\, U_n = 4\sigma_1^2 n^{-1} + o(n^{-1})$.

Now write $\phi(S) = \psi(X_i, X_j) - \theta$ for $S = \{i, j\}$, and denote by $S_2^{(\nu)}$ the set of triples $(S_1, S_2, S_3)$ of elements of $S_{n,2}$ such that $S_1 \cup S_2 \cup S_3$ has exactly $\nu$ elements. Then

$$E(\sum_{(n,2)} \phi(S))^3 = \sum_{\nu=2}^{6} \sum_{S_2^{(\nu)}} E\{\phi(S_1)\phi(S_2)\phi(S_3)\}, \qquad (9)$$

and in view of (8) and the fact that $S_2^{(\nu)}$ has $O(n^\nu)$ elements, to approximate $\kappa_{3,n}$ up to $O(n^{-1})$ we need only consider the terms in (9) for $\nu = 4, 5$ and 6.

108

For $\nu = 5$ and 6, for any choice of $S_1, S_2$ and $S_3$ in $S_2^{(\nu)}$ at least one of the sets is disjoint from the others and so $E\phi(S_1)\phi(S_2)\phi(S_3)$ is zero by the usual independence arguments. For $\nu = 4$, the only arrangements of $S_1, S_2$ and $S_3$ for which independence arguments do not lead to $E\phi(S_1)\phi(S_2)\phi(S_3) = 0$ are of the types

$$(i,j),(j,k)(k,l) \tag{10}$$

or

$$(i,j),(i,k)(i,l) \tag{11}$$

with $i, j, k, l$ distinct.

By the $H$-decomposition,

$$\psi(X_i, X_j) - \theta = h^{(1)}(X_i) + h^{(1)}(X_j) + h^{(2)}(X_i, X_j)$$

so using the independence argument again, we find that for the arrangement (10) we get

$$\begin{aligned}
E\phi(S_1)\phi(S_2)\phi(S_3) &= Eh^{(2)}(X_1, X_2)h^{(2)}(X_2, X_3)h^{(2)}(X_3, X_4) \\
&\quad + 3Eh^{(1)}(X_1)h^{(2)}(X_1, X_2)h^{(2)}(X_2, X_3).
\end{aligned}$$

However, by the conditioning argument used in the proof of Theorem 3 of Section 1.6 the two terms on the right hand side are both zero, since $E\{h^{(2)}(X_1, X_2)|X_1\} = 0$. Thus we need only consider the arrangement (11). For this type

$$\begin{aligned}
E\{\phi(S_1)\phi(S_2)\phi(S_3)\} &= E\{h^{(1)}(X_1)^3\} \\
&\quad + 3E\{h^{(1)}(X_1)h^{(1)}(X_2)h^{(2)}(X_1, X_2)\} \\
&= K_3\sigma_1^3 \quad \text{say.}
\end{aligned}$$

There are $24\binom{n}{4}$ such arrangements, so that by (8) and (9) we get

$$\kappa_{3,n} = K_3 n^{\frac{1}{2}} + o(n^{-1}). \tag{12}$$

A similar but more complicated argument not given here shows that

$$\kappa_{4,n} = K_4 n^{-1} + o(n^{-1})$$

where

$$K_4\sigma_1^4 = E\{h^{(1)}(X_1)^4\} - 3\sigma_1^4 + 12E\{h^{(1)}(X_1)^2 h^{(1)}(X_2)h^{(2)}(X_1,X_2)\}$$
$$+ 12E\{h^{(1)}(X_1)h^{(1)}(X_2)h^{(2)}(X_1,X_3)h^{(2)}(X_2,X_3)\}.$$

For more general formulae, see Withers (1988). Denote the d.f. of $\sqrt{n}(U_n - \theta)/(s.d.(U_n))$ by $F_n$. Then in the present case the formal expansion (7) takes the form

$$F_n(x) = \Phi(x) - \phi(x) \left[ \tfrac{1}{6}K_3(x^2 - 1)n^{-\frac{1}{2}} \right.$$
$$\left. + \left\{ \tfrac{1}{24}K_4(x^3 - 3x) + \tfrac{1}{72}K_3^2(x^5 - 10x^3 + 15x) \right\} n^{-1} \right] + o(n^{-1}) \quad (13)$$

The validity of this formal expansion up to three terms has been studied by Callaert, Janssen and Veraverbeke (1980) Bickel, Gotze and Van Zwet (1986) and Korolyuk and Borovskikh (1986) who give sufficient conditions for the validity of the expansion.

The first two papers impose natural fourth moment conditions on the kernel sufficient to ensure the finiteness of $K_3$ and $K_4$, and also a condition on the characteristic function of $h^{(1)}(X)$.

Callaert et. al. also impose a complicated technical condition on the characteristic function of a certain sum involving the kernel, while Bickel et. al. require a more easily verified condition, namely that there exist sufficiently many distinct eigenvalues of the kernel $h^{(2)}$ with respect to the common distribution function of the $X's$. Korolyuk and Borovskikh impose a Cramér condition on the characteristic function of $h^{(1)}(X)$. We refer the reader requiring more information to these papers. See also the survey paper by Gupta and Panchapakesan (1982), and Takahashi (1988).

Edgeworth expansions for sequences of statistics having a non-normal limit limit distribution are also possible. Gotze (1979) considers asymptotic expansions for the case $k = 2$. His results are in fact for von Mises functionals, but apply broadly also to $U$-statistics. He also considers expansions for multivariate U-statistics (Gotze (1987)).

110

## 3.4 The strong law of large numbers for $U$-statistics

This section treats the generalisation of the strong law of large numbers to $U$-statistic sequences, and associated rate of convergence results. Since the proofs involve a heavy use of martingale results, we begin by collecting together a few elementary facts about martingales. Readers familiar with this topic may wish to skip Section 3.4.1.

### 3.4.1. Martingales

We now turn to a review of basic martingale theory. No proofs are given for the most standard results, instead the reader is referred to the many texts available. Billingsley (1979) and Breiman (1968) are particularly lucid on this topic.

### Basic definitions

A *martingale* is a sequence $\{X_n\}_{n=1}^{\infty}$ of random variables defined on a probability space $(\Omega, \mathcal{F}, P)$ together with a sequence of $\sigma$-fields $\{\mathcal{F}_n\}_{n=1}^{\infty}$, satisfying the following:

(a) The $\sigma$-fields $\mathcal{F}_n$ are an increasing sequence of sub $\sigma$-fields of $\mathcal{F}$: $\mathcal{F}_1 \subseteq \mathcal{F}_2 \subseteq \ldots \subseteq \mathcal{F}$.

(b) Each random variable $X_n$ is $\mathcal{F}_n$-measurable. (The usual jargon is to say that the $X_n$ are "adapted" to the $\mathcal{F}_n$).

(c) The random variables $X_n$ are all integrable: i.e. $E|X_n| < \infty$ for all $n$.

(d) $E(X_m|\mathcal{F}_n) = X_n$ almost surely for all $m, n$ with $m \geq n$.

The last condition expresses the notion of a "fair game"; if $X_n$ represents a gambler's winnings after time $n$, then the gambler's expected future gain, given his present gain, coincides with his present winnings. The classic example of a martingale is that of a sequence of sums of independent zero mean random variables.

A *submartingale* is a sequence satisfying all the above requirements except (d); a submartingale instead satisfies the weaker condition

(d') $E(X_m|\mathcal{F}_n) \geq X_n$ for $m \geq n$.

111

If $\{X_n\}$ is a martingale and $\phi(x)$ is a convex function, then $\{\phi(X_n)\}$ is a submartingale provided $E\phi(X_n)$ exists for all $n$. Thus $\{|X_n|\}$ and $\{|X_n|^r\}$ for $r > 1$ are submartingales if $\{X_n\}$ is a martingale.

A related concept, more useful for applications to $U$-statistic theory, is that of a reverse martingale. Here the associated $\sigma$-fields $\mathcal{F}_n$ form a decreasing sequence, and in place of (d) a reverse martingale satisfies $E(X_m|\mathcal{F}_n) = X_n$ for $n \geq m$. A reverse submartingale is similar, but instead, satisfies $E(X_m|\mathcal{F}_n) \geq X_n$ for $n \geq m$.

## Convergence results

The r.v.s of a martingale sequence all have the same expectation, and reverse martingale sequences have particularly nice convergence properties. We have

**Theorem 1.** *Let $\{X_n\}$ be a reverse martingale. Then there is a random variable $X$ such that*
*(a) $X$ is measurable with respect to the $\sigma$-field $\mathcal{F}_\infty = \cap_{n=1}^{\infty} \mathcal{F}_n$;*
*(b) $X_n$ converges to $X$ almost surely and in $L_1$;*
*(c) $\lim_{n \to \infty} E(X_1|\mathcal{F}_n) = X$ a.s.*

Since it is true that $\lim E(Z|\mathcal{F}_n) = E(Z|\mathcal{F}_\infty)$ for any decreasing sequence of $\sigma$-fields and any integrable random variable $Z$, the limit in (c) above is in fact $E(X_1|\mathcal{F}_\infty)$. We state this fact for the record as Theorem 2:

**Theorem 2.** *Let $\{X_n\}$ be a reverse martingale. Then $X_n$ converges to $E(X_1|\mathcal{F}_\infty)$ almost surely and in $L_1$.*

## Martingale inequalities

The following results will be useful in the sequel.

**Theorem 3.** *Let $\{X_n\}$ be a reverse martingale, with associated $\sigma$-fields $\mathcal{F}_n$, and suppose that $E(|X_n|^r) < \infty$ for each $n$, where $r > 1$. Then for every $c > 0$ and $m > 1$,*

$$Pr(\sup_{n \geq m} |X_n| > c) \leq c^{-r} E|X_m|^r.$$

**Proof.** First note that in view of the remarks above, $|X_n|^r$ is a reverse submartingale since $X_n$ is a reverse martingale. Next, define the sets $A_j = \cap_{l=j+1}^{n} \{|X_l|^r \leq c^r\} \cap \{|X_j|^r > c^r\}$ for $j = m, \ldots, n$. The $A_j$'s are disjoint and

$$\left\{ \sup_{n \geq j \geq m} |X_j|^r > c^r \right\} = \bigcup_{j=m}^{n} A_j.$$

Moreover, $A_j$ is in $\mathcal{F}_j$ for each $j$, since $\mathcal{F}_j \supseteq \mathcal{F}_{j+1} \supseteq \cdots \supseteq \mathcal{F}_n$. Hence

$$c^r Pr \left( \sup_{m \geq j \geq m} |X_j| > c \right) = c^r Pr \left( \sup_{n \geq j \geq m} |X_j|^r > c^r \right)$$

$$= \sum_{j=m}^{n} c^r Pr(A_j)$$

$$< \sum_{j=m}^{n} \int_{A_j} |X_j|^r dP$$

$$\leq \sum_{j=m}^{n} \int_{A_j} E(|X_m|^r | \mathcal{F}_j) dP$$

$$= \sum_{j=m}^{n} \int_{A_j} |X_m|^r dP$$

$$\leq E|X_m|^r$$

using the facts that $|X_j|^r$ is a submartingale and that $A_j$ is in $\mathcal{F}_j$. Hence

$$Pr \left( \sup_{n \geq j \geq m} |X_j| > c \right) \leq c^{-r} E|X_m|^r \tag{1}$$

and the proof is completed by noting that the probabilities on the left of (1) form an increasing sequence whose limit is $Pr \left( \sup_{n \geq m} |X_n| > c \right)$.

**Theorem 4. (The Hájek-Rényi inequality).** *Let* $\{X_n\}$ *be a forward martingale with* $E(X_n^2)$ *finite for each* $n$, *and let* $c_n$ *be a decreasing sequence of positive constants. Then if* $X_0 = 0$, *we have*

$$Pr \left( \max_{1 \leq j \leq n} |c_j X_j| \geq \varepsilon \right) \leq \varepsilon^{-2} \sum_{j=1}^{n} c_j^2 E(X_j^2 - X_{j-1}^2).$$

113

The proof depends on the concept of a stopping time and several auxiliary lemmas, which we prove for completeness. The proofs are simplified versions of results proved in more generality in Chapter 7 of Chow and Teicher (1978). Note that Ahmad (1980) has a similar result for $U$-statistics.

A *stopping time* is an integer-valued r.v. $T$ such that $\{T = n\}$ is in $\mathcal{F}_n$ for each $n$. If the set of possible values of $T$ is bounded, $T$ is called a bounded stopping time.

**Lemma A.** *Let $\{X_n\}$ be a forward martingale and $T$ a bounded stopping time. Then assuming all expectations exist, we have $EX_T = EX_1$.*

**Proof.** Let $N$ be a bound on $T$ (i.e. $Pr(T \leq N) = 1$). The events $\{T > n\}$ are in $\mathcal{F}_n$ since $\{T > n\}' = \cup_{\nu=1}^{n} \{T = \nu\}$ which is in $\mathcal{F}_n$ because $\{T = \nu\}$ is in $\mathcal{F}_\nu$, the $\sigma$-fields are increasing and $\sigma$-fields are closed under the formation of complements. Now write

$$
\begin{aligned}
EX_1 &= \int_{\{T=1\}} X_1 dP + \int_{\{T>1\}} X_1 dP \\
&= \int_{\{T=1\}} X_1 dP + \int_{\{T>1\}} E(X_2|\mathcal{F}_1) \, dP \\
&= \int_{\{T=1\}} X_T \, dP + \int_{\{T=2\}} X_T \, dP + \int_{\{T>2\}} X_2 dP
\end{aligned}
$$

using the martingale property and the definition of conditional expectation. We can obviously repeat the argument on the last summand and so

$$
\begin{aligned}
E(X_1) &= \sum_{n=1}^{N} \int_{\{T=n\}} X_T \, dP \\
&= \int X_T \, dP \\
&= EX_T.
\end{aligned}
$$

**Lemma B.** *Let $\{X_n, \mathcal{F}_n\}_{n=0}^{\infty}$ be a non-negative (forward) submartingale with $X_0 = 0$ and $\mathcal{F}_0 = \mathcal{F}_1$, and let $T$ be a bounded stopping time with $Pr(T \leq N) = 1$. Let $v_n$ be a decreasing sequence of positive constants. Then*

$$
E(v_T X_T) \leq \sum_{n=1}^{N} v_n E(X_n - X_{n-1}).
$$

114

**Proof.** Define $Y_n = v_n X_n - E \sum_{\nu=1}^{n} \{v_\nu E(X_\nu - X_{\nu-1}|\mathcal{F}_{\nu-1}) + (v_\nu - v_{\nu-1})X_{\nu-1}\}$. Using elementary properties of conditional expectations, it follows easily that $\{Y_n, \mathcal{F}_n\}$ is a forward martingale with $E(Y_1) = 0$, and so by Lemma A, $E(Y_T) = 0$. Thus

$$E(v_T X_T) = E \sum_{\nu=1}^{T} [E\{v_\nu(X_\nu - X_{\nu-1})|\mathcal{F}_{\nu-1}\} + (v_\nu - v_{\nu-1})X_{\nu-1}]$$

$$\leq E \sum_{\nu=1}^{T} E\{v_\nu(X_\nu - X_{\nu-1})|\mathcal{F}_{\nu-1}\} \tag{2}$$

since $E\{(v_\nu - v_{\nu-1})\}X_{\nu-1} \leq 0$ by hypothesis. Because the sequence $\{X_n\}$ is a submartingale, $E(X_\nu - X_{\nu-1}|\mathcal{F}_{\nu-1}) \geq 0$ and each summand in (2) is positive, so

$$E(v_T X_T) \leq E \sum_{\nu=1}^{N} E\{v_\nu(X_\nu - X_{\nu-1})|\mathcal{F}_{\nu-1}\}$$

$$= \sum_{\nu=1}^{N} v_\nu E(X_\nu - X_{\nu-1}).$$

Now we can prove Theorem 4.

Let $A$ be the event that at least one of the r.v.s $c_j^2 X_j^2$, $j = 1, \ldots, n$ is greater than or equal to $\varepsilon^2$, and define a r.v. $T$ by

$$T = \begin{cases} \min\{j : c_j^2 X_j^2 \geq \varepsilon^2\} & \text{on } A, \\ n & \text{on } A'. \end{cases}$$

Then $\{T = j\}$ is in $\mathcal{F}_j$, so $T$ is a bounded stopping time. Now $\{X_n^2\}$ is a submartingale, and on $A$, $c_T^2 X_j^2 \geq \varepsilon^2$ so that

$$\varepsilon^2 Pr\left(\max_{1 \leq j \leq n} |c_j X_j| \geq \varepsilon\right) = \varepsilon^2 Pr\left(\max_{1 \leq j \leq n} c_j^2 X_j^2 \geq \varepsilon^2\right)$$

$$\leq \int_A c_T^2 X_T^2 dP$$

$$\leq E(c_T^2 X_T^2)$$

$$\leq \sum_{j=1}^{n} c_j^2 E(X_j^2 - X_{j-1}^2)$$

115

by Lemma B.

**Corollary 1.** *If $\{X_n\}$ is a reverse martingale sequence, then $X_n, \ldots, X_1$ is a forward martingale sequence, and so by Theorem 4, if $\{c_n\}$ is an increasing sequence and $X_n$ a reverse martingale, then*

$$Pr\left(\max_{1 \le j \le n} |c_j X_j| \ge \varepsilon\right) \le \varepsilon^{-2} \sum_{j=1}^{n} c_j^2 E(X_j^2 - X_{j+1}^2)$$

*where $X_{n+1} = 0$.*

**Corollary 2.** *Let $\{X_n\}$ be a reverse martingale, and $\{c_n\}$ an increasing sequence such that $\sum_{n=1}^{\infty} c_n^2 (E(X_n^2) - E(X_{n+1}^2)) < \infty$. Then*

$$Pr\left(\sup_{n \ge m} |c_n X_n| > \varepsilon\right) \le \varepsilon^{-2} \sum_{n=m}^{\infty} c_n^2 (E(X_n^2) - E(X_{n+1}^2)).$$

**Proof.** The sets $\{\sup_{N > n \ge m} |c_n X_n| > \varepsilon\}$ increase with $N$, and their limit is the set $\{\sup_{n \ge m} |c_n X_n| > \varepsilon\}$. Letting $N \to \infty$ in both sides of

$$Pr\left(\max_{N \ge n \ge m} |c_n X_n| > \varepsilon\right) \le \varepsilon^{-2} \sum_{n=m}^{N} c_n^2 (E(X_n^2) - E(X_{n+1}^2))$$

proves the corollary.

**Corollary 3.** *Let $\{X_n\}$ be a forward martingale and $a \ge 1$. Then*

$$Pr\left(\max_{1 \le j \le n} |X_j| \ge \varepsilon\right) \le \varepsilon^{-a} E|X_n|^a.$$

**Proof.** The proof is almost identical to that of Theorem 4 and is omitted.

Our next result is a martingale inequality due to Dharmadhikari, Fabian and Jogdeo (1968).

**Theorem 5.** *Let $\{X_n\}$ be a forward martingale, with $X_0 = 0$ and define $\gamma_{nr}$ by $\gamma_{nr} = E|X_n - X_{n-1}|^r$ and $\beta_{nr}$ by $\beta_{nr} = n^{-1} \sum_{j=1}^{n} \gamma_{jr}$. Then there is a constant $C_r$ depending only on $r$ such that for all $r \ge 2$ and $n = 1, 2, \ldots$*

$$E(|X_n|^r) \le C_r \beta_{nr} n^{r/2}.$$

116

For a proof, which is elementary but quite lengthy, see the reference.

Our final result is an inequality for reverse martingales due to Chatterji (1969), which is used in our proof of the Berry-Esseen Theorem in Section 3.3.2.

**Theorem 6.** Let $\{X_n, \mathcal{F}_n\}$ be a reverse martingale satisfying $E|X_n|^p < \infty$ for $1 \leq p \leq 2$, and with $X_{n+1} = 0$. Then

$$E|X_1|^p \leq 2(\sum_{j=1}^{n} E|X_j - X_{j+1}|^p).$$

**Proof.** The theorem is proved by induction. For $n = 2$ the result is trivial, so suppose that the result is true for $n - 1$. Consider the inequality

$$|1 + y|^p - 1 - py \leq 2|y|^p$$

which is valid for all $y$ and $1 \leq p \leq 2$. Replacing $y$ by $y/x$ and multiplying by $|x|^p$ gives

$$|x + y|^p \leq |x|^p + 2p|x|^{p-1}\text{sgn}(x)y + 2|y|^p$$

and hence

$$|X_1|^p \leq E|X_2|^p + 2pE\{|X_2|^{p-1}\text{sgn}(X_2)(X_1 - X_2)\} + 2E|X_1 - X_2|^p. \quad (3)$$

The induction step will be completed and the theorem proved if we can show that the middle term of the right hand side of (3) is zero. This follows from

$$E\{|X_2|^{p-1}\text{sgn}(X_22)(X_1 - X_2)\} = E\left\{E\{|X_2|^{p-1}\text{sgn}(X_2)(X_1 - X_2)|\mathcal{F}_2\}\right\}$$

which is zero since the reverse martingale property entails

$$E\{|X_2|^{p-1}\text{sgn}(X_2)(X_1 - X_2)|\mathcal{F}_2\}$$
$$= |X_2|^{p-1}\text{sgn}(X_2)E(X_1|\mathcal{F}_2) - |X_2|^p\text{sgn}(X_2)X_2$$
$$= 0.$$

### 3.4.2  *U*-statistics as martingales and the SLLN

The relevance of Section 3.4.1 to our theme is simply this: *U*-statistics are martingales, in fact are both forward and reverse martingales, as our next two results show. The first, due to Hoeffding (1961), represents $U_n$ as a forward martingale. The second, essentially simpler, was discovered by Berk (1966) and is a reverse martingale representation.

**Theorem 1.** *Let $U_n$ be a sequence of U-statistics based on a kernel $\psi$ satisfying $E|\psi(X_1,\ldots,X_k)| < \infty$, and let $\mathcal{F}_n = \sigma(X_1,\ldots,X_n)$. Then $\{\binom{n}{c} H_n^{(c)}\}_{n=c}^{\infty}$ is a martingale adapted to the $\mathcal{F}_n$ for $c = 1,\ldots,k$.*

**Proof.** Since the $\sigma$-fields $\mathcal{F}_n$ clearly increase and the r.v.s $H_n^{(c)}$ are clearly integrable and measurable $\mathcal{F}_n$, it remains only to prove (d) in the definition of Section 3.4.1. We have

$$E\left\{\binom{n+1}{c} H_{n+1}^{(c)} \,\middle|\, \mathcal{F}_n\right\} = \sum_{(n+1,c)} E\{h^{(c)}(X_{i_1},\ldots,X_{i_c})|\mathcal{F}_n\},$$

and $E\{h^{(c)}(X_{i_1},\ldots,X_{i_c})|\mathcal{F}_n\} = h^{(c)}(X_{i_1},\ldots,X_{i_c})$ provided no index $i_j$ equals $n+1$, and is zero otherwise by the properties of the $H$-decomposition. Hence

$$E\left\{\binom{n+1}{c} H_{n+1}^{(c)} \,\middle|\, \mathcal{F}_n\right\} = \sum_{(n,c)} h^{(c)}(X_{i_1},\ldots,X_{i_c})$$
$$= \binom{n}{c} H_n^{(c)}$$

as required.

**Theorem 2.** *With the hypotheses of Theorem 1, $\{U_n\}_{n=k}^{\infty}$ is a reverse martingale adapted to the $\sigma$-fields $\mathcal{F}_n = \sigma(U_n, U_{n+1},\ldots)$*

**Proof.** Clearly the properties (a) - (c) of the definition of a martingale are satisfied. For property (d), note that because of the inherent symmetry involved, we can write

$$E(\psi(X_{i_1},\ldots,X_{i_k})|\mathcal{F}_n) = E(\psi(X_1,\ldots,X_k)|\mathcal{F}_n)$$

for every subset $\{i_1, \ldots, i_k\}$, of $\{1, 2, \ldots, n\}$ and hence

$$U_n = E(U_n|\mathcal{F}_n) = \binom{n}{k}^{-1} \sum_{(n,k)} E(\psi(X_{i_1}, \ldots, X_{i_k})|\mathcal{F}_n)$$

$$= E(\psi(X_1, \ldots, X_k)|\mathcal{F}_n).$$

To verify (d), let $n \geq m$. Then

$$E(U_m|\mathcal{F}_n) = E\{E(\psi(X_1, \ldots, X_k)|\mathcal{F}_m) \mid \mathcal{F}_n\}$$

$$= E\{\psi(X_1, \ldots, X_k)|\mathcal{F}_n\}$$

$$= U_n$$

since $\mathcal{F}_n \subseteq \mathcal{F}_m$.

An alternative characterisation of $U_n$ as a reverse martingale is possible: For our i.i.d. sequence $X_1, X_2, \ldots$, let $\mathbf{X}_{(n)}$ denote the $n$-vector of order statistics from the sample $X_1, \ldots, X_n$, and let $(R_1, \ldots, R_n)$ denote the corresponding vector of ranks. Knowledge of $\mathbf{X}_{(n+1)}$ and $R_{n+1}$ implies knowledge of $\mathbf{X}_{(n)}$ and $X_{n+1}$, so that if we set $\mathcal{F}_n = \sigma(\mathbf{X}_{(n)}, X_{n+1}, X_{n+2}, \ldots)$, then $\mathcal{F}_{n+1} = \sigma(\mathbf{X}_{(n+1)}, X_{n+2}, \ldots) \subseteq \sigma(\mathbf{X}_{(n+1)}, R_{n+1}, X_{n+2}, \ldots) = \sigma(\mathbf{X}_{(n)}, X_{n+1}, X_{n+2}, \ldots) = \mathcal{F}_n$ so that the $\mathcal{F}_n$'s form a decreasing sequence of $\sigma$-fields. Moreover, $U_n$ has the representation

$$U_n = E(\psi(X_1, \ldots, X_k)|\mathcal{F}_n)$$

since

$$E(\psi(X_1, \ldots, X_k)|\mathcal{F}_n) = E(\psi(X_1, \ldots, X_k)|\mathbf{X}_{(n)})$$

$$= \sum_{(n)} E(\psi(X_1, \ldots, X_k)|\mathbf{X}_{(n)}, \mathbf{R}_{(n)} = \mathbf{r}^{-1})/(n!)$$

$$= \sum_{(n)} \frac{k!(n-k)!}{n!} \psi(X_{r_1}, \ldots, X_{r_k})$$

$$= U_n$$

where $\mathbf{R}_{(n)}$ is the vector of ranks and $\mathbf{r}^{-1}$ denotes the permutation which is the inverse of $\mathbf{r}$. Thus $U_n$ is $\mathcal{F}_n$ measurable, and as before,

$$E(U_n|\mathcal{F}_{n+1}) = E\{E(\psi(X_1, \ldots, X_k)|\mathcal{F}_n)|\mathcal{F}_{n+1}\}$$

$$= E(\psi(X_1, \ldots, X_k)|\mathcal{F}_{n+1})$$

$$= U_{n+1}$$

119

so that $U_n$ is a reverse martingale.

As an example of the use of the martingale property of $U$-statistics, we derive a bound for the central moments of $U$-statistics that is more precise than that given in Section 1.5.

**Theorem 3.**   *Let $X_1, \ldots, X_n$, be independent and identically distributed r.v.s with d.f. $F$ and suppose that $U_n$ is a $U$-statistic with kernel $\psi$ of degree $k$ based on the $X's$. Let $\gamma = E|\psi(X_1, \ldots, X_k) - \theta|^r$ and further suppose that $\gamma$ is finite for some $r \geq 2$. Then there is a constant $C_r$ depending only on $r$ such that*

$$E|U_n - \theta|^r \leq C_r \gamma n^{-r/2}.$$

**Proof.** We give a proof for the case $k = 2$. An extension to general $k$ may be found in Janssen (1981).

Let $U_n - \theta = 2H_n^{(1)} + H_n^{(2)}$ be the $H$-decomposition of $U_n$. We first derive a bound for $E|H_n^{(1)}|^r$. Note that

$$\int |h^{(1)}(x)|^r dF = \int \left| \int (\psi(x_1, x_2) - \theta) \, dF(x_1) \right|^r dF(x_2)$$

$$\leq \int \int |\psi(x_1, x_2) - \theta|^r dF(x_1) \, dF(x_2)$$

so that $E|h^{(1)}(X_1)|^r < \gamma$. Also, $nH_n^{(1)}$ is a forward martingale by Theorem 1, so by Theorem 5 of Section 3.4.1, there exists a constant $C_r$ depending only on $r$ such that

$$E|H_n^{(1)}|^r \leq C_r E|h^{(1)}(X_1)|^r n^{-r/2}, \tag{1}$$

and so $E|H_n^{(1)}|^r \leq C_r \gamma n^{-r/2}$. (Here and in the rest of the proof we use $C_r$ to denote a generic constant depending only on $r$. The actual value of $C_r$ changes according to context.)

Now we find an analogous bound for $H_n^{(2)}$. Define r.v.s $\xi_j$ by $\xi_j = \sum_{i=1}^{j-1} h^{(2)}(X_i, X_j)$ for $j > 1$. Then $\binom{n}{2} H_n^{(2)} = \xi_2 + \cdots + \xi_n$ and $\{\binom{n}{2} H_n^{(2)}\}$ is a forward martingale by Theorem 1. Applying Theorem 5 of Section 3.4.1, we obtain

$$E\left|\binom{n}{2} H_n^{(2)}\right|^r \leq C_r \max_{1 < j \leq n} E|\xi_j|^r n^{r/2}. \tag{2}$$

120

To bound $E|\xi_j|^r$, consider the sequence

$$W_k = \sum_{i=1}^{k} h^{(2)}(X_i, X_j)$$

for $k = 1, 2, \ldots, j - 1$. It is easy to see that $\{W_k\}$ is a martingale adapted to the $\sigma$-fields $\sigma(X_1, \ldots, X_k, X_j)$, and so by using Theorem 5 of Section 3.4.1 again we get

$$E|\xi_j|^r = E|W_{j-1}|^r \leq C_r E|h^{(2)}(X_1, X_2)|^r j^{r/2}.$$

Thus

$$\max_{1 < j \leq n} E|\xi_j|^r \leq C_r E|h^{(2)}(X_1, X_2)|^r n^{r/2}$$

and so from (2) we get

$$E|H_n^{(2)}|^r \leq C_r E|h^{(2)}(X_1, X_2)|^r n^{-r}. \qquad (3)$$

Since $h^{(2)}(x_1, x_2) = \psi(x_1, x_2) - h^{(1)}(x_1) - h^{(2)}(x_2) - \theta$, it follows by Minkowski's inequality that $E|h^{(2)}(X_1, X_2)|^r \leq C_r \gamma$ and so (3) becomes

$$E|H_n^{(2)}|^r \leq C_r \gamma n^{-r}.$$

Combining (1) and (2) and using Minkowski's inequality again yields the result.

A theorem to cover the case $1 \leq r < 2$ has been proved by Chen (1980). He shows that in this case

$$E|U_n - \theta|^r \leq C E|\psi(X_1, \ldots, X_k) - \theta|^r n^{1-r}.$$

For the degenerate case, with $\sigma_1 = 0$, he also shows that $n^{1-r}$ can be changed to $n^{2(1-r)}$ in the above estimate. More refined asymptotic moment formulae are given in Ronzhin (1982). See also Korolyuk and Borovskikh (1986a).

Next we use the results developed above to address the question of the strong consistency of $U$-statistics. Note that Theorem 2 implies that $U_n$ converges almost surely to $E(\psi(X_1, \ldots, X_k)|\mathcal{F}_\infty)$.

It turns out that this random variable is a constant, and hence is equal to its expectation $\theta = E\psi(X_1, \ldots, X_n)$, so that $U_n$ converges a.s. to $\theta$ or in other words obeys the strong law of large numbers (i.e. is strongly consistent). This was first proved by Hoeffding, using forward martingale arguments. Proofs using the reverse representation is also possible, being similar to that used in the usual "martingale" proof of the SLLN for i.i.d. sequences given by Doob (see e.g. Doob (1953) p341). We state the theorem and given proofs of both types. We also give a proof based on the Hewitt-Savage zero-one law.

**Theorem 3.** *Suppose $E|\psi(X_1, \ldots, X_k)| < \infty$. Then $U_n$ converges a.s. to $\theta$.*

### First Proof (Hoeffding (1961))

Hoeffding's proof is contained in an unpublished technical report so we shall accordingly give a full account of it. Although elementary, it is quite lengthy so we give a summary of the proof before getting down to the details. For simplicity we consider only the case $k = 2$.

Because of the $H$-decomposition and the strong law for i.i.d. summands, we need only prove that, if $S_n$ denotes the sum

$$S_n = \sum_{(n,2)} h^{(2)}(X_i, X_j)$$

then $n^{-2} S_n \to 0$ almost surely. For notational convenience we drop the superscript from $h^{(2)}(x, y)$, and write $h$ for $h^{(2)}$. Define

$$h_{(j)}(x, y) = \begin{cases} h(x, y) & \text{if } |h(x,y)| \leq j^2 \text{ ;} \\ 0 & \text{otherwise,} \end{cases} \qquad (4)$$

and define $S'_n = \sum_{1 \leq i < j \leq n} h_{(j)}(X_i, X_j)$. We then prove

$$n^{-2}(S_n - S'_n) \to 0 \quad a.s. \qquad (5)$$

and

$$n^{-2} S'_n \to 0 \quad a.s. \qquad (6)$$

122

The result (5) is proved using ad hoc arguments, while (6) is proved using martingale arguments, in particular those of Lemma B below.

We begin by establishing two auxiliary results, the second of which is due to Hoeffding.

**Lemma A.** Let $\{X_k\}$ be a sequence of random variables, and suppose that $\sup_{m>n} |X_m| \xrightarrow{P} 0$. Then $X_n \to 0$ a.s.

**Proof.** First note that the set $\{X_n \to 0\}$ can be written

$$\bigcap_{k=1}^{\infty} \bigcup_{n=1}^{\infty} \bigcap_{m=n}^{\infty} \left\{ |X_m| < \frac{1}{k} \right\}.$$

Further, the sets $\left\{ |X_m| < \frac{1}{k} \right\}$ decrease with $k$, so that if we define sets $B_k$ by $B_k = \bigcup_{n=1}^{\infty} \bigcap_{m=n}^{\infty} \left\{ |X_m| < \frac{1}{k} \right\}$ then $\{B_k\}$ is a decreasing sequence of sets and hence

$$Pr(X_n \to 0) = Pr\left( \bigcap_k B_k \right) = \lim_k Pr(B_k). \tag{7}$$

But for all $n$

$$Pr(B_k) \geq Pr\left( \bigcap_{m>n} \left\{ |X_m| < \frac{1}{k} \right\} \right) = Pr\left( \sup_{m>n} |X_m| < \frac{1}{k} \right)$$

so letting $n \to \infty$ we obtain $Pr(B_k) = 1$ and hence the result follows from (7).

**Lemma B.** Let $X_n$ be a forward martingale and suppose that $E|X_n|^a < \infty$ for all $n$ and some $a \geq 1$. Further suppose that for some $b > 0$

$$\sum_{\nu=1}^{\infty} 2^{-ab\nu} E|X_{2^\nu}|^a < \infty. \tag{8}$$

Then $n^{-b} X_n \to 0$ a.s.

**Proof.** By Corollary 3 of Section 3.4.1, for every integer $\nu$ it follows that

$$Pr\left( \max_{2^{\nu-1} \leq m \leq 2^\nu} |X_m| > \varepsilon \right) \leq \varepsilon^{-a} E|X_{2^\nu}|^a. \tag{9}$$

123

Also

$$\left\{ \sup_{m \geq n} \left| m^{-b} X_m \right| > \varepsilon \right\} \subseteq \bigcup_{\nu \geq [\log_2 n + 1]} \left\{ \max_{2^{\nu-1} \leq m < 2^\nu} \left| m^{-b} X_m \right| > \varepsilon \right\}$$

so that

$$Pr \left( \sup_{m \geq n} \left| m^{-b} X_m \right| > \varepsilon \right) \leq \sum_{\nu \geq [\log_2 n + 1]} Pr \left( \max_{2^{\nu-1} \leq m < 2^\nu} \left| m^{-b} X_m \right| > \varepsilon \right)$$

$$\leq \sum_{\nu \geq [\log_2 n + 1]} Pr \left( \max_{2^{\nu-1} \leq m < 2^\nu} \left| X_m \right| > 2^{(\nu-1)b} \varepsilon \right)$$

$$\leq \varepsilon^{-a} 2^{ba} \sum_{\nu \geq [\log_2 n + 1]} (1/2)^{\nu ba} E |X_{2^\nu}|^a$$

by (9). In view of (8), as $n \to \infty$ we get $\sup_{m \geq n} \left| m^{-b} X_m \right| \xrightarrow{p} 0$, and hence $n^{-b} X_n \to 0$ a.s. by Lemma A.

**Proof of Theorem 3 (Continued)** Our first task is to establish (5). To this end, let $Y_{ij} = h(X_i, X_j)$ and $Y'_{ij} = h_{(j)}(X_i, X_j)$. Then if $m < n$,

$$n^{-2}(S_n - S'_n) = n^{-2}(S_m - S'_m) + \sum_{i=1}^{m} n^{-2} \sum_{j=m+1}^{n} (Y_{ij} - Y'_{ij})$$

$$+ n^{-2} \sum_{m < i < j \leq n} (Y_{ij} - Y'_{ij}). \qquad (10)$$

The first term in (10) obviously converges to zero a.s. for fixed $m$, as $n \to \infty$. To demonstrate the convergence of the second term, it is enough to prove that $n^{-2} \sum_{j=2}^{n} (Y_{1j} - Y'_{1j}) \to 0$ a.s. Define

$$g^{(j)}(x, y) = \begin{cases} h(x, y) & \text{if } h(x, y) > j^2 ; \\ 0 & \text{otherwise,} \end{cases}$$

and set $g_1^{(j)}(x) = E g^{(j)}(X_1, x)$. Then writing $V_n = \sum_{j=2}^{n} \{ g^{(j)}(X_1, X_j) - g_1^{(j)}(X_1) \}$ we have

$$n^{-2} \sum_{j=2}^{n} (Y_{1j} - Y'_{1j}) = n^{-2} V_n + n^{-2} \sum_{j=2}^{n} g_1^{(j)}(X_1). \qquad (11)$$

124

It is easily seen that $V_n$ is a forward martingale. Also, because of the inequalities $E|g^{(j)}(X_1, X_2)| \leq E|h(X_1, X_2)|$ and $E|g_1^{(j)}(X)| \leq E|h(X_1, X_2)|$, it follows that $E|V_n| = O(n)$. (The integrability of $h$ follows from the assumption that $E|\psi(X_1, X_2)|$ exists). Thus we may apply Lemma B with $b = 2, a = 1$ to conclude that $n^{-2}V_n \to 0$ a.s. For the second term in (11), note that the function $g_1^{(j)}$ is bounded by the non-negative function g defined by

$$g(x) = \int |h(x, y)| dF(y)$$

which is finite a.e. and so

$$|n^{-2} \sum_{j=2}^{n} g_1^{(j)}(X_1)| \leq n^{-1} g(X_1).$$

Thus the second term in (11) converges to zero a.s., and we see from (11) that the second term of (10) converges to zero a.s.

For the third term in (10), note that

$$Pr\left(\bigcap_{m<i<j<\infty} \{Y_{ij} = Y_{ij}'\}\right) = 1 - Pr\left(\bigcup_{m<i<j<\infty} \{Y_{ij} \neq Y_{ij}'\}\right)$$

$$\geq 1 - \sum_{m<i<j<\infty} Pr(Y_{ij} \neq Y_{ij}')$$

and that

$$Pr(Y_{ij} \neq Y_{ij}') = \int_{|h(x,y)|>j^2} dF(x)dF(y)$$

$$\leq \sum_{\nu=j^2}^{\infty} \nu^{-1} a_{\nu+1}$$

where

$$a_\nu = \int_{\nu-1<|h(x,y)|\leq\nu} |h(x, y)| dF(x) dF(y).$$

Note that the sequence $\{a_\nu\}$ satisfies $\sum_{\nu=1}^{\infty} a_\nu = E|h(X_1, X_2)| < \infty$.

We can write

$$\sum_{m<i<j<\infty} Pr[Y_{ij} \neq Y'_{ij}] \leq \sum_{m<i<j<\infty} \sum_{\nu=j^2}^{\infty} \nu^{-1} a_{\nu+1}$$

$$\leq \sum_{\nu=(m+2)^2}^{\infty} \nu^{-1} \sum_{m+2\leq j\leq\nu^{\frac{1}{2}}} (j-m-1)a_{\nu+1}$$

$$\leq \sum_{\nu=(m+2)^2}^{\infty} a_{\nu+1},$$

and since the series $\sum_{\nu=1}^{\infty} a_\nu$ converges, we get

$$\lim_{m\to\infty} Pr\left(\bigcap_{m<i<j<\infty} \{Y_{ij} = Y'_{ij}\}\right) = 1.$$

Now

$$\bigcap_{m<i<j<\infty} \{Y_{ij} = Y'_{ij}\} \subseteq \bigcap_{m<i<j<\infty} \{|Y_{ij} - Y'_{ij}| < \varepsilon/2\}$$

$$\subseteq \left\{\sup_{m<i<j<\infty} |Y_{ij} - Y'_{ij}| < \varepsilon\right\}$$

so that

$$\lim_{m\to\infty} Pr\left(\sup_{m<i<j<\infty} |Y_{ij} - Y'_{ij}| < \varepsilon\right) = 1$$

for all $\varepsilon > 0$. By Lemma A, it follows that $Y_{ij} - Y'_{ij} \to 0$ a.s. as $i,j \to \infty$ with $i < j$. Since the convergence of a sequence $x_n$ to zero implies the convergence of $n^{-1}\sum_{i=1}^n x_i$ to zero, we must have $n^{-2}\sum_{1\leq i<j\leq n} Y_{ij}-Y'_{ij} \to 0$ a.s.

We now turn to the proof of (6). For fixed $j$, let

$$h_{(j)}(x,y) = h_{(j)}^{(1)}(x) + h_{(j)}^{(2)}(y) + h_{(j)}^{(2)}(x,y) = \theta_j$$

be a (slightly modified) $H$-decomposition of $h_{(j)}$, where

$$\theta_j = \int\int h_{(j)}(x,y)dF(x)dF(y)$$

126

and

$$h_{(j)}^{(1)}(x) = \int h_{(j)}(x,y)dF(y). \tag{12}$$

Then

$$n^{-2}S_n' = n^{-2} \sum_{1 \le i < j \le n} \{h_{(j)}^{(1)}(X_i) + h_{(j)}^{(1)}(X_j)\} + n^{-2} \sum_{1 \le i < j \le n} h_{(j)}^{(2)}(X_i, X_j)$$

$$- n^{-2} \sum_{j=2}^{n} (j-1)\theta_j. \tag{13}$$

The functions $h_{(j)}(x,y)$ converge to zero as $j \to \infty$ and are dominated by the integrable function $|h(x,y)|$, so that $\theta_j \to 0$ by the dominated convergence theorem. Thus by a standard result, $n^{-1}\sum_{j=1}^{n}\theta_j \to 0$ and the third sum in (13) converges to zero since

$$\left| n^{-2} \sum_{j=2}^{n} (j-1)\theta j \right| \le n^{-1} \sum_{j=2}^{n} |\theta_j|.$$

Now consider the first sum in (13). The functions $h_{(j)}^{(1)}(x)$ defined by (12) satisfy

$$h_{(j)}^{(1)}(x) = - \int_{|h(x,y)|>j^2} h(x,y)dF(y)$$

since $Eh(X_1, x) = 0$, and so

$$|h_j^{(1)}(x)| \le \int_{|h(x,y)|>j^2} |h(x,y)|dF(y).$$

If we denote this last integral by $f_j(x)$, we see that the $f_j$'s form a decreasing sequence of non-negative functions converging to zero and are dominated by the integrable function $E|h(X_1, x)|$. Again using the dominated convergence theorem we see that $\int f_j(x)dF(x) \to 0$.

Now write for fixed $m < n$

$$\left| n^{-2} \sum_{1 \le i < j \le n} \{h_j^{(1)}(X_i) + h_j^{(1)}(X_j)\} \right|$$

$$\le n^{-2} \sum_{1 \le i < j \le n} \{f_j(X_i) + f_j(X_j)\}$$

127

$$\leq n^{-2} \sum_{1 \leq i < j \leq m} (f_0(X_i) + f_0(X_j)) + n^{-2} \sum_{i=1}^{m} \sum_{j=m+1}^{n} \{f_0(X_i) + f_m(X_j)\}$$

$$+ n^{-2} \sum_{m < i < j \leq n} \{f_m(X_i) + f_m(X_j)\}$$

$$= n^{-2}(n-1) \sum_{i=1}^{m} f_0(X_i) + n^{-2}(n-1) \sum_{i=m+1}^{n} f_m(X_i). \tag{14}$$

Clearly the first term of (14) converges a.s. to zero, while by the strong law of large numbers for i.i.d. summands the second converges a.s. to $Ef_m(X_i) = \int f_m(x)dF(x)$, which can be made arbitrarily small by choosing $m$ sufficiently large. Hence the first sum in (13) converges a.s. to zero.

For the second sum in (13), let $T_n = \sum_{1 \leq i < j \leq n} h_{(j)}^{(2)}(X_i, X_j)$. Then $T_n$ is a forward martingale, and using Theorem 3(ii) of Section 1.6 we can deduce that $Cov(h_{(j)}^{(2)}(X_i, X_j), h_{(j)}^{(2)}(X_\mu, X_\nu)) = 0$ unless $\{\mu, \nu\} = \{i, j\}$. Thus, using the fact that $E|h_{(j)}^{(2)}(X_1, X_2)|^2 \leq E|h_{(j)}(X_1, X_2)|^2$, we can write

$$E(T_n^2) = \sum_{1 \leq i < j \leq n} \int (h_{(j)}^{(2)}(x, y))^2 dF(x)dF(y)$$

$$\leq \sum_{j=2}^{n}(j-1) \sum_{\nu=1}^{j^2} \nu a_\nu$$

$$\leq n^2 \sum_{\nu=1}^{n^2} \nu a_\nu.$$

Hence $\sum_{l=1}^{\infty} 2^{-4l} E|T_{2^l}|^2 \leq 2 \sum_{\nu=1}^{\infty} a_\nu < \infty$ and so, using Lemma B with values of $a$ and $b$ equal to 2, we obtain $n^{-2}T_n \to 0$ a.s. and so all terms in (13) converge a.s. to 0. The theorem is proved.

**Second Proof.** Since $U_n$ is a reverse martingale adapted to the $\sigma$-field $\mathcal{F}_n$, we have by Theorem 2 of Section 3.4.1

$$\lim_n U_n = \lim_n E(\psi(X_1, \ldots, X_k)|\mathcal{F}_n)$$

$$= E(\psi(X_1, \ldots, X_k)|\mathcal{F}_\infty)$$

where $\mathcal{F}_\infty$ is the $\sigma$-field $\bigcap_{n=k}^{\infty} \mathcal{F}_\infty$. Hence

$$E(\lim_n U_n) = E\{E(\psi(X_1, \ldots, X_k)|\mathcal{F}_\infty)\}$$

$$= E\psi(X_1, \ldots, X_k),$$

128

by Theorem 3.4.4 of Billingsley (1979).

Now let $\mathcal{G}_n = \sigma(X_n, X_{n+1}, \ldots)$, and denote by $\mathcal{G}_\infty$ the tail $\sigma$-field $\bigcap_{n=k}^\infty \mathcal{G}_n$. By the Kolmogorov Zero-One law (see e.g. Billingsley, Theorem 2.2.1), the events of $\mathcal{G}_\infty$ have probability 0 or 1, and $\mathcal{G}_\infty$-measurable functions are almost surely constant. Hence it is enough to prove that $\lim U_n$ is $\mathcal{G}_\infty$-measurable, for then $\lim U_n$ is constant, and hence must equal its expectation $E\psi(X_1, \ldots, X_k)$.

To prove that $\lim U_n$ is $\mathcal{G}_\infty$-measurable, we must show that $\lim U_n$ is $\mathcal{G}_m$-measurable for some arbitrary $m > k$. Write

$$U_n = \binom{n}{k}^{-1} S_n + \binom{n}{k}^{-1} T_n \tag{15}$$

where $T_n$ is the sum of all terms $\psi(X_{i_1}, \ldots, X_{i_k})$ for which at least one index $i_j$ is $< m$, and $S_n$ the sum of all terms all of whose indices are greater than or equal to $m$. Then $\binom{n}{k}^{-1} S_n$ is $\mathcal{G}_m$-measurable for all $n \geq m$, and thus so is $\lim_n \binom{n}{k}^{-1} S_n$, provided the limit exists.

We claim that $\lim_n \binom{n}{k}^{-1} T_n$ is zero. To see this, note that $T_n$ is a sum of $k-1$ terms $T_n^{(1)}, \ldots, T_n^{(k-1)}$, the $p^{th}$ of which consists of $\psi$ summed over sets having exactly $p$ indices $\geq m$, and $(k-p)$ indices $< m$. Thus $T_n^{(p)}$ is a sum of $\binom{m-1}{k-p}$ terms of the form

$$\sum_{m \leq i_1 < \cdots < i_p \leq n} \psi(X_{j_1}, \ldots, X_{j_{k-p}}, X_{i_1}, \ldots, X_{i_p})$$

where $j_1, \ldots, j_{k-p}$ are fixed integers with $1 \leq j_1 < \ldots < j_{k-p} < m$. Now for fixed $j_1, \ldots, j_{k-p}$, the sum above when divided by $\binom{n-m+1}{p}$ is a reverse martingale, as may be seen by copying the proof of Theorem 2 and hence converges to an integrable r.v. Thus $\binom{n-m+1}{p}^{-1} T_n^{(p)}$ converges to an integrable r.v. for $p = 1, 2, \ldots, k-1$, and so

$$\lim_n \binom{n}{k}^{-1} T_n = \lim_n \binom{n}{k}^{-1} \sum_{p=1}^{k-1} T_n^{(p)}$$

$$= \sum_{p=1}^{k-1} \lim_n \binom{n-m+1}{p} \binom{n}{k}^{-1} \binom{n-m+1}{p}^{-1} T_n^{(p)}$$

$$= 0$$

129

since $\binom{n-m+1}{p}\binom{n}{k}^{-1} = O(n^{-1})$. Hence from (15) we see that $\lim_n \binom{n}{k}^{-1} S_n$ exists and in fact equal $\lim_n U_n$, so that $\lim_n U_n$ is $\mathcal{G}_m$-measurable. Since $m$ is arbitrary, $\lim_n U_n$ is $\mathcal{G}_\infty$-measurable and hence is constant a.s.

**Third proof.** A third proof, due to Arvesen (1969), is based on the Hewitt-Savage zero-one law. We need a little background on symmetric sets. Consider an arbitrary set $S$ in $\sigma(X_1, X_2, \ldots)$. By standard theory, there exists an infinite-dimensional Borel set $B_\infty$ such that

$$S = X^{-1}(B_\infty) = \{(X_1, X_2, \ldots) \epsilon B_\infty\}$$

The set $S$ is called symmetric if it is invariant under any finite permutation of the labels of the X's; i.e. if

$$S = \{(X_{i_1}, \ldots, X_{i_n}, X_{n+1}, \ldots) \epsilon B_\infty\}$$

where $i_1, \ldots, i_n$ is any permutation of $1, 2, \ldots, n$.

Now let $\epsilon > 0$ and consider the set $U(\epsilon) = \{U_n > \theta + \epsilon$ infinitely often$\}$, which can be written $\bigcap_{n=1}^\infty \bigcup_{m=n}^\infty \{U_n > \theta + \epsilon\}$. Since the set $U(\epsilon)$ can also be written in the form $U(\epsilon) = \bigcap_{n=N}^\infty \bigcup_{m=n}^\infty \{U_n > \theta + \epsilon\}$ for arbitrary $N$, it is clear that $U(\epsilon)$ is symmetric since for sufficiently large $n$, $U_n$ is invariant under any particular finite permutation of the $X's$. By the Hewitt-Savage zero-one law (see e.g. Breiman (1969) p.63), all symmetric sets have probability zero or one when the sequence $X_1, X_2, \ldots$ is i.i.d. The fact that $E(U_n) = \theta$ means that $Pr(U(\epsilon))$ cannot be one, so it must be zero. By a similar argument, the probability of the set $L(\epsilon) = \{U_n < \theta - \epsilon$ infinitely often$\}$ is zero. Now the set where $U_n$ does not converge to $\theta$ can be written as $\bigcup_{k=1}^\infty L(1/k) \cup U(1/k)$ so that

$$Pr(U_n \text{ does not converge to } \theta) \leq \sum_{k=1}^\infty Pr(U(1/k)) + Pr(L(1/k)) = 0$$

and hence $U_n$ converges to $\theta$ with probability one.

If the assumption that $E|\psi(X_1, X_2)| < \infty$ in Theorem 3 is strengthened to $E|\psi(X_1, X_2)|^2 < \infty$, a very simple proof of Theorem 3 is possible: since $\{U_n\}$ is a reverse martingale, we have

$$Pr\left(\sup_{n \leq m} |U_m - \theta| > \epsilon\right) \leq \epsilon^{-2} E|U_n - \theta|^2 = O(n^{-1})$$

130

using Corollary 3 of Section 3.4.1. The conclusion then follows by Lemma A.

A rate of convergence for the strong law of large numbers can also be established. Our next theorem is due to Grams and Serfling (1973). A preliminary result due to Baum and Katz is needed. We state it without proof in a form reported in Hanson (1970).

**Lemma C.** *Let* $Y_1, Y_2, \ldots$ *be independently and identically distributed with mean* $\mu$, *and suppose that* $Pr(|Y_i| > n) = O(n^{-r})$ *where* $r > 1$. *Denote* $n^{-1} \sum_{i=1}^{n} Y_i$ *by* $\overline{Y}_n$. *Then for each* $\varepsilon > 0$,

$$Pr\left( \sup_{m>n} |\overline{Y}_m - \mu| > \varepsilon \right) = O(n^{1-r}).$$

**Theorem 4.** *Suppose that* $E|\psi(X_1, \ldots, X_k)|^{2r} < \infty$ *for some positive integer* $r$. *Then for each* $\varepsilon > 0$,

$$Pr\left( \sup_{m \geq n} |U_m - \theta| > \varepsilon \right) = O(n^{1-2r}).$$

**Proof.**  Using the $H$-decomposition

$$U_n = \theta + kH_n^{(1)} + R_n$$

we can write

$$Pr\left( \sup_{m \geq n} |U_m - \theta| > \varepsilon \right) \leq Pr\left( \sup_{m \geq n} k|H_m^{(1)}| > \frac{\varepsilon}{2} \right)$$
$$+ Pr\left( \sup_{m \geq n} |R_n| > \frac{\varepsilon}{2} \right). \qquad (16)$$

Using the Markov inequality we can write

$$Pr(k|h^{(1)}(X_i)| > n) \leq k^{2r} E|h^{(1)}(X_i)|^{2r} n^{-2r} = O(n^{-2r}),$$

so that by Lemma C the first term on the right of (16) is $O(n^{1-2r})$. For the second term, note that $R_n$ is a $U$-statistic (see the discussion after Theorem 1 of Section 1.6) and so is a reverse martingale. Moreover the kernel of $R_n$,

$\rho$ say, inherits the integrability properties of $\psi$, so that $E|\rho(X_1, \ldots, X_k)|^{2r}$ and hence $E|R_n|^{2r}$ are both finite. Thus by Theorem 3 of Section 3.4.1,

$$Pr\left(\sup_{m \geq n} |R_m| > \varepsilon\right) \leq \varepsilon^{-2r} E(|R_n|^{2r}).$$

Now $R_n$ is a $U$-statistic with first order degeneracy, so by Theorem 2 of Section 1.5, $E(|R_n|^{2r}) = O(n^{-2r})$, and the theorem is proved.

Lin (1981) improves this result by showing that for $\alpha > 1/2$

$$\sum_{n=k}^{\infty} n^{r(1+\alpha)-2} Pr\left(\sup_{m \geq n} m^{-\alpha} |U_m - \theta| > \varepsilon\right) < \infty$$

which implies that

$$Pr\left(\sup_{m \geq n} |U_m - \theta| > \varepsilon\right) = o(n^{1-2r}).$$

Kokic (1987) extends Lin's result to degenerate $U$-statistics having $\sigma_{d-1}^2 = 0$, $\sigma_d^2 > 0$ and obtains the result

$$\sum_{n=k}^{\infty} n^{r(d+1+2\alpha)-\gamma} Pr\left(\sup_{m \geq n} m^{-\alpha} |U_m - \theta| > \varepsilon\right) < \infty$$

for $\alpha > -(d+1)/2$, $\gamma > 1$. This implies that

$$Pr\left(\sup_{m \geq n} |U_m - \theta| > \varepsilon\right) = O(n^{-(d+1)r}).$$

Note that this last result can also be derived by using the method of Theorem 4.

## 3.5 The law of the iterated logarithm for $U$-statistics

We now deal with the generalisation to $U$-statistics of the last of our four classical results for i.i.d. summands, the law of the iterated logarithm (LIL). This theorem gives a rate of convergence for the strong law of large numbers, and in the form due to Hartman and Winter (1941), states that if $S_n = X_1 + \cdots + X_n$ for i.i.d. summands $X_1, \ldots, X_n$, having mean zero and variance $\sigma^2$, then

$$\limsup_n S_n / \sqrt{2\sigma^2 n \log \log n} = 1 \quad a.s.$$

Since non-degenerate $U$-statistics are "almost" a sum of i.i.d. summands, we might suspect that a version of the LIL also holds in this context, and this is in fact the case. We have

**Theorem 1.** *Let $\{U_n\}$ be a sequence of $U$-statistics based on a non-degenerate kernel $\psi$ of order $k$ satisfying $E|\psi(X_1, \ldots, X_k)|^2 < \infty$ and hence having asymptotic variance $k^2 \sigma_1^2 / n$. Then the LIL holds for $U_n$: we have*

$$\limsup_n \frac{n(U_n - \theta)}{\sqrt{2k^2 \sigma_1^2 n \log \log n}} = 1 \quad a.s.$$

**Proof.** Using the $H$-decomposition, we can write

$$n(U_n - \theta) = k \sum_{i=1}^{n} h^{(1)}(X_i) + n R_n$$

where $R_n$ is a $U$-statistic having a first-order degeneracy, and so $Var R_n = O(n^{-2})$. The result will follow from the LIL for i.i.d. summands if we can show that

$$n R_n / \sqrt{n \log \log n} \to 0 \quad a.s.$$

By direct computation, it follows that $E(R_n^2 - R_{n+1}^2) = O(n^{-3})$. The $U$-statistic $R_n$ is a reverse martingale, and $n^{\frac{1}{2}} / \sqrt{\log \log n}$ is increasing in $n$ so that by Corollary 2 of Section 3.4.1

$$Pr \left( \sup_{n \geq m} |n R_n| / \sqrt{n \log \log n} > \varepsilon \right) \leq \varepsilon^{-2} \sum_{n=m}^{\infty} n (\log \log n)^{-1} E(R_n^2 - R_{n+1}^2)$$

$$\leq C \sum_{n=m}^{\infty} n^{-2} / \log \log n$$

for some constant $C$. Since this last series is convergent, the result follows by Lemma A of Section 3.4.2.

A LIL for degenerate $U$-statistics have been proved by Dehling, Denker and Philipp (1986), and a functional LIL by Dehling (1989a). See also Dehling (1989b).

## 3.6 Invariance principles

Let $X_1, X_2, \ldots$ be a sequence of independent and identically distributed random variables defined on some probability space $\Omega$, and assume that the $X_i$ have mean zero and variance $\sigma^2$.

The sequences of partial sums $S_n = X_1 + \cdots + X_n$ are the subject of the classical results of probability theory discussed in Section 3.1. A further generalisation of these results is possible by means of identifying partial sum sequences with random functions and developing so-called *invariance principles* to describe the limiting behaviour of the resulting sequences of random functions.

A random function $X(t)$ indexed by $[0, 1]$ is a family of random variables defined on some probability space $(\Omega, \mathcal{F}, P)$, and can be regarded in two ways:

(i) For each $t, X(t) = X(t, .)$ is a random variable (i.e. a function defined on $\Omega$ which is measurable with respect to the $\sigma$-field $\mathcal{F}$),

(ii) For fixed $\omega$ in $\Omega, X(\cdot, \omega)$ is a function on $[0, 1]$.

If for almost all $\omega$ in $\Omega, X(\cdot, \omega)$ is a continuous function on $[0,1]$ we say that $X(t)$ has *continuous sample paths*; the random functions in this section will be of this type.

Denote by $C[0, 1]$ the set of all continuous functions on $[0, 1]$. $C[0, 1]$ is a normed linear space with norm

$$||x - y|| = \sup_{t \in [0,1]} |x(t) - y(t)|. \tag{1}$$

Any random function with continuous sample paths can be identified with a probability measure on $C[0, 1]$ in the following way. Consider the $\sigma$-field of Borel subsets of $C[0, 1]$ i.e. the $\sigma$-field generated by the open (with respect to the norm (1)) subsets of $C[0, 1]$. If $X(t)$ is a random function, with all the r.vs defined on a probability space $(\Omega, \mathcal{F}, P)$, then we may define a probability measure $\mu$ by

$$\mu = PX^{-1}$$

i.e. for any Borel subset $B$ of $C[0, 1]$ we define

$$\mu(B) = P\{\omega : X(\cdot, \omega) \epsilon B\}.$$

There is thus a duality between random functions on $[0, 1]$ with continuous sample paths and probability measures on $C[0, 1]$.

## Weak invariance principles

A sequence of random functions $X_n(t)$ with corresponding measures $\mu_n$ is said to *converge weakly* to a measure $\mu$ (written $\mu_n \Rightarrow \mu$) if $\mu_n(A) \to \mu(A)$ for all Borel subsets $A$ with $\mu(\partial(A)) = 0$, where $\partial(A)$ denotes the boundary of $A$.

Numerous equivalent characterisations of weak convergence are known; for these and additional detail on the above matters the reader is referred to Billingsley (1968).

A random function of particular interest is the so-called *Brownian motion* $W(t)$ which has the properties

(i) For a fixed $t$, the r.v. $W(t)$ is normally distributed with mean 0 and variance $t$;

(ii) If $t_1 < \ldots < t_r$ are points in $[0, 1]$ then $W(t_2) - W(t_1), W(t_3) - W(t_2), \ldots, W(t_r) - W(t_{r-1})$ are all independent;

(iii) The random function $W(t)$ has continuous sample paths.

This random function can be used to model the displacement of a microscopic particle suspended in water and undergoing Brownian motion. The measure on $C[0, 1]$ corresponding to this random function is called the Wiener measure and will also be denoted in the sequel by $W$.

Now consider the partial sum sequences $S_n = X_1 + \cdots + X_n$. Define a random function $S_n(t)$ by

$$S_n(t) = \begin{cases} 0, & t = 0; \\ S_j/\sigma n^{\frac{1}{2}}, & t = j/n, j = 1, 2, \ldots n; \end{cases}$$

and by linear interpolation for other values of $t$, so that

$$S_n(t) = \{S_{[nt]} + (nt - [nt])X_{[nt]+1}\}/\sigma n^{\frac{1}{2}} \tag{2}$$

for $t$ in [0,1]. Obviously $S_n(t)$ has continuous sample paths, so $S_n(t)$ corresponds to a measure on $C[0, 1]$ which we also denote by $S_n$. The convergence behaviour of $S_n(t)$ is given by the following result, which is due to Donsker (1951). See also Billingsley (1968).

**Theorem 1.** *Let* $X_1, X_2, \ldots$ *be a sequence of i.i.d. random variables with mean 0 and finite variance* $\sigma^2$. *Let* $S_n(t)$ *be defined by (2). Then* $S_n \Rightarrow W$ *where* $W$ *is Wiener measure.*

Theorem 1 is an analogue of the central limit theorem, but is more useful. As an example of its utility, suppose that $g : C[0,1] \rightarrow \mathbb{R}$ is a continuous function defined on $C[0,1]$. Then (see Billingsley (1968) p31) the random variable $g(S_n(t))$ converges in distribution to $g(W(t))$. Take $g(x) = \sup_{t \in [0,1]} x(t)$, then $g$ is continuous and we can use this result to find the distribution of $\max_{1 \leq j \leq n} S_j / \sigma n^{\frac{1}{2}} = \sup_{t \in [0,1]} S_n(t)$, which must converge in distribution in the light of the above remarks to $\sup_{t \in [0,1]} W(t)$ where $W(t)$ is a Brownian motion.

Theorem 1 does not require the $X's$ to have any particular distribution: the (weak) limit is invariant (hence the name "weak invariance principle") under this choice as long as the $X$'s have finite variance. By choosing a particular special case, the limit distribution (i.e. the distribution of $\sup_t W(t)$) can be computed, and in fact for $\alpha > 0$

$$Pr\left(\sup_{t \in [0,1]} W(t) \leq \alpha\right) = \frac{2}{\sqrt{2\pi}} \int_0^\alpha e^{-u^2/2} \, du.$$

For details the reader is referred to Billingsley (1968) Section 10. For an excellent survey of invariance principles, see Heyde (1981).

Theorem 1 has been generalised to $U$-statistic sequences by Miller and Sen (1972):

**Theorem 2.** *Let* $\{U_n\}$ *be a sequence of* $U$-*statistics based on a non-degenerate kernel* $\psi$ *of degree* $k$, *and suppose that* $E|\psi(X_1, \ldots, X_k)|^2 < \infty$. *Define for* $t \in [0,1]$

$$U_n(t) = \begin{cases} 0 & t = j/n, \ j = 0, 1, \ldots k-1, \\ j(U_j - \theta)/k\sigma_1 n^{\frac{1}{2}} & t = j/n, \ j = k, \ldots, n, \end{cases}$$

136

*and by linear interpolation between these points, so that for $\frac{i}{n} < t < \frac{i+1}{n}$,*

$$U_n(t) = U_n(j/n) + (nt - j)\left\{U_n\left(\frac{j+1}{n}\right) - U_n\left(\frac{j}{n}\right)\right\}.$$

*Then $U_n \Rightarrow W$.*

**Proof.** Let

$$U_n - \theta = kH_n^{(1)} + R_n$$

be the $H$-decomposition of $U_n$, where the random variables $H_n^{(1)}$ and $R_n$ have the properties $EH_n^{(1)} = 0$, $Var\, kH_n^{(1)} = k^2\sigma_1^2/n$, $H_n^{(1)}$ is uncorrelated with $R_n$, and $R_n$ is a $U$-statistic with $ER_n = 0$ and $Var\, R_n = O(n^{-2})$. Further, by direct calculation it is easy to see that $Var\, R_n - Var\, R_{n+1} = O(n^{-3})$. Following (2), define a random function $H_n^{(1)}(t)$ by $H_n^{(1)}(0) = 0$ and

$$H_n^{(1)}(t) = \{[nt]H_{[nt]}^{(1)} + (nt - [nt])(h^{(1)}(X_{[nt]+1}))\}/\sigma_1 n^{\frac{1}{2}}$$

so that by Theorem 1, $H_n^{(1)}(t) \Rightarrow W$. By Theorem 4.1 of Billingsley (1968), it is thus enough to prove that

$$\sup_{t\epsilon[0,1]} |U_n(t) - H_n^{(1)}(t)|$$

converges in probability to zero. Consider

$$\sup_{t\epsilon[0,1]} |U_n(t) - H_n^{(1)}(t)| = \max_{1 \le j \le n} |U_n(j/n) - H_n^{(1)}(j/n)|$$

$$\le \max_{1 \le j < k} |jH_j^{(1)}|/\sigma_1 n^{\frac{1}{2}} + \max_{k \le j \le n} |jR_j|/k\sigma_1 n^{\frac{1}{2}} \quad (3)$$

The first term of (3) converges to zero in probability as $n \to \infty$, since

$$Pr\left(\max_{1 \le j < k} |jH_j^{(1)}| > \varepsilon n^{\frac{1}{2}}\right) \le \varepsilon^{-2} n^{-1}\, Var(kH_k^{(1)})$$

by Kolmogorov's inequality (see e.g. Billingsley (1979) Theorem 22.2). For the second term of (3), note that $R_j$ is a reverse martingale, so that by Corollary 2 of Section 3.4.1,

$$Pr\left(\max_{k \le j \le n} |jR_j| > n^{\frac{1}{2}}\varepsilon\right) \le \varepsilon^{-2} n^{-1} \sum_{j=k}^{n} j^2\{Var R_j - Var R_{j+1}\}$$

$$\le C\, \varepsilon^{-2} n^{-1} \sum_{j=1}^{n} j^{-1}$$

$$\le C\, \varepsilon^{-2} n^{-1} \log n$$

for some constant $C$. Thus the second term of (3) converges in probability to zero and the theorem is proved.

A parallel theory exists, which substitutes the space $\mathcal{D}[0,1]$ of all right-continuous functions defined on $[0,1]$ for the space $C$ above. Our theorems are readily adapted to this new context, so we skip the details.

A different type of random function constructed from $U_n$ has been considered by Loynes (1971). He defines for $n \leq m$

$$S_n^*(t) = \begin{cases} U_m/\{VarU_n\}^{\frac{1}{2}} & \text{for } t = Var\,U_m/Var\,U_n, \\ 0 & \text{for } t = 0, \end{cases}$$

and between these points by linear interpolation. The random function $S_n^*(t)$ thus depends on $U_n, U_{n+1}, \ldots$ rather than $U_k, \ldots, U_n$ as in the last theorem. However the conclusion remains unchanged : $S_n^* \Rightarrow W$. Loynes' proof does not depend on the $H$-decomposition but rather uses the martingale structure of $U_n$ directly. We refer the reader to his paper for details. Weak invariance principles for degenerate $U$-statistics are treated by Neuhaus (1977) and Ronzhin (1986). A dual invariance principle covering both the degenerate and non-degenerate cases has been given by Hall (1979). A different approach using stochastic integrals is taken by Denker, Grittenberger and Keller (1985). Csörgő and Horváth (1988) consider approximating $U$-statistics by Gaussian processes both weakly and "in probability", and apply their results to detecting changepoints.

## Strong invariance principles

Invariance principles exist for modes of convergence stronger than weak convergence. In particular we may consider *strong invariance principles* which describe the almost sure asymptotic behaviour of random sequences. In formulating strong invariance principles, it is convenient to consider random functions indexed by $[0, \infty)$ rather than $[0,1]$.

The classic result for i.i.d. sequences is the following, due to Strassen:

**Theorem 3.** *Let $\{X_n\}$ be a sequence of i.i.d. random variables with $EX_n = 0$ and $Var\,X_n = \sigma^2$. Define a random function $S(t)$ on $[0, \infty)$*

138

by $S(n) = \sum_{i=1}^{n} X_i$ and by linear interpolation otherwise. Then there is a Brownian motion $W(t)$ such that

$$S(t) = \sigma W(t) + O((tf(t))^{\frac{1}{4}} \log t) \quad a.s.$$

where $f$ is a function satisfying the conditions

(i) $f(t)$ is increasing on $[0, \infty)$;

(ii) $t^{-1} f(t)$ is decreasing on $[0, \infty)$;

(iii) $\sum_{n \geq 1} f(n\sigma^2)^{-1} \int_{x^2 > f(n\sigma^2)} x^2 dF(x) < \infty$

where $F$ is the common distribution function of the $X$'s.

Theorem 3 is a special case of a result of Strassen who in fact proves Theorem 3 for martingales, subject to similar conditions, so we omit the proof. The reader is referred to Theorem 4.4 of Strassen (1967).

To extend Theorem 3 to $U$-statistics is straightforward: we use the $H$-decomposition, and apply Theorem 3 to the projection and the Hájek-Rényi inequality to the remainder.

**Theorem 4.** *(Sen (1974)). Let $\{U_n\}$ be a sequence of $U$-statistics based on a non-degenerate kernel $\psi$ of degree $k$. Define a random function on $[0, \infty)$ by*

$$S(t) = \begin{cases} 0; & t = 0, 1, 2, \ldots, k-1; \\ n(U_n - \theta); & t = n, n \geq k; \end{cases}$$

*and by linear interpolation elsewhere. Suppose $f$ is a function satisfying (i) and (ii) of Theorem 3 and*

(iii)' $\sum_{n=1}^{\infty} f(cn) \int_{[h^{(1)}(x)]^2 > f(cn)} [h^{(1)}(x)]^2 dF < \infty$ *for all $c > 0$.*

*Then there is a Brownian motion $W(t)$ on $[0, \infty)$ such that*

$$S(t) = k\sigma_1 W(t) + O((tf(t))^{\frac{1}{4}} \log t) \quad a.s.$$

*as $t \to \infty$.*

**Proof.** Write $U_n - \theta = kH_n^{(1)} + R_n$, where $H_n^{(1)}$ and $R_n$ are as in the proof of Theorem 2. Then

$$S(t) = H^{(1)}(t) + R(t) \tag{3}$$

139

where $H^{(1)}(t) = knH_n^{(1)}$ for $n = t$ and is defined by linear interpolation elsewhere, and $R(t)$ similarly. In view of Theorem 3, we need only prove

$$nR_n/(nf(n))^{\frac{1}{4}} \log n \to 0 \quad a.s.$$

If $c_j = n/(nf(n))^{\frac{1}{4}} \log n$, then because of Lemma A of Section 3.4.1, it is enough to prove $\lim_n Pr\left(\sup_{j \geq n} |c_j R_j| > \varepsilon\right) = 0$. Since $c_j$ is increasing for $n \geq 8$, we can apply the Hájek-Rényi equality in our usual manner and obtain

$$Pr(\sup_{j \geq n} |c_j R_j| > \varepsilon) \leq \varepsilon^{-2} \sum_{j=n}^{\infty} c_j^2(E(R_j^2) - E(R_{j+1}^2))$$

$$\leq C \ \varepsilon^{-2} \sum_{j=n}^{\infty} c_j^2 j^{-3}$$

$$= O(n^{-\frac{1}{2}})$$

hence proving the theorem.

Invariance principles in which the U-statistic is approximated almost surely by an integral with respect to a Kiefer process have been considered by Dehling, Denker and Philipp (1984). See de Wet (1987) for a summary of their results. A strong invariance principle and LIL for degenerate U-statistics have been proved by Dehling, Denker and Philipp (1986).

## 3.7 Asymptotics for $U$-statistic variations

In this section we treat the asymptotic properties of the $U$-statistic variations introduced in Chapter 2. We begin with generalised $U$-statistics.

### 3.7.1 Asymptotics for generalised $U$-statistics

Using the notation of Theorem 3 of Section 2.2, any generalised $U$-statistic $U_{n_1,n_2}$ with $m = 2$ can be written as

$$U_{n_1,n_2} = \theta + k_1 H_{n_1,n_2}^{(1,0)} + k_2 H_{n_1,n_2}^{(0,1)} + R_n \tag{1}$$

where $H_{n_1,n_2}^{(1,0)} = n_1^{-1} \sum_{i=1}^{n_1} h^{(1,0)}(X_i)$, $H_{n_1,n_2}^{(0,1)} = n_2^{-1} \sum_{j=1}^{n_2} h^{(0,1)}(Y_j)$ and $Var R_n = o(n^{-1})$. Thus $n_1^{\frac{1}{2}} H_{n_1,n_2}^{(1,0)}$ and $n_2^{\frac{1}{2}} H_{n_1,n_2}^{(0,1)}$ both converge to normal distributions with mean zero and variances $\delta_{1,0}^2$ and $\delta_{0,1}^2$ respectively. Let $N = n_1 + n_2$ and suppose that $n_1$ and $n_2 \to \infty$ in such a way that $p_n = n_1 N^{-1} \to p$, where $0 < p < 1$. Then since

$$N^{\frac{1}{2}}(U_{n_1,n_2} - \theta) = p_N^{-\frac{1}{2}} n_1^{\frac{1}{2}} k_1 H_{n_1,n_2}^{(1,0)} + (1 - p_N)^{-\frac{1}{2}} n_2^{\frac{1}{2}} k_2 H_{n_1,n_2}^{(0,1)} + o_p(1)$$

and since $H_{n_1,n_2}^{(1,0)}$ and $H_{n_1,n_2}^{(0,1)}$ are independent, it follows that $N^{\frac{1}{2}}(U_{n_1,n_2} - \theta)$ converges in distribution to a normal r.v. with zero mean and variance $p^{-1} k_1^2 \delta_{1,0}^2 + (1 - p)^{-1} k_2^2 \delta_{0,1}^2$. Lehmann (1951) also proves the result by a different method. We state it as a formal theorem:

**Theorem 1.** *Let $U_{n_1,n_2}$ be a generalised $U$-statistic based on two independent samples $X_1, \ldots, X_{n_1}$ and $Y_1, \ldots, Y_{n_2}$, and suppose that both $\delta_{0,1}^2$ and $\delta_{1,0}^2$ are positive, where*

$$\delta_{c,d}^2 = Var\, h^{(c,d)}(X_1, \ldots, X_c; Y_1, \ldots, Y_d)$$

*and the functions $h^{(c,d)}$ are defined by equation (3) of Section 2.2.1. Then if $N = n_1 + n_2$, and if $p_n = n_1 N^{-1} \to p$, where $0 < p < 1$, we have*

$$N^{\frac{1}{2}}(U_{n_1,n_2} - \theta) \xrightarrow{D} N(0, N(0, p^{-1} k_1^2 \delta_{1,0}^2 + (1 - p)^{-1} k_2^2 \delta_{0,1}^2)$$

*as $N \to \infty$.*

**Example 1.**

Consider the problem of testing if two absolutely continuous distributions having the same known median are in fact equal. Let $F$ and $G$ be two distribution functions both having median zero, and let $X_1, \ldots, X_{n_1}$ and $Y_1, \ldots, Y_{n_2}$ be random samples from $F$ and $G$ repectively. Sukhatme (1957) proposes a test for the hypothesis $F = G$ based on the statistic

$$T_N = \frac{1}{n_1 n_2} \sum_{i=1}^{n_1} \sum_{j=1}^{n_2} K(X_i, Y_j)$$

where

$$K(x,y) = \begin{cases} 1, & \text{if } 0 < x < y \text{ or } y < x < 0, \\ 0, & \text{otherwise.} \end{cases}$$

Under the hypothesis that $F = G$, $ET = \frac{1}{4}$, and the functions $h^{(1,0)}$ and $h^{(0,1)}$ are given by

$$h^{(1,0)} = \begin{cases} \frac{3}{4} - F(x), & 0 < x, \\ F(x) - \frac{1}{4}, & x < 0; \end{cases}$$

and $h^{(0,1)} = h^{(1,0)}$. Hence the quantitites $\delta_{1,0}$ and $\delta_{0,1}$ are both equal to

$$\int_{-\infty}^{0} (F(x) - \tfrac{1}{2})^2\, dx + \int_{0}^{\infty} (\tfrac{1}{2} - F(x))^2\, dx = \frac{1}{12},$$

and the asymptotic distribution of $N^{\frac{1}{2}}(T_N - \frac{1}{4})$ under the null hypothesis is normal with variance $(p^{-1} + (1-p)^{-1})/12$. The test is thus asymptotically distribution-free.

Our next theorem extends Theorem 1 to a vector of generalised $U$-statistics, in a form that will be used in Section 6.2.5.

**Theorem 2.** *Let $U^{(j)}_{n_1,\ldots,n_m}$, $j = 1, \ldots, m$ be generalised $U$-statistics, based on kernels*

$$\psi^{(j)}(x_{1,1}, \ldots, x_{1,k_1}; \ldots; x_{m,1}, \ldots, x_{m,k_m})$$

*and a common set $X_{i,j}$, $j = 1, \ldots, n_i$, $i = 1, \ldots, m$ of independent random variables, where $X_{i,j}$ is assumed to have distribution function $F_i$. Suppose that the expectation of $U^{(j)}_{n_1,\ldots,n_m}$ is $\theta_j$, and (using a slightly different notation than that of Section 2.2) let $h_l^{(j)}$ be the set of first kernel functions in the $H$-decomposition of $U^{(j)}_{n_1,\ldots,n_m}$, i.e.*

$$h_l^{(j)}(x) = E\psi^{(j)}(X_{1,1}, \ldots, X_{1,k_1}; \ldots; x, X_{l,2}, \ldots, x_{l,k_l}; \ldots) - \theta_j.$$

*Denote by $\Sigma_i$ the covariance matrix of the random vector*

$$\mathbf{Y}_{i,j} = (h_i^{(1)}(X_{i,j}), \ldots, h_i^{(m)}(X_{i,j}))$$

*and suppose that $n_1, \ldots, n_m \to \infty$ in such a way that $n_j N^{-1} \to p_j > 0$, where $N = n_1 + \cdots + n_m$. Also let $\mathbf{U}_N = (U^{(1)}_{n_1,\ldots,n_m}, \ldots, U^{(m)}_{n_1,\ldots,n_m})$ and $\boldsymbol{\theta} = (\theta_1, \ldots, \theta_m)$. Then $N^{\frac{1}{2}}(\mathbf{U}_N - \boldsymbol{\theta})$ converges in distribution to*

142

a multivariate normal distribution with mean vector zero and covariance
matrix $p_1^{-1} k_1^2 \Sigma_1 + \cdots + p_m^{-1} k_m^2 \Sigma_m$.

**Proof.** The $H$-decomposition of $U_{n_1,\ldots,n_m}^{(j)}$ is

$$U_{n_1,\ldots,n_m}^{(j)} - \theta_j = \sum_{l=1}^{m} \frac{k_l}{n_l} \sum_{r=1}^{n_l} h_l^{(j)}(X_{l,r}) + o_p(N^{-\frac{1}{2}})$$

and so

$$N^{\frac{1}{2}}(\mathbf{U}_N - \boldsymbol{\theta}) = \sum_{l=1}^{m} p_l^{-\frac{1}{2}} k_l n_l^{-\frac{1}{2}} \sum_{r=1}^{n_l} \mathbf{Y}_{l,r} + o_p(1). \tag{2}$$

Since the vectors $\mathbf{Y}_{l,r}$ are all independent, and $n_l^{-\frac{1}{2}} \sum_{r=1}^{n_l} \mathbf{Y}_{l,r}$ converges in
distribution to $MN(0, \Sigma_i)$, the result follows from (2).

The asymptotic normality result above has been supplemented by a
weak invariance principle for generalised $U$-statistics by Sen (1974). The
martingale results of Section 3.4.1 cannot be applied directly since there
is no natural ordering of pairs of integers. However, with the aid of the
$H$-decomposition (1), it is only necessary to prove the remainder is negligible in probability, as martingale arguments can then be applied to $H_{n_1,n_2}^{(1,0)}$
and $H_{n_1,n_2}^{(0,1)}$ which are $U$-statistics rather than generalised $U$-statistics. Sen
(1977) also treats the strong convergence of generalised $U$-statistics.

### 3.7.2 The independent, non-identically distributed case

We prove the following result due to Hoeffding (1948a), which establishes the asymptotic normality under standard conditions for $U$-statistics
based on independent but not identically distributed random variables.

**Theorem 1.** Let $X_1, X_2, \ldots$ be independent r.v.s with distribution functions $F_1, F_2, \ldots$. Define for $i = 1, 2, \ldots, n$

$$\psi_{1,i}(x) = \binom{n-1}{k-1}^{-1} \sum_{(n-1,k-1)}^{(-i)} \psi_{1,i;i_2,\ldots i_k}(x)$$

where the sum is taken over all $(k-1)$-subsets $\{i_2, \ldots, i_k\}$ of $\{1, 2, \ldots, n\}$
that do not contain $i$, and

$$\psi_{1;i;i_2,\ldots,i_k}(x) = E\psi(x, X_{i_2}, \ldots, X_{i_k}) - \theta\{i, i_2, \ldots, i_k\}.$$

143

Suppose that

(i) $Var\,\psi(X_{i_1}, \ldots, X_{i_k}) < A$ for some constant $A$ and all $i_1, \ldots, i_k$;

(ii) For some $\delta > 0$, $E|\psi_{1,i}(X_i)|^{2+\delta} < \infty$ for all $i$, and

$$\lim_{n \to \infty} \sum_{i=1}^{n} E|\psi_{1,i}(X_i)|^{2+\delta} \Big/ \left(\sum_{i=1}^{n} E|\psi_{1,i}(X_i)|^2\right)^{(2+\delta)/2} = 0.$$

Then $(U_n - E(U_n))/(Var U_n)^{\frac{1}{2}} \xrightarrow{D} N(0,1)$.

**Proof.** Let $S_n$ denote the sum

$$S_n = \frac{k}{n} \sum_{i=1}^{n} \psi_{1,i}(X_i) \tag{1}$$

By the Liapounov central limit theorem (see, e.g. Chow and Teicher (1978) p293), the conditions (ii) ensure that $S_n/(Var S_n)^{\frac{1}{2}} \xrightarrow{D} N(0,1)$, so we need only prove that

$$\frac{S_n}{(Var S_n)^{\frac{1}{2}}} - \frac{U_n - E(U_n)}{(Var U_n)^{\frac{1}{2}}} \xrightarrow{D} 0. \tag{2}$$

Since the r.v. in (2) has zero mean, it suffices to prove that its variance converges to zero, which will be the case if and only if

$$\frac{(Cov(S_n, U_n))^2}{Var S_n Var U_n} \to 1. \tag{3}$$

Now

$$
\begin{aligned}
Cov(U_n, S_n) &= \frac{k}{n} \binom{n}{k}^{-1} \sum_{(n,k)} \sum_{i=1}^{n} Cov(\psi(X_{i_1}, \ldots, X_{i_k}), \psi_{1,i}(X_i)) \\
&= \frac{k}{n} \binom{n}{k}^{-1} \sum_{i=1}^{n} \sum_{(n-1,k-1)}^{(-i)} Cov(\psi(X_i, X_{i_2}, \ldots, X_{i_k}), \psi_{1,i}(X_i)) \\
&= \frac{k}{n} \binom{n}{k}^{-1} \sum_{i=1}^{n} \binom{n-1}{k-1} Var\,\psi_{1,i}(X_i) \\
&= \frac{k^2}{n^2} \sum_{i=1}^{n} Var\,\psi_{1,i}(X_i) \\
&= Var\,S_n \tag{4}
\end{aligned}
$$

144

and

$$Var\, S_n = \frac{k^2}{n^2} \sum_{i=1}^{n} Var\, \psi_{1,i}(X_i)$$

$$= \frac{k^2}{n^2} \sum_{i=1}^{n} \binom{n-1}{k-1}^{-2} \sum_{(n-1,k-1)}^{(-i)} \sum_{(n-1,k-1)}^{(-i)}$$
$$Cov\big(\psi(X_i, X_{i_2}, \ldots, X_{i_k}), \psi(X_i, X_{j_2}, \ldots, X_{j_k})\big) \qquad (5)$$

There are $n\binom{n-1}{k-1}\binom{n-k}{k-1}$ terms of (5) for which the sets $\{i_2, \ldots, i_k\}$ and $\{j_2, \ldots, j_k\}$ are disjoint, and $O(n^{2k-2})$ other terms. Moreover, every pair of $k$-subsets $S$ and $T$ for which $|S \cap T| = 1$ corresponds to exactly one term of (5), and the covariances in (5) are uniformly bounded by assumption (1). It follows that we can write (5) as

$$\frac{k^2}{n^2} \binom{n-1}{k-1}^{-2} \sum_{|S \cap T|=1} \sigma^2(S, T) + o(n^{-1})$$

$$= \binom{n}{k}^{-2} \left\{ \binom{n}{k} \binom{k}{1} \binom{n-k}{k-1} \bar{\sigma}_{1,n}^2 \right\} + o(n^{-1})$$

$$= k\bar{\sigma}_{1,n}^2 n^{-1} + o(n^{-1}),$$

using the notation of Section 2.3. The proof is completed by using (3) of that Section and (4) and (5).

To complement the CLT above, the reader may wish to consult Sen (1984), who proves weak and strong invariance principles, Ghosh and Dasgupta (1982) for Berry-Esseen theorems and Dasgupta (1984) who proves a result on large deviation probabilities.

### 3.7.3 Asymptotics for $U$-statistics based on stationary sequences

We begin by introducing another form of weak dependence for stationary sequences that is weaker than the notion of absolute regularity discussed in Section 2.4.2.

The *strong mixing coefficient* of a stationary sequence $X_t$ is the number $\alpha(\tau), \tau > 0$ defined by

$$\alpha(\tau) = \sup_{\substack{A \in \mathcal{M}(-\infty, t) \\ B \in \mathcal{M}(t+\tau, \infty)}} |P(A \cap B) - P(A)P(B)|$$

and the sequence is called *strongly mixing* if $\alpha(\tau) \to 0$ as $\tau \to 0$. Note that some authors use the term *completely regular* for such sequences. Strong mixing is a weaker condition than absolute regularity, as our first result shows.

**Theorem 1.** *Every absolutely regular process is strongly mixing. Moreover, if $X_t$ is strongly mixing, and $\phi$ is a Borel measurable function, then the process $\phi(X_t)$ is strongly mixing.*

**Proof.** Let $A$ and $B$ be sets in $\mathcal{M}(-\infty, t)$ and $\mathcal{M}(t + \tau, \infty)$ respectively for some $t$ and $\tau$. Then by elementary properties of conditional expectations,

$$|P(A \cap B) - P(A)P(B)| = |\int_A (P(B|\mathcal{M}(-\infty, t)) - P(B))dP|$$

$$\leq \int_A \sup_{B \in \mathcal{M}(t+\tau, \infty)} |P(B|\mathcal{M}(-\infty, t)) - P(B)|dP$$

$$\leq E \sup_{B \in \mathcal{M}(t+\tau, \infty)} |P(B|\mathcal{M}(-\infty, t)) - P(B)|$$

so that $\alpha(\tau) \leq \beta(\tau)$ for all $\tau > 0$, proving the first assertion. For the second, let $\mathcal{M}_\phi(a, b)$ denote the $\sigma$-field generated by the collection $\mathcal{C}_\phi$ of sets of the form $\{(\phi(X_{t_1}), \ldots, \phi(X_{t_k})) \in B_k\}$ where $k$ is a positive integer, $B_k$ ranges over all $k$-dimensional Borel sets, and $t_1, \ldots, t_k$ are integers between $a$ and $b$. Define the function $\phi_{(k)} : \mathbb{R}_k \to \mathbb{R}_k$ by

$$\phi_{(k)}(x_1, \ldots, x_k) = (\phi(x_1), \ldots, \phi(x_k)).$$

Since $\{(\phi(X_{t_1}), \ldots, \phi(X_{t_k})) \in B_k\} = \{(X_{t_1}, \ldots, X_{t_k}) \in \phi_{(k)}^{-1}(B_k)\}$, and since $\phi_{(k)}$ is Borel measurable, it follows that every set in $\mathcal{C}_\phi$ is of the form $\{(X_{t_1}, \ldots, X_{t_k}) \in C_k\}$ for a Borel set $C_k$ and so $\mathcal{M}_\phi(a, b) \subseteq \mathcal{M}(a, b)$, where $\mathcal{M}(a, b)$ is as defined in Section 2.4.2. If $\alpha$ and $\alpha_\phi$ denote the strong mixing coefficients of the processes $X_t$ and $\phi(X_t)$, we have

$$\alpha_\phi(\tau) = \sup_{\substack{A \in \mathcal{M}_\phi(-\infty, t) \\ B \in \mathcal{M}_\phi(\tau+t, \infty)}} |P(A \cap B) - P(A)P(B)|$$

$$\leq \sup_{\substack{A \in \mathcal{M}(-\infty, t) \\ B \in \mathcal{M}(t+\tau, \infty)}} |P(A \cap B) - P(A)P(B)|$$

$$= \alpha(\tau),$$

which proves the second assertion since $X_t$ is strongly mixing.

We can now prove our central limit theorem, which is due to Yoshihara (1976):

**Theorem 2.** *Let $X_t$ be a stationary sequence satisfying the conditions of Theorem 1 of Section 2.4.2, and let $U_n$ be a $U$-statistic of degree two based on $X_1, \ldots, X_n$. Then provided $\sigma^2 > 0$, we have $n^{\frac{1}{2}}(U_n - \theta)/2\sigma \xrightarrow{D} N(0, 1)$.*

**Proof.** As in the proof of Theorem 1 of Section 2.4.2, let $U_n = \theta + 2H_n^{(1)} + H_n^{(2)}$ be the $H$-decomposition of $U_n$, so that

$$n^{\frac{1}{2}}(U_n - \theta)/2\sigma = n^{-\frac{1}{2}}\sum_{t=1}^{n} h^{(1)}(X_t)/\sigma + n^{\frac{1}{2}}H_n^{(2)}/2\sigma.$$

In a similar manner, we see that $n^{\frac{1}{2}}EH_n^{(2)} \to 0$ and $nVar\, H_n^{(2)} \to 0$ so that $n^{\frac{1}{2}}H_n^{(2)} \xrightarrow{D} 0$. Accordingly, to prove the theorem we need only show that $n^{-\frac{1}{2}}\sum_{t=1}^{n} h^{(1)}(X_t)/\sigma \xrightarrow{D} N(0, 1)$. This follows from the standard CLT for strongly mixing sequences. For example, by Theorem 18.5.3 of Ibragimov and Linnik (1971) the desired result will follow if we can show that

(i) $\sup_t E|h^{(1)}(X_t)|^{\delta/(2+\delta)} < \infty$

and

(ii) $\sum_t (\alpha_{h^{(1)}}(t))^{\delta/(2+\delta)} < \infty$.

The condition (i) has been assumed as part of the statement of the theorem, and (ii) follows from the assumption that $\alpha(n) = O(n^{-(2+\delta')/\delta'})$ and the fact that $\alpha_{h^{(1)}}(t) \leq \alpha(t)$. The theorem is proved.

Recall that an $m$-dependent stationary process has a complete regularity coefficient $\beta$ which satisfies $\beta(\tau) = 0$ for $\tau > m$, so that provided an $m$-dependent sequence $X_t$ satisfies the moment conditions of the above theorem, the other conditions are met and the $U$-statistic based on $X_t$ will be asymptotically normal.

Several authors have considered other asymptotic results for $U$-statistics based on stationary sequences, including central limit theorems for stronger kinds of weak dependence, such as uniform mixing and *-mixing. (A process is uniformly mixing if it satisfies the condition

$$\phi(\tau) = \sup_{\substack{A \in \mathcal{M}(-\infty, t) \\ B \in \mathcal{M}(t+\tau, \infty)}} |P(A \cap B) - P(A)P(B)|/P(A) \to 0$$

147

as $\tau \downarrow 0$, and *-mixing if the above condition is true, and also the condition obtained by replacing $P(A)$ by $P(B)$ in the denominator.

Thus *-mixing is equivalent to uniform mixing of both the process and the time reversed process. Sen (1972) deals with the CLT, the LIL and the weak invariance principle for *-mixing processes. Yoshihara (1976) covers the same ground for absolutely regular processes, as do Denker and Keller (1983), who also consider uniformly and *-mixing processes. Eagleson (1979) considers both a CLT and a limit theorem for degenerate $U$-statistics based on uniformly mixing processes. Yoshihara (1984) proves a Berry-Esseen theorem for absolutely regular processes. Malevich and Abdalimov (1983) give a similar theorem for $U$-statistics based on $m$-dependent sequences.

### 3.7.4 Asymptotics for $U$-statistics based on finite population sampling

This section is concerned with the asymptotic normality of $U$-statistics based on simple random sampling without replacement from some finite population. As in the i.i.d. case, asymptotic normality is proved by the usual projection technique: we show that the normalised $U$-statistic is asymptotically equivalent to a sum of exchangeable r.v.s, and apply the classical finite-population CLT.

Specifically, let $\{P_N\}$ be a sequence of populations, each of size $N$, and let $x_1, \ldots, x_N$ be the population labels. (The $x_i$ depend on $N$, but for notational simplicity we do not denote this explicitly). Let $X_1, \ldots, X_n$ be a simple random sample of size $n$ chosen from $x_1, \ldots, x_N$. Again, $X_1, \ldots, X_N$ depend on $N$ but we do not make this notationally explicit. We assume that as $n$ and $N$ increase, $nN^{-1} \to \alpha$ where $0 < \alpha < 1$.

Now let $U_n$ be a $U$-statistic with a kernel $\psi$ of degree $k$ based on $X_1, \ldots, X_n$. Write

$$U_n = U_N + \frac{k}{n} \sum_{i=1}^{n} (\psi_1(X_i) - U_N) + R_n \qquad (1)$$

It is clear from (1) that $ER_n = 0$ and we show below in Lemma A that $nVar\,R_n \to 0$, so that the asymptotic behaviours of $n^{\frac{1}{2}}(U_n - U_N)$ and $k \sum_{i=1}^{n} (\psi_1(X_i) - U_N)/n^{\frac{1}{2}}$ are identical.

Many authors have considered conditions that will ensure the asymptotic normality of $\sum_{i=1}^{n}(\psi_1(X_i) - U_N)/\sqrt{n}$. Since the finite population central limit theorem is equivalent to a central limit theorem for linear rank statistics, this problem has been extensively studied in both the sample survey and non-parametric literature. We use a version due to Chernoff and Teicher (1958). They prove (their Corollary 4) the following result: Let $Y_1, \ldots, Y_N$ be exchangeable r.vs with means $\mu_N$, having the properties

(i) $Cov(Y_1, Y_2) = -\sigma/(N-1) + o(N^{-1})$;

(ii) $\max_{1 \leq i \leq N} |Y_i - \mu_N|/N^{\frac{1}{2}} \xrightarrow{P} 0$;

(iii) $N^{-1} \sum_{i=1}^{N} |Y_i - \mu_N|^2 \xrightarrow{P} \sigma^2$;

where $\sigma > 0$. Then

$$\sum_{i=1}^{n}(Y_i - \mu_N)/n^{\frac{1}{2}} \xrightarrow{D} N(0, \sigma^2(1-\alpha)).$$

To apply this result to our setting, set $Y_i = \psi_1(X_i), i = 1, 2, \ldots, N$ so that for the moment, we are sampling the entire population. We will check (i)–(iii) above. By direct calculation,

$$Cov(\psi_1(X_1), \psi_1(X_2)) = -\bar{\sigma}_{1,N}^2/(N-1)$$

so that provided $\bar{\sigma}_{1N}^2 \to \sigma^2$, (i) is true. For (ii), note that

$$\max_{1 \leq i \leq N} |Y_i - \mu_N| = \max_{1 \leq i \leq N} |\psi_1(x_i) - U_N|,$$

so that in our setting (ii) reduces to $\lim_{N \to \infty} \max_{1 \leq i \leq N} |\psi_1(x_i) - U_N|/N^{\frac{1}{2}} = 0$. The condition (iii) also reduces to $\bar{\sigma}_{1,N}^2 \to \sigma^2$ in our setting, since

$$\frac{1}{N} \sum_{i=1}^{N} |Y_i - \mu_N|^2 = \frac{1}{N} \sum_{i=1}^{N} |\psi_1(x_i) - U_N|^2 = \bar{\sigma}_{1,N}^2.$$

Thus, in view of the discussion below, we have proved the following.

**Theorem 1.** *Suppose that*

(i) $\max_{1 \leq i \leq N} |\psi_1(x_i) - U_N|/N^{\frac{1}{2}} \to 0$;

(ii) $\bar{\sigma}_{1,N}^2 \to \sigma^2 > 0$

*as* $N \to \infty$. *Then as* $N$ *and* $n \to \infty$, *with* $nN^{-1} \to \alpha$,

$$n^{\frac{1}{2}}(U_n - U_N) \xrightarrow{D} N(0, k^2\sigma^2(1-\alpha)).$$

Note that a condition that implies (i) is that for some $\delta > 0$, $E|\psi_1(X_1) - U_N|^{2+\delta}$ be uniformly bounded as $N \to \infty$. To see this, consider

$$Pr\left(\max_{1 \leq i \leq N} |\psi_1(X_i) - U_N|/N^{\frac{1}{2}} \geq \epsilon\right) \leq Pr\left(\bigcup_{i=1}^{N}\{|\psi_1(X_i) - U_N| \geq \epsilon N^{\frac{1}{2}}\}\right)$$

$$\leq NPr(|\psi_1(X_1) - U_N| \geq \epsilon N^{\frac{1}{2}})$$

$$\leq \frac{E|\psi_1(X_1) - U_N|^{2+\delta}}{N^{\delta/2}\epsilon^{2+\delta}}$$

by the Markov inequality. Thus, letting $N \to \infty$, we obtain

$$\lim_{N \to \infty} Pr\left(\max_{1 \leq i \leq N} |\psi_1(X_i) - U_N|/N^{\frac{1}{2}} \geq \epsilon\right) = 0.$$

However, the r.v. $\max_{1 \leq i \leq N} |\psi_1(X_i) - U_N|$ is constant, so that for all sufficiently large $N$,

$$\max_{1 \leq i \leq N} |\psi_1(x_i) - U_N|/N^{\frac{1}{2}} < \epsilon$$

proving (i). The condition that $E|\psi_1(X_1)|^{2+\delta}$ be uniformly bounded is that employed by Nandi and Sen (1963) in their version of the theorem.

It remains only to prove

**Lemma A.** *Let* $R_n$ *be defined by (1). Then* $Var\, R_n = o(n^{-1})$.

**Proof.** We have

$$Var\, R_n = Var\, U_n - \frac{2k}{n}Cov(U_n, \sum_{i=1}^{n}(\psi_1(X_i) - U_N))$$

$$+ \frac{k^2}{n^2}Var\left((\sum_{i=1}^{n}(\psi(X_i) - U_N)\right). \tag{2}$$

The covariance in (2) can be written

$$Cov(U_n, \sum_{i=1}^{n}(\psi_1(X_i) - U_N)) = nCov(U_n, \psi_1(X_1))$$

$$= n\binom{n}{k}^{-1}\sum_{(n,k)} Cov(\psi(X_{i_1}, \ldots, X_{i_k}), \psi_1(X_1)). \tag{3}$$

150

Now $Cov(\psi(X_{i_1}, \ldots, X_{i_k}), \psi_1(X_1)) = Cov(\psi(X_1, \ldots, X_k), \psi_1(X_1))$ if the set $\{i_1, \ldots, i_k\}$ contains 1, and $Cov(\psi(X_1, \ldots, X_k), \psi_1(X_{k+1}))$ otherwise, and $\binom{n-1}{k-1}$ of the sets of $S_{n,k}$ contain 1 and $\binom{n-1}{k}$ do not, so that (3) equals

$$kCov(\psi(X_1, \ldots, X_k), \psi_1(X_1))$$
$$+ (n-k)Cov(\psi(X_1, \ldots, X_k), \psi_1(X_{k+1})). \quad (4)$$

Now

$$
\begin{aligned}
Cov(\psi(X_1, \ldots, X_k), \psi_1(X_1)) &= E\{E(\psi(X_1, \ldots, X_k)\psi_1(X_1)|X_1)\} - U_N^2 \\
&= E\{\psi_1(X_1)E(\psi(X_1, \ldots, X_k)|X_1)\} - U_N^2 \\
&= E\psi_1^2(X_1) - U_N^2 \\
&= \bar{\sigma}_{1,N}^2 \quad (5)
\end{aligned}
$$

and

$$Cov(\psi(X_1, \ldots, X_k), \psi_1(X_{k+1}))$$
$$= E\{\psi_1(X_{k+1})E(\psi(X_1, \ldots, X_k)|X_{k+1})\} - U_N^2. \quad (6)$$

To compute $E(\psi(X_1, \ldots, X_k)|X_{k+1})$, write

$$
\begin{aligned}
E(\psi(X_1, \ldots, X_k)|X_{k+1} = x_j) &= \{(N-1)_{(k)}\}^{-1} \sum_{i_1, \ldots, i_k \neq j} \psi(x_{i_1}, \ldots, x_{i_k}) \\
&= \{(N-1)_{(k)}\}^{-1} T_j \text{ say,}
\end{aligned}
$$

where

$$\sum_{i_1, \ldots, i_k} \psi(x_{i_1}, \ldots, x_{i_k}) = T_j + k \sum_{i_2, \ldots, i_k \neq j} \psi(x_j, x_{i_2}, \ldots, x_{i_k})$$

so that

$$T_j = N_{(k)}U_N - k\{(N-1)_{(k-1)}\}\psi_1(x_j)$$

and

$$E(\psi(X_1, \ldots, X_k)|X_{k+1}) = \frac{N}{N-k}(U_N - \frac{k}{N}\psi_1(X_{k+1})).$$

Substituting this in (6), we get

$$Cov(\psi(X_1, \ldots, X_k), \psi_1(X_{k+1})) = \frac{-k\bar{\sigma}_{1,N}^2}{N-K}. \quad (7)$$

151

Substituting (5) and (7) into (4), we get

$$Cov(U_n, \frac{k}{n} \sum_{i=1}^{n} (\psi_1(X_i) - U_N)) = \frac{k^2(N-n)}{n(N-k)} \bar{\sigma}_{1,n}^2. \qquad (8)$$

Similarly

$$Var \sum_{i=1}^{n} (\psi_1(X_i) - U_N) = n(n-1)Cov(\psi_1(X_1), \psi_1(X_2)) + nVar\psi_1(X_1),$$

and

$$Cov(\psi_1(X_1)\psi_2(X_2)) = \frac{1}{N(N-1)} \sum_{i \neq j} \psi_1(x_i)\psi_1(x_j) - U_N^2$$

$$= \frac{1}{N(N-1)} \left( \sum_{i=1}^{N} \sum_{j=1}^{N} \psi_1(x_i)\psi_1(x_j) - \sum_{i=1}^{N} \psi_1^2(x_i) \right) - U_N^2$$

$$= \frac{N}{N-1} U_N^2 - \frac{1}{N-1} E\psi_1^2(X_1) - U_N^2$$

$$= -\bar{\sigma}_{1,N}^2/(N-1)$$

so that

$$Var \sum_{i=1}^{n} (\psi_1(X_i) - U_N) = \frac{n(N-n)\bar{\sigma}_{1,N}^2}{N-1}. \qquad (9)$$

The result now follows from (2) using equation (15) of Section 2.5, (8) and (9).

This asymptotic result is complemented in the literature by both a Berry-Esseen theorem and an invariance principle. The former is due to Zhao and Chen (1987), who prove the following result for kernels of degree two: let $U_n$ have kernel $\psi$ of degree two, let $\alpha_N = nN^{-1}$, and suppose that $E|\psi_1(X_1)|^3 < \infty$. Then there is a constant $C$ depending neither on $N$, $\psi$ or the $x$'s such that

$$\sup_x \left| Pr\left( \frac{n^{\frac{1}{2}}(U_n - U_N)}{2\bar{\sigma}_{1,N}(1-\alpha_N)^{\frac{1}{2}}} \le x \right) - \Phi(x) \right|$$

$$< \frac{C}{\sqrt{n}} \left( \frac{E|\psi(X_1, X_2)|^2}{\bar{\sigma}_{1,N}^2} + \frac{E|\psi_1(X_1)|^3}{\bar{\sigma}_{1,N}^3} \right).$$

152

Milbrodt (1987) proves an invariance principle, using the Nandi-Sen assumption of the uniform boundedness of $E|\psi_1(X_1)|^{2+\delta}$ for some $\delta > 0$.

### 3.7.5 Asymptotics for weights and generalised $L$-statistics

We begin the section with a theorem giving conditions under which a weighted $U$-statistic is asymptotically normal. Recall that a weighted $U$-statistic is one of the form

$$W_n = \sum_{(n,k)} w(S)\psi(S) \tag{1}$$

where $\psi(S) = \psi(X_{i_1}, \dots, X_{i_k})$ and the weights $w(S)$ satisfy $\sum_{(n,k)} w(S) = 1$.

**Theorem 1.** *Let $W_n$ be a weighted $U$-statistic of the form (1) and define*

$$w_{i,n} = \sum_{S:i \in S} w(S), \quad W_n^* = \sum_{i=1}^{n} w_{i,n}\psi_1(X_i),$$

*the notation reflecting the fact that the weights may depend on $n$. Suppose that*

(i) $\max\limits_{1 \leq i \leq n} |w_{i,n}|^2 \big/ \sum_{i=1}^n w_{i,n}^2 \to 0$,

(ii) $Var\, W_n^*/W_n \to 0$ as $n \to \infty$

and

(iii) $E|\psi_1(X_1)|^{2+\delta} < \infty$ for some $\delta > 0$.

Then $(W_n - \theta)/(Var\, W_n)^{\frac{1}{2}} \xrightarrow{D} N(0,1)$.

**Proof.** The usual strategy works here: we apply a classical result (in this case Liapounov's central limit theorem) to $W_n^*$, and show that the difference between $W_n^*$ and $W_n$, properly normalised, is asymptotically negligible.

We assume without loss of generality that $\theta = 0$, so that $EW_n = EW_n^* = 0$. Consider the r.v.

$$R_n = W_n/(Var W_n)^{\frac{1}{2}} - W_n^*/(Var W_n^*)^{\frac{1}{2}}. \tag{1}$$

The mean of $R_n$ is clearly zero, and its variance is

$$2(1 - Cov(W_n, W_n^*)/\sqrt{Var W_n Var W_n^*}), \tag{2}$$

153

so that to prove $R_n \xrightarrow{P} 0$, it is enough in view of (ii) to show that $Var\, W_n^* = Cov(W_n, W_n^*)$. This is accomplished by noting that

$$Cov(W_n, W_n^*) = \sum_{i=1}^{n} w_{i,n} Cov(W_n, \psi_1(X_i))$$

and

$$Cov(W_n, \psi_1(X_i)) = \sum_{(n,k)} w(S) Cov(\psi(S), \psi_1(X_i))$$
$$= w_{i,n} \sigma_1^2$$

so that

$$Cov(W_n, W_n^*) = \sum_{i=1}^{n} w_{i,n}^2 \sigma_1^2$$
$$= Var\, W_n^*.$$

Now set $Y_{i,n} = w_{i,n} \psi_1(X_i)$. The r.v.s $Y_{1,n}, \ldots, Y_{n,n}$ are independent and satisfy the condition $E|Y_{in}|^{2+\delta} < \infty$ by (iii), so by the Liapounov CLT we will obtain $W_n^* / (Var\, W_n^*) \xrightarrow{D} N(0,1)$ if we can show that

$$\sum_{i=1}^{n} E|Y_{n,i}|^{2+\delta} \bigg/ \left( \sum_{i=1}^{n} Var\, Y_{n,i} \right)^{(2+\delta)/2} \to 0$$

or equivalently, that

$$\sum_{i=1}^{n} |w_{i,n}|^{2+\delta} \bigg/ \left( \sum_{i=1}^{n} w_{i,n}^2 \right)^{(2+\delta)/2} \to 0. \tag{3}$$

But

$$\sum_{i=1}^{n} |w_{i,n}|^{2+\delta} \le \max_{1 \le i \le n} |w_{i,n}|^\delta \sum_{i=1}^{n} w_{i,n}^2$$

so that

$$\sum_{i=1}^{n} |w_{i,n}|^{2+\delta} \bigg/ \left( \sum_{i=1}^{n} w_{i,n}^2 \right)^{(2+\delta)/2} \le \left\{ \frac{\max\limits_{1 \le i \le n} |w_{1,n}|^2}{(\sum_{i=1}^{n} w_{i,n}^2)} \right\}^{\delta/2}$$

and (3) and hence the theorem follows from assumption (i).

Note that, in the case $k = 2$, we have

$$Var\, W_n = \sum_{|S_1 \cap S_2| = 1} w(S_1)w(S_2)\sigma_1^2 + \sum_{(n,2)} w^2(S)\sigma_2^2$$

$$= \sum_{i=1}^{n} w_{i,n}^2 \sigma_1^2 + \sum_{(n,2)} w^2(S)(\sigma_2^2 - 2\sigma_1^2)$$

so that (ii) is implied by $\sum_{(n,2)} w^2(S) / \sum_{i=1}^{n} w_{i,n}^2 \to 0$.

Next we turn to the asymptotics of generalised $L$-statistics. Because this limit theory uses differential approximation techniques not used in the rest of the book, we give only a brief sketch. The reader wishing a fuller account is referred to Serfling (1984) and the references therein.

Using the notation of Section 2.7, a generalised $L$-statistic based on a symmetric kernel of degree $k$ is one of the form

$$\sum_{i=1}^{N} C_{n,i} W_{i:n}$$

where the quantities $W_{i:n}$ are the ordered kernel values $\psi(S)$ and $N = \binom{n}{k}$. Assuming that the constants $C_{n,i}$ are given by (3) of 2.7, we can write

$$\sum_{i=1}^{N} C_{n,i} W_{i:n} = \int_0^1 H_n^{-1}(t) J(t) dt = T(H_n)$$

where $T(H) = \int_0^1 H^{-1}(t) J(t) dt$. Also

$$T(H_n) = T(H) + \binom{n}{k}^{-1} \sum_{(n,k)} IC(\psi(X_{i_1}, \dots, X_{i_k})) + R_n$$

and, using the methodology of the differential approximation, it can be shown that $R_n$ is asymptotically negligible, and that

$$n^{\frac{1}{2}}(T(H_n) - T(H)) \quad \text{and} \quad n^{\frac{1}{2}} \binom{n}{k}^{-1} \sum_{(n,k)} IC(\psi(X_{i_1}, \dots, X_{i_k}))$$

155

have similar asymptotic distributions, which will be normal by standard $U$-statistic theory provided the kernel $IC(\psi(x_1,\ldots,x_k),H,T)$ is non degenerate. Similarly, if the generalised $L$-statistic is of the form

$$T(H_n) = \sum_{i=1}^{n} c_i H_n^{-1}(p_i)$$

the same result holds true.

Recall that trimmed $U$-statistics, where the extreme kernel values are discarded, are a special case of the generalised $L$-statistics considered above. Another type of trimming, where the extreme $X's$ are discarded before the $U$-statistic is calculated, is considered by Janssen, Serfling and Veraverbeke (1987).

### 3.7.6   Random $U$-statistics

Let $N_n$ be a sequence of r.v.s taking values $k$, $k+1\ldots$ independently of the $X$'s, and consider the $U$-statistic $U_{N_n}$ based on a random number $N_n$ of the $X$'s. The asymptotic behaviour of $U_{N_n}$ is of interest in the problems of sequential estimation discussed briefly in Section 6.4, so we give a brief account here, without proofs.

An early theorem of this type was proved by Sproule (1974), who shows that if $N_n/n \xrightarrow{p} 1$, then $U_{N_n}$ is asymptotically normal. This result remains true if instead we have $N_n/n - \lambda \xrightarrow{p} 0$, where $\lambda$ is some positive r.v. having a discrete distribution. This result has been supplemented by various Berry-Esseen theorems; see Ahmad(1980) and Csenki (1981). Recent results have been obtained by Aerts and Callaert (1986), who prove the following theorem:

**Theorem 1.** *Let $U_n$ be a non-degenerate $U$-statistic based on a sequence $X_1,\ldots,X_n$ of i.i.d. random variables and having kernel $\psi$ of degree $k$. Assume that for some $\delta > 0$ and $t > (4+\delta)/3$*

$$E|\psi_1(X_1) - \theta|^{2+\delta} < \infty \quad \text{and} \quad E|\psi(X_1,\ldots,X_k)|^t < \infty.$$

*Also let $\epsilon_n$ be a sequence of positive numbers converging to zero such that $n^{-\delta} \leq \epsilon_n$ for all sufficiently large $n$. Let $N_n$ be a sequence of positive r.vs, and $\tau$ a positive r.v. satisfying for some constants $c_1$ and $c_2$*

(i) $Pr(|[n\tau]^{-1}N_n - 1| > c_1\epsilon_n) = O(\epsilon_n^{\frac{1}{2}})$,

156

(ii) $Pr(\tau < c_1 n^{-1} \epsilon_n^{-1/\delta}) = O(\epsilon_n^{\frac{1}{2}})$

and

(iii) *The r.v. $\tau$ is independent of the $X$'s.*

*Then*

$$\sup_x |Pr(N_n^{\frac{1}{2}}(U_{N_n} - \theta)/k\sigma_1 \leq x) - \Phi(x)| = O(\epsilon_n).$$

Note that if we assume in addition that the $N_n$ are also independent of $\tau$ and the $X$'s, then condition (i) may be replaced by the weaker condition $Pr([n\tau]^{-1} N_n < 1 - \alpha) = O(\epsilon_n^{1/2})$.

We also note that Horváth (1985) has considered a strong law of large numbers for the present situation.

## 3.8 Kernels with estimated parameters

Often we must deal with a statistic that is "almost" a $U$-statistic except that the kernel contains some unknown parameter that must be estimated from the data. For example, a $U$-statistic to estimate the variance in the case when the mean $\mu$ is known is $n^{-1} \sum_{i=1}^{n}(X_i - \mu)^2$. If $\mu$ is unknown, it must be replaced by an estimate, and we use instead the familiar estimate $(n-1)^{-1} \sum_{i=1}^{n}(X_i - \bar{X})^2$.

For the general case, consider a $U$-statistic $U_n(\lambda)$ based on a kernel $\psi(x_1, \ldots, x_k; \lambda)$ which depends on $m$ unknown parameters $\lambda_j$ that are functionals of $F$, the common distribution function of the $X$'s. Let $\hat{\lambda}$ be an estimate of this (vector) parameter based on $X_1, \ldots, X_n$. The question of how and when the asymptotic distributions of $U_n(\lambda)$ and $U_n(\hat{\lambda})$ differ has been addressed in the $U$-statistic context by Sukhatme (1958), Randles (1983) and Randles and de Wet (1987), and we now give a brief account of their results, without proofs.

First suppose that the $U$-statistic is non-degenerate, and let $\theta(\gamma) = E\psi(X_1, \ldots, X_k; \gamma)$. Heuristically, if the function $\psi(X_1, \ldots, X_k; \gamma)$ is differentiable as a function of $\gamma$, then we may expand this function about $\lambda$ and obtain

$$\psi(X_1, \ldots, X_k; \hat{\lambda}) = \psi(X_1, \ldots, X_k; \lambda) + (\hat{\lambda} - \lambda)\frac{\partial}{\partial \lambda}\psi(X_1, \ldots, X_k; \lambda^*)$$

for some $\lambda^*$ near $\lambda$. Summing over all $k$-subsets we get

$$n^{\frac{1}{2}}(U_n(\hat{\lambda}) - \theta(\lambda)) = n^{\frac{1}{2}}(U_n(\lambda) - \theta(\lambda)) + n^{\frac{1}{2}}(\hat{\lambda} - \lambda)U_n'(\lambda) + o_p(1)$$

157

where $U_n'$ is the vector of $U$-statistics based on $\frac{\partial}{\partial \gamma_j} \psi(X_1, \ldots, X_k; \lambda)$. Assume that $n^{\frac{1}{2}}(\hat{\lambda} - \lambda)$ and $n^{\frac{1}{2}}(U_n(\lambda) - \theta(\lambda))$ are jointly asymptotically normal, with asymptotic covariance matrix $\Sigma$, and suppose that $EU_n' = (\frac{\partial}{\partial \gamma_1}\theta(\lambda), \ldots, \frac{\partial}{\partial \gamma_m}\theta(\lambda)) = \theta'(\lambda)$, say. If $\theta'(\lambda) = 0$, then because of the SLLN for $U$-statistics, we must have $U_n' \xrightarrow{P} 0$ and the asymptotic distributions of $n^{\frac{1}{2}}(U_n(\hat{\lambda}) - \theta(\lambda))$ and $n^{\frac{1}{2}}(U_n(\lambda) - \theta(\lambda))$ will be the same. If $\theta'(\lambda) \neq 0$, the asymptotic distribution of $n^{\frac{1}{2}}(U_n(\hat{\lambda}) - \theta(\lambda))$ will be normal with zero mean and asymptotic variance $(1, \theta'(\lambda))^T \Sigma (1, \theta'(\lambda))$.

If $\psi$ is not differentiable in $\lambda$, the same conclusions may hold, provided $\theta$ is differentiable. The paper by Randles cited above gives conditions under which this happens.

**Example 1. The sample variance.**

Here $\psi(x; \mu) = (x - \mu)^2$, and $\theta'(\mu) = 0$, so replacing $\mu$ by $\bar{X}$ does not change the asymptotic distribution.

**Example 2. Testing if populations differ only in location.**

Suppose we want to test if two distribution functions $F$ and $G$ differ only in location, i.e. we want to test the hypothesis that $F(x - \xi_1) = G(x - \xi_2)$ for all $x$, where $\xi_1$ and $\xi_2$ are the medians of $F$ and $G$. Sukhatme (1958) proposes a modification of the test described in Example 1 of Section 3.7.1. If $\xi_1$ and $\xi_2$ are known, a suitable test statistic is

$$T_N(\xi_1, \xi_2) = \frac{1}{n_1 n_2} \sum_{i=1}^{n_1} \sum_{j=1}^{n_2} K(X_i - \xi_1, Y_j - \xi_2)$$

where $K$ is the kernel defined in Example 1 of Section 3.7.1. Replacing $\xi_1$ and $\xi_2$ by sample medians $\widetilde{X}$ and $\widetilde{Y}$ gives

$$T_N(\widetilde{X}, \widetilde{Y}) = \frac{1}{n_1 n_2} \sum_{i=1}^{n_1} \sum_{j=1}^{n_2} K(X_i - \widetilde{X}, Y_j - \widetilde{Y}).$$

We show below that this substitution does not effect the asymptotic distribution of the statistic, provided we assume that $F$ and $G$ have densities symmetric about their respective medians. The kernel $K$ is not differentiable, but the function

$$\theta(\gamma_1, \gamma_2) = EK(X - \gamma_1, Y - \gamma_2)$$

158

satisfies all the conditions needed to apply the result described above. We need to show that the partial derivatives of $\theta$ are zero at the medians.

The function $\theta(\gamma_1, \gamma_2)$ is given by

$$\theta(\gamma_1, \gamma_2) = Pr(0 < X - \gamma_1 < Y - \gamma_2) + Pr(Y - \gamma_2 < X - \gamma_1 < 0)$$

$$= \int_{\gamma_1}^{\infty} f(u)(1 - G(u + \gamma_2 - \gamma_1))\, du$$

$$+ \int_{-\infty}^{\gamma_1} f(u)(1 - G(u + \gamma_2 - \gamma_1))\, du,$$

so

$$\frac{\partial}{\partial \gamma_1} \theta(\gamma_1, \gamma_2) = -f(\gamma_1)\{1 - 2G(\gamma_2)\} + \int_{-\infty}^{\gamma_1} f(u)g(u + \gamma_2 - \gamma_1)\, du$$

$$- \int_{\gamma_1}^{\infty} f(u)g(u + \gamma_2 - \gamma_1)\, du$$

and

$$\frac{\partial}{\partial \gamma_2} \theta(\gamma_1, \gamma_2) = \int_{-\infty}^{\gamma_1} f(u)g(u + \gamma_2 - \gamma_1)\, du - \int_{\gamma_1}^{\infty} f(u)g(u + \gamma_2 - \gamma_1)\, du.$$

Putting $\gamma_1 = \xi_1$, $\gamma_2 = \xi_2$ we get

$$\frac{\partial \theta}{\partial \gamma_1} = \frac{\partial \theta}{\partial \gamma_2} = \int_{-\infty}^{0} f(u+\xi_1)g(u+\xi_2-\xi_1)\, du - \int_{0}^{\infty} f(u+\xi_1)g(u+\xi_2-\xi_1)\, du$$

which is zero in view of the symmetry of $f$ and $g$ about $\xi_1$ and $\xi_2$.

De Wet and Randles (1987) have considered the case when the basic $U$-statistic is degenerate of order one. Assume that $k = 2$ and that the kernel $h(x, y; \lambda)$ is of the form

$$h(x, y; \lambda) = \int_{-\infty}^{\infty} g(x, t; \lambda)g(y, t; \lambda)\, dM(t)$$

where $M$ is some positive finite measure, and $\lambda$ is an $m$-vector of parameters. Also assume that for some (vector-valued) function $\alpha$,

$$\hat{\lambda} = \lambda + n^{-1} \sum_{i=1}^{n} \alpha(X_i) + o_p(n^{-\frac{1}{2}})$$

159

so that $n^{-\frac{1}{2}}(\hat{\lambda} - \lambda)$ is asymptotically normal with covariance matrix $\Sigma$ equal to the covariance matrix of the random vector $\alpha(X)$. Let $\mu(t;\gamma) = Eg(X,t;\gamma)$ and suppose that $\mu(t;\lambda) = 0$ which will typically be the case for degenerate kernels. Denote the vector of partial derivatives of $\mu(t;\gamma)$ with respect to $\gamma_1,\ldots,\gamma_m$ by $\partial\mu(t;\gamma)$ and define a kernel $h^*$ by

$$h^*(x,y;\lambda) = \int_{-\infty}^{\infty} (g(x,t;\lambda) + \partial\mu(t;\lambda)\alpha(x))(g(y,t;\lambda) + \partial\mu(t;\lambda)\alpha(y))\,dM(t).$$

Let $\{\delta_\nu^*\}$ and $\{\delta_\nu\}$ be the sequences of eigenvalues of the linear operators associated with the kernels $h^*$ and $h$. Then de Wet and Randles show that under certain conditions, the asymptotic distribution of $n(U_n - \theta)$ is that of

$$\sum_{\nu=1}^{\infty} (\delta_\nu^* Z_i^2 - \delta_\nu)$$

where the $Z_i$'s are independent $N(0,1)$ random variables. Thus the vanishing of the partial derivatives at $\lambda$ once again implies that the asymptotic distribution is unaffected by the estimation of unknown parameters.

## Example 3. The Cramér-von Mises statistic.

If we want to test if an unknown distribution function $F$ equals some specified distribution function $F_0$, we can use the Cramér-von Mises statistic, which takes the form

$$\omega_n^2 = n \int_{-\infty}^{\infty} (F_n(x) - F_0(x))^2\,dF_0(x)$$

where $F_n$ is the empirical distribution function of a sample $X_1,\ldots,X_n$ distributed as $F$. Rearranging the above expression gives

$$\omega_n^2 = n^{-1} \sum_{i=1}^{n} \sum_{j=1}^{n} h(X_i, X_j)$$

where the kernel $h$ is given by

$$h(x,y) = \int_{-\infty}^{\infty} \left(I\{x \le t\} - F_0(t)\right)\left(I\{y \le t\} - F_0(t)\right) dF_0(t)$$

This statistic is an example of a so-called $V$-statistic; such statistics are discussed in Section 4.2. An asymptotically equivalent statistic is the $U$-statistic based on the same kernel $h$. Under the hypothesis that $F = F_0$, this $U$-statistic is readily seen to have zero mean and is in fact degenerate of order one. The eigenvalues are $\{\delta_\nu\} = (\nu\pi)^{-2}$; for an indication of how these are derived, see de Wet (1987).

Often, however, we might want to test that $F$ belongs to a specific family of distribution functions, so we might want to test that $F(x) = F_0((x - \xi)/\sigma)$, where $F_0$ is known, and has density $f_0$ say. If $\xi$ and $\sigma$ are known, the usual Cramér-von Mises statistic is

$$\omega^2(\xi, \sigma) = n \int_{-\infty}^{\infty} (F_n(x\sigma + \xi) - F_0(x))^2 \, dF_0(x)$$

which has kernel

$$h(x, y; \xi, \sigma) = \int_{-\infty}^{\infty} \big(I\{x \leq \xi + \sigma t\} - F_0(t)\big)\big(I\{y \leq \xi + \sigma t\} - F_0(t)\big) \, dF_0(t)$$

However, the statistic needs to be modified if the location and scale parameters are unknown. We can use instead the statistic $\omega^2(\hat{\xi}, \hat{\sigma})$ where $\hat{\xi}$ and $\hat{\sigma}$ are suitable estimates. In this case the function $\mu$ takes the form

$$\mu(t; \gamma_1, \gamma_2) = F_0((\gamma_1 + \gamma_2 t - \mu)/\sigma) - F_0(t)$$

and the partial derivatives are

$$\frac{\partial}{\partial \gamma_1} \mu(t; \gamma) \bigg|_{\substack{\gamma_1 = \xi \\ \gamma_2 = \sigma}} = \sigma^{-1} f_0(t), \quad \frac{\partial}{\partial \gamma_2} \mu(t; \gamma) \bigg|_{\substack{\gamma_1 = \xi \\ \gamma_2 = \sigma}} = \sigma^{-1} t f_0(t),$$

and so the asymptotic distributions of the unmodified and modified statistics are different, and the latter depends on the actual estimates of $\xi$ and $\sigma$ that are chosen.

Note that this statistic, with or without estimated parameters, is one of a family of statistics for testing independence and goodness-of-fit. Other statistics of this type include the Anderson-Darling statistic and the statistic due to Hoeffding described in Section 6.2.4.

## 3.9    Bibliographic details

The basic asymptotic normality result in Section 3.2.1 is due to Hoeffding (1948a), while the theorem on first-order degeneracy asymptotics is due to Serfling (1980). For different proofs, see Gregory (1977), Eagleson (1979) and Hall (1979). The discussion in Section 3.2.3 is adapted from Rubin and Vitale (1980), and the material on Poisson convergence is taken from Babour and Eagleson (1984). The proof of the Berry-Esseen theorem in Section 3.3.2 is from Friedrich (1989), and Bickel (1974) provided the basis for Section 3.3.3.

The basic facts on martingales in Section 4.3.1 are taken from Billingsley (1979) and Chow and Teicher (1978), and the proofs of Theorem 3 of Section 3.4.2 are based on Hoeffding's 1961 technical report and Arvesen (1968). Basic material on invariance principles was adapted from Strassen (1967), Billingsley (1968) and Heyde (1981). The proofs of Theorems 2 and 4 are due to Miller and Sen (1972) and Sen (1974b).

The proofs of the asymptotic results of Section 3.7 are taken from Hoeffding (1948b) for the non-identically distributed case, Yoshihara (1976) for weakly dependent sequences, Nandi and Sen (1963) for $U$-statistics based on finite population sampling, Nowicki and Wierman (1987) for the weighted case and Aerts and Callaert (1986) for random $U$-statistics.

The material in Section 3.8 is taken from Randles (1983), de Wet (1987) and de Wet and Randles (1987).

# CHAPTER FOUR

## Related Statistics

### 4.1 Introduction

This chapter is concerned with three classes of statistics related to $U$-statistics. In the present section the general class of symmetric statistics (statistics invariant under relabelling of the sample random variables) is introduced. All $U$-statistics are symmetric, and some $U$-statistic results carry over to the general case. A nice characterisation of $U$-statistics in the class of symmetric statistics is presented in Theorem 2. Section 4.1.2 discusses asymptotic results.

Section 4.2 deals with von Mises statistical functionals or $V$-statistics, and discusses the connection between $V$-statistics and $U$-statistics. Examples are given illustrating how the asymptotic behaviour of $V$-statistics may be deduced from that of the corresponding $U$-statistics.

Incomplete $U$-statistics are the subject of the last part of the chapter. The question of choice of design for incomplete $U$-statistics and the related asymptotics are covered in some detail.

### 4.1.1. Symmetric statistics: basics

If $X_1, \ldots, X_n$ are independently and identically distributed with d.f. $F$, and $\theta$ is some parameter depending on $F$, it is natural to estimate $\theta$ by means of a symmetric function of the $X$'s. There is a considerable literature on the subject of symmetric statistics, and in this section we discusss the connections between this theory and the theory of $U$-statistics, which are of course symmetric statistics in their own right.

Specifically, consider a sequence of *symmetric functions* $S_n(x_1, \ldots, x_n)$ where each $S_n$ is a function of $n$ arguments invariant under permutations of those arguments.

We begin by considering a generalisation of the $H$-decomposition (For a further generlisation, see for example, Efron and Stein (1981).) As in Section 1.6, define for $c = 1, 2, \ldots, n$

$$s_n^{(c)}(x_1, \ldots, x_c) = E(S_n(X_1, \ldots, X_n)|X_1 = x_1, \ldots, X_c = x_c)$$

$$-\sum_{j=0}^{c-1} \sum_{(c,j)} s_n^{(j)}(x_{i_1}, \ldots, x_{i_c}) \qquad (1)$$

and let $s_n^{(0)} = ES_n(X_1, \ldots, X_n)$. Then we have

$$S_n(X_1, \ldots, X_n) = \sum_{j=0}^{n} \sum_{(n,j)} s_n^{(j)}(X_{i_1}, \ldots, X_{i_j}) \qquad (2)$$

which follows by the arguments of Theorem 1 of Section 1.6. For $j > 0$, the quantities $s_n^{(j)}(X_1, \ldots, X_j)$ have zero mean, and as in Theorem 3 of 1.6,

$$Cov(s_n^{(j)}(X_{i_1}, \ldots, X_{i_j}), s_n^{(j')}(X_{i'_1}, \ldots, X_{i'_{j'}})) = 0$$

unless $j = j'$ and the two sets $\{i_1, \ldots, i_j\}$ and $\{i'_1, \ldots, i'_{j'}\}$ coincide.

Note that the functions $s_n^{(j)}, j \le n$, depend on $n$ as well as $j$, for nothing in the above formulation prevents $S_n$ from being a a completely arbitrary sequence of symmetric functions. However, in applications they will be a sequence of estimators (based on sample size $n$) estimating some parameter $\theta$.

If the functions $s_n^{(j)}$ are identically zero for $j \ge k$ and $n \ge k$, the sequence of symmetric statistics is said to have *finite order* $k$. Under these circumstances, we can write it as a $U$-statistic, albeit with a kernel $\psi_n$ depending on the sample size:

**Theorem 1.** *Let $S_n$ be a sequence of symmetric statistics of finite order $k$. Then $S_n$ is a $U$-statistic of degree $k$ with kernel $\psi_n$ (depending on $n$) given by*

$$\psi_n(x_1, \ldots, x_k) = \sum_{j=0}^{k} \binom{n}{j} \binom{k}{j}^{-1} \sum_{(k,j)} s_n^{(j)}(x_{i_1}, \ldots, x_{i_j}). \qquad (3)$$

**Proof.** If $S_n$ is of finite order $k$, then from (2) we have

$$S_n = \sum_{j=0}^{k} \sum_{(n,j)} s_n^{(j)}(X_{i_1}, \ldots, X_{i_j}). \qquad (4)$$

164

Define $S_n^{(j)}(x_1, \ldots, x_k) = \sum_{(k,j)} s_n^{(j)}(x_{i_1}, \ldots, x_{i_j})$. Then using the identities (c.f. Section 1.6)

$$\sum_{(n,k)} S_n^{(j)}(x_{i_1}, \ldots, x_{i_k}) = \binom{n-j}{k-j} \sum_{(n,j)} s_n^{(j)}(x_{i_1}, \ldots, x_{i_j})$$

and

$$\binom{n}{k}\binom{n-j}{k-j}^{-1} = \binom{n}{j}\binom{k}{j}^{-1},$$

the right hand side of (4) can be written

$$\sum_{j=0}^{k} \binom{n-j}{k-j}^{-1} \sum_{(n,k)} S_n^{(j)}(X_{i_1}, \ldots, X_{i_k})$$

$$= \binom{n}{k}^{-1} \sum_{j=0}^{k} \binom{n}{j}\binom{k}{j}^{-1} \sum_{(n,k)} S_n^{(j)}(X_{i_1}, \ldots, X_{i_k})$$

$$= \binom{n}{k}^{-1} \sum_{(n,k)} \psi_n(X_{i_1}, \ldots, X_{i_k}), \tag{5}$$

proving the theorem.

**Example 1. Sample variance.**

If $S_n = (n-1)^{-1} \sum_{i=1}^{n} (X_i - \bar{X})^2$, then straightforward calculations yield

$$s_n^{(1)}(x_1) = n^{-1}\{(x_1 - \mu) - \sigma^2\},$$

$$s_n^{(2)}(x_1, x_2) = -2\{n(n-1)\}^{-1}(x_1 - \mu)(x_2 - \mu)$$

and

$$s_n^{(j)}(x_1, \ldots, x_j) = 0 \quad \text{for} \quad j \geq 3.$$

Using formula (3), the kernel of the corresponding $U$-statistic is $\psi_n(x_1, x_2) = \frac{1}{2}(x_1 - x_2)^2$. Since $\psi_n$ does not depend on $n$ in this case, the sample variance is a true $U$-statistic.

From Theorem 1, we obtain a characterisation of $U$-statistics as symmetric statistics of finite order for which the kernels (3) do not depend on $n$.

A more easily verifiable condition is contained in the next theorem, which is due to Lenth (1983).

**Theorem 2.** *Let $S_n$ be a sequence of symmetric statistics, and suppose that for $n > k$,*

$$S_n(X_1, \ldots, X_n) = n^{-1} \sum_{i=1}^{n} S_{n-1}(X_1, \ldots, X_{i-1}, X_{i+1}, \ldots, X_n). \quad (6)$$

*Then $S_n$ is a U-statistic of degree $k$. Conversely, any U-statistic of degree $k$ satisfies (6).*

**Proof.** We will show that $S_n$ is a $U$-statistic of degree $k$ with kernel $\psi(x_1, \ldots, x_k) = S_k(x_1, \ldots, x_k)$. The proof is by induction on $n$. By (6), the theorem is trivially true for $n = k + 1$. Assume it is true for $n$, so that $S_n$ is a $U$-statistic with kernel $\psi$. We will show that $S_{n+1}$ is a $U$-statistic with kernel $\psi$. Using the induction hypothesis and (6), we get

$$S_{n+1} = (n+1)^{-1} \sum_{i=1}^{n+1} S_n(X_1, \ldots, X_{i-1}, X_{i+1}, \ldots, X_{n+1})$$

$$= (n+1)^{-1} \sum_{i=1}^{n+1} \binom{n}{k}^{-1} \sum_{T \in \mathcal{S}_i} \psi(T) \quad (7)$$

where $\mathcal{S}_i$ is the class of subsets of $\{1, 2, \ldots, n+1\}$ not containing $i$. Now each subset $T$ of $\mathcal{S}_{n+1,k}$ is contained in exactly $n - k + 1$ of the classes $\mathcal{S}_i$ since $T \in \mathcal{S}_i$ for every $i$ for which $i$ is not a member of $T$. Hence (7) reduces to

$$(n+1)^{-1}(n-k+1)\binom{n}{k}^{-1} \sum_{(n+1,k)} \psi(X_{i_1}, \ldots, X_{i_k})$$

$$= \binom{n+1}{k}^{-1} \sum_{(n+1,k)} \psi(X_{i_1}, \ldots, X_{i_k})$$

showing that $S_{n+1}$ is a $U$-statistic of order $k$.

For the converse, the above argument reverses to show that every $U$-statistic of degree $k$ satisfies (6).

166

The "leave-one-out" condition (6) is reminiscent of the *jackknife*, which is discussed in greater detail in Chapter 5; see also Efron (1982). In fact we can prove

**Theorem 3.** *Let $S_n$ be a symmetric statistic, and let $S_n(JACK)$ denote its jackknifed version,*

$$S_n(JACK) = n^{-1} \sum_{i=1}^{n} \{nS_n - (n-1)S_{n-1}(X_1, \ldots, X_{i-1}, X_{i+1}, \ldots, X_n)\}$$

*Then $S_n$ is a U-statistic if and only if $S_n(JACK) = S_n$.*

**Proof.** The condition $S_n(JACK) = S_n$ is equivalent to (6) since

$$S_n(JACK) - S_n$$

$$= (n-1)\left(S_n - n^{-1} \sum_{i=1}^{n} S_{n-1}(X_1, \ldots, X_{i-1}, X_{i+1}, \ldots, X_n)\right).$$

**Example 2. L-statistics.**

Let $X_{1:n} < \cdots < X_{n:n}$ be the order statistics of the sample and let $S_n = \sum_{i=1}^{n} C_{n,i} X_{i:n}$ be an $L$-statistic (c.f. Section 2.7). Then because of the symmetry of $S_n$,

$$S_{n-1}(X_1, \ldots, X_{i-1}, X_{i+1}, \ldots, X_n) = \sum_{j=1}^{i-1} C_{n-1,j} X_{j:n} + \sum_{j=i+1}^{n} C_{n-1,j-1} X_{j:n}$$

and so $n^{-1} \sum_{i=1}^{n} S_{n-1}(X_1, \ldots, X_{i-1}, X_{i+1}, \ldots X_n) = \sum_{i=1}^{n} b_{n,i} X_{i:n}$ where

$$nb_{n,1} = (n-1)C_{n-1,1},$$

$$nb_{n,j} = (j-1)C_{n-1,j-1} + (n-j)C_{n-1,j}, \quad 2 \leq j \leq n-1,$$

and

$$nb_{n,n} = (n-1)C_{n-1,n-1}.$$

Thus a necessary and sufficient condition for an $L$-statistic to be a $U$-statistic is

$$nC_{n,1} = (n-1)C_{n-1,1};$$

$$nC_{n,j} = (j-1)C_{n-1,j-1} + (n-j)C_{n-1,j}, 2 \leq j \leq n-1;$$

and

$$nC_{n,n} = (n-1)C_{n-1,n-1}.$$

As an example, consider the $L$-statistic

$$L_n = \sum_{i=1}^{n} C_{n,i} X_{i:n}$$

where $C_{n,i} = 2(2i - n - 1) / n(n-1)$. It is routine to verify the above conditions, so that $L_n$ is a $U$-statistic. To find the kernel, we find an integer $k$ for which (6) is satisfied, and set $\psi = S_k$. A little trial and error shows that $L_2 = X_{2:2} - X_{1:2}$ and $L_3 = \frac{2}{3}(X_{3:3} - X_{1:3})$ so that we can write

$$L_3 = \tfrac{1}{3}\big\{ X_{2:3} - X_{1:3}) + (X_{3:3} - X_{1:3}) + (X_{3:3} - X_{2:3}) \big\}$$

and hence $L_n$ is a $U$-statistic with kernel $\psi(x_1, x_2) = |x_1 - x_2|$. This statistic is commonly known as *Gini's mean difference*.

Our last result specialises the decomposition (2) to the case of $U$-statistics.

**Theorem 4.** *If $S_n$ is a $U$-statistic of degree $k$, then the functions $s_n^{(j)}$ in (1) are given by*

$$s_n^{(j)}(x_1, \ldots, x_j) = \begin{cases} \binom{k}{j}\binom{n}{j}^{-1} h^{(j)}(x_1, \ldots, x_j) & \text{if } j \leq k; \\ 0 & \text{otherwise,} \end{cases}$$

*where $h^{(1)}, \ldots, h^{(k)}$ are the functions appearing in the $H$-decomposition of $S_n$ defined in Section 1.6.*

**Proof.** By induction on $j$. For $j = 1$,

$$s_n^{(1)}(x_1) = E(S_n | X_1 = x_1) - \theta$$
$$= \binom{n}{k}^{-1} \sum_{(n,k)} E(\psi(X_{i_1}, \ldots, X_{i_k}) | X_1 = x_1) - \theta, \qquad (8)$$

and $E(\psi(X_{i_1}, \ldots, X_{i_k}) | X_1 = x_1) - \theta$ equals $\psi_1(x_1) - \theta$ if 1 is in the set $\{i_1, \ldots, i_k\}$ and zero otherwise. Hence, formula (8) reduces to $\binom{n}{k}^{-1}\binom{n-1}{k-1}$

168

$(\psi_1(x_1) - \theta) = kn^{-1}h^{(1)}(x_1)$, proving the theorem for $j = 1$. Now assume the theorem is true for $1, 2, \ldots, j-1$; we will prove it true for $j$ when $j \leq k$. Using the $H$-decomposition for $U$-statistics, we can write

$$E(S_n|X_1, \ldots, X_j) = \theta + \sum_{c=1}^{k} \binom{k}{c} E(H_m^{(c)}|X_1, \ldots, X_j) \qquad (9)$$

and

$$E(H_n^{(c)}|X_1, \ldots, X_j) = \binom{n}{c}^{-1} \sum_{(n,c)} E(h^{(c)}(X_{i_1}, \ldots, X_{i_c})|X_1, \ldots, X_j). \quad (10)$$

By Theorem 2 of Section 1.6, $E(h^{(c)}(X_{i_1}, \ldots, X_{i_c})|X_1, \ldots, X_j)$ is zero unless $\{i_1, \ldots, i_c\}$ is a subset of $\{1, 2, \ldots, j\}$, in which case the conditional expectation equals $h^{(c)}(X_{i_1}, \ldots, X_{i_c})$. Hence by (9) and (10)

$$E(S_n|X_1, \ldots, X_j) = \theta + \sum_{c=1}^{j} \binom{k}{c}\binom{n}{c}^{-1} \sum_{(j,c)} h^{(c)}(X_{i_1}, \ldots, X_{i_c})$$

$$= \binom{k}{j}\binom{n}{j}^{-1} h^{(j)}(X_1, \ldots, X_j) + \sum_{c=1}^{j-1} \sum_{(j,c)} s_n^{(c)}(X_{i_1}, \ldots, X_{i_c})$$

using the induction hypothesis. This last equation in conjunction with (1) proves the result for $j \leq k$. Finally, note that, by the above reasoning, if $j > k$ and assuming the theorem true for $1, 2, \ldots, j-1$, we have

$$E(S_n|X_1, \ldots, X_j) = \sum_{c=0}^{k} \binom{k}{c}\binom{n}{c}^{-1} \sum_{(j,c)} h^{(c)}(X_{i_1}, \ldots, X_{i_c})$$

$$= \sum_{c=0}^{k} \sum_{(j,c)} s_n^{(c)}(X_{i_1}, \ldots, X_{i_c}) \qquad (11)$$

Using the induction hypothesis, we see that (11) and (1) imply $s_n^{(j)} = 0$.

**Example 3. The sample variance.**

From previous examples, we have

$$s_n^{(1)}(x_1) = n^{-1}\{(x_1 - \mu) - \sigma^2\} = 2n^{-1}h^{(1)}(x)$$

and

$$s_n^{(2)}(x_1, x_2) = -\binom{n}{2}^{-1}(x_1 - \mu)(x_2 - \mu) = \binom{n}{2}^{-1}h^{(2)}(x_1, x_2).$$

### 4.1.2 Asymptotic behaviour of symmetric statistics

Let $S_n(X_1, \ldots, X_n)$ be a sequence of symmetric statistics. The results of the last section made no assumptions about the functions $S_n$, but in practice the sequence of statistics are estimators of some parameter for different sample sizes and have certain convergence properties. The restriction that this places on the sequence has been made precise by different authors in different ways.

For example, Efron and Stein (1981), in Sections 3 and 4 consider symmetric functions for which the functions $s_n^{(c)}(x_1, \ldots, x_c)$ are of the form

$$s_n^{(c)}(x_1, \ldots, x_c) = n^{-c}\alpha_n^{(c)}(x_1, \ldots, x_c)$$

where $\alpha_n^{(c)}$ converges pointwise to some $\alpha^{(c)}(x_1, \ldots x_c)$, and in particular consider the case where $\alpha_n^{(c)}$ is independent of $n$. Note that by Theorem 4 of the last section, $U$-statistics are of this form with

$$s_n^{(c)} = n^{-c}\alpha_n^{(c)}$$

and

$$\alpha_n^{(c)} = c!\binom{k}{c}h^{(c)} + o(1).$$

Dynkin and Mandelbaum (1983), using a different notation, consider sequences where the functions $s_n^{(j)}$ do not depend on $n$. Rubin and Vitale (1980) consider symmetric statistics of finite order, but allow the $s_n^{(j)}$ to depend on $n$, subject to certain convergence requirements. Their approach has been discussed in Section 3.2.3 in the $U$-statistic context, but the discussion given there is easily extended to cover the case of symmetric statistics. Recall that we may write

$$s_n^{(c)}(x_1, \ldots, x_c) = \sum_{i_1=1}^{\infty} \cdots \sum_{i_c=1}^{\infty} \left\langle s_n^{(c)}, e_{i_1} \ldots e_{i_c} \right\rangle e_{i_1}(x_1) \ldots e_{i_c}(x_c)$$

where the functions $e_1, e_2, \ldots$ along with the constant function $e_0 = 1$ form an orthonormal basis of $\mathcal{L}_2(F)$, the space of all functions square integrable with respect to $F$, the common d.f. of the $X$'s. The inner product above is that of $L_2(F^k)$, the space of functions square integrable with respect to the $k$-dimensional product measure $dF(x_1)\ldots dF(x_k)$. Then we have

$$\sum_{(n,c)} s_n^{(c)}(X_{j_1},\ldots,X_{j_c})$$

$$= \sum_{i_1=1}^{\infty}\cdots\sum_{i_c=1}^{\infty} \left\langle s_n^{(c)}, e_{i_1}\ldots e_{i_c}\right\rangle \sum_{(n,c)} e_{i_1}(X_{j_1})\ldots e_{i_c}(X_{j_c})$$

and in the term $\sum_{(n,c)} e_{i_1}(X_{j_1})\ldots e_{i_c}(X_{j_c})$ we may group like terms together and denote by $r_m(\mathbf{i})$ the number of times the index $m$ appears in the index set $\mathbf{i} = (i_1,\ldots,i_c)$. Then according to the results of Section 2.3.2,

$$n^{\frac{c}{2}}\binom{n}{c}^{-1} \sum_{(n,c)} e_{i_1}(X_{j_1})\ldots e_{i_c}(X_{j_c}) \xrightarrow{\mathcal{D}} \prod_{m=1}^{\infty} H_{r_m(\mathbf{i})}(Z_m)$$

where $Z_1, Z_2, \ldots$ is a sequence of independent $N(0,1)$ r.vs, and $H_k$ is the $k$th Hermite polynomial.

If we assume that $\binom{n}{c}n^{-\frac{c}{2}}s_n^{(c)}$ converges to $\alpha^{(c)}$ in $L_2(F^c)$, then it follows as in Section 3.2.3 using a standard truncation argument that the random variable $S_n(X_1,\ldots,X_n) - s_n^{(0)}$ converges in distribution to

$$\sum_{c=1}^{k}\sum_{i_1=1}^{\infty}\cdots\sum_{i_c=1}^{\infty} \left\langle \alpha^{(c)}, e_{i_1}\ldots e_{i_c}\right\rangle \prod_{m=1}^{\infty} H_{r_m(\mathbf{i})}(Z_m).$$

Rubin and Vitale prove this result using a Lindeberg condition for a triangular array of r.v.s, rather than a sequence $X_1, X_2, \ldots$ of i.i.d. r.v.s. Readers are referred to their paper for details.

**Example 1. Asymptotic distribution of $U$-statistics.**

Let $S_n = \sqrt{n}(U_n - \theta)$. Then by Theorem 4 of the previous section,

$$s_n^{(c)} = n^{1/2}\binom{k}{c}\binom{n}{c}^{-1} h^{(c)}(x_1,\ldots,x_c)$$

so that

$$\lim_n n^{-c/2}s_n^{(c)} = \begin{cases} kh^{(c)}, & c = 1; \\ 0, & c > 1; \end{cases}$$

171

and the limit distribution of $S_n = \sqrt{n}(U_n - \theta)$ is that of

$$\sum_{i=1}^{\infty} \langle kh^{(1)}, e_i \rangle \prod_{m=1}^{\infty} H_{r_m(i)}(Z_m) = \sum_{i=1}^{\infty} k \langle h^{(1)}, e_i \rangle Z_i$$

where the $e_i$'s, together with the constant function 1, are an arbitrary orthonormal basis of $L_2(F)$. We may take a basis containing the function $e_1 = h^{(c)}/\sigma_1$; for this basis we have

$$\langle kh^{(1)}, e_i \rangle = \begin{cases} k\sigma_1, & i = 1; \\ 0, & i \neq 1; \end{cases}$$

and so the limit distribution is that of $k\sigma_1 Z_1$, i.e. $N(0, k^2\sigma_1^2)$.

The approach of Dynkin and Mandelbaum employs different techniques: namely the device of Poisson randomisation of the sample size together with basic Hilbert space theory. The resulting limit distributions are characterised in terms of multiple Wiener integrals.

Every sequence of symmetric statistics based on a sequence $X_1, X_2, \ldots$ of i.i.d. random variables having d.f. F (with equivalent probability measure $\mu$) and for which the functions $s_n^{(j)}$ of (2) do not depend on $n$ can be identified with a sequence of functions $s^{(0)}, s^{(1)}, \ldots$ for which

(i) $s^{(j)}$ is a symmetric function with $j$ arguments, and

(ii) $Es^{(j)}(x_1, \ldots, x_{j-1}, X_j) = 0$.

The set of all such sequences is a vector space under obvious definitions of addition and scalar multiplication, and if we confine attention to sequences satisfying the additional requirement

(iii) $\sum_{j=0}^{\infty}(j!)^{-1}E(s^{(j)}(X_1, \ldots, X_j))^2 < \infty$

then equipped with the inner product

$$\langle s, t \rangle_H = \sum_{j=0}^{\infty}(j!)^{-1}E\big(s^{(j)}(X_1, \ldots, X_j)t^{(j)}(X_1, \ldots, X_j)\big)$$

the resulting space is a Hilbert space $H$.

The Dynkin-Mandelbaum result characterises the limiting distribution of the sequence

$$Y_n(s) = \sum_{j=0}^{n} n^{-\frac{j}{2}} S_n^{(j)}(s^{(j)}) \tag{1}$$

172

where $S_n^{(j)}(s^{(j)})$ is given by

$$S_n^{(j)}(s^j) = \sum_{(n,j)} s^{(j)}(X_{i_1}, \ldots, X_{i_j}) \tag{2}$$

for $s = (s^{(0)}, s^{(1)}, \ldots)$ in $H$.

A type of sequence basic to this approach, as indeed it is to the Rubin-Vitale approach, is the sequence $s^\phi$ whose elements are

$$(s^\phi)^{(j)}(x_1, \ldots, x_j) = \prod_{i=1}^{j} \phi(x_i),$$

where $E\phi(X_1) = 0$ and $E\phi^2(X_1) < \infty$. Note that such sequences are in $H$ since (i) and (ii) are clearly satisfied and

$$\sum_{j=0}^{\infty} (j!)^{-1} E\{(s^\phi)^{(j)}\}^2 = \sum_{j=0}^{\infty} (j!)^{-1} \{E\phi^2(X_1)\}^j = e^{E\phi^2} < \infty.$$

The proof of the Dynkin-Mandelbaum result revolves around two families of random variables, the first obtained from (1) by randomising the sample size, and the second based on a certain Gaussian family.

Denote by $\mathcal{E}$ all functions $\phi$ of one variable having finitely many values and satisfying $E\phi(X_1) = 0$. Call these functions *elementary*, and note that they are dense in the metric of $L_2(F)$ in the set of functions used in the definition of the $s^\phi$ above. Also define a family of random variables $\{\varepsilon(\phi), \phi \in \mathcal{E}\}$ by

$$\varepsilon(\phi) = e^{G(\phi) - \frac{1}{2}\mu(\phi^2)}$$

where $G(\phi)$ is a family of Gaussian random variables satisfying the conditions $EG(\phi) = 0$ and $Cov(G(\phi), G(\psi)) = \mu(\phi\psi)$, and where the notation $\mu(f)$ for a function $f$ means $Ef(X_1)$. Note that these conditions imply that $G(t\phi) = tG(\phi)$ for all real $t$ and that $G(\phi_1 + \phi_2) = G(\phi_1) + G(\phi_2)$.

Also define for $s \in H$,

$$Z_\lambda(s) = \sum_{j=0}^{N_\lambda} \lambda^{-\frac{j}{2}} S_{N_\lambda}^{(j)}(s^{(j)}) \tag{3}$$

where $N_\lambda$ is a Poisson variate with mean $\lambda$, and is independent of the $X_i's$. Denote by $\mathcal{F}_\lambda$ and $\mathcal{F}$ the families of random variables

$$\mathcal{F}_\lambda = \{Z_\lambda(s^\phi) : \phi \in \mathcal{E}\} \quad \text{and} \quad \mathcal{F} = \{\varepsilon(\phi) : \phi \in \mathcal{E}\}.$$

Both are subspaces of the space of all square integrable random variables (we may suppose that all r.v.s are defined on the same probability space) and consequently admit inner products of the form $\langle U, V \rangle = EUV$.

Moreover, we show below that

$$E\,Z_\lambda(s^\phi)Z_\lambda(s^\psi) = E\,\varepsilon(\phi)\varepsilon(\psi) \tag{4}$$

and so the mapping $I_\lambda : \mathcal{F}_\lambda \to \mathcal{F}$ defined by $I_\lambda(Z_\lambda(s^\phi)) = \varepsilon(\phi)$ may be extended to an isometry between the Hilbert spaces $\mathcal{H}_\lambda$ and $\mathcal{H}$ generated by $\mathcal{F}_\lambda$ and $\mathcal{F}$ respectively.

In fact, the space $\mathcal{H}_\lambda$ turns out to be $\{Z_\lambda(s) : s \in H\}$ and if we define

$$W(s) = I_\lambda(Z_\lambda(s)), \tag{5}$$

the space $\mathcal{H}$ is just $\{W(s) : s \in H\}$. Next we note that, given $\phi_1, \ldots, \phi_m$ in $\mathcal{E}$, the random vector $(Z_\lambda(s^{\phi_1}), \ldots, Z_\lambda(s^{\phi_m}))$ converges in distribution to $(\varepsilon(\phi_1), \ldots, \varepsilon(\phi_m))$ as $\lambda \to \infty$. Linear combinations of the former vector must thus converge to linear combinations of the latter, showing that elements of $\mathcal{F}_\lambda$ converge to the corresponding elements of $\mathcal{F}$. This correspondence is simply extended to $\mathcal{H}_\lambda$ and $\mathcal{H}$, so that $Z_\lambda(s)$ converges to $W(s)$ for all $s$ in $H$. Since $Z_n(s)$ and $Y_n(s)$ differ by a negligible amount in probability, this shows that $Y_n(s)$ converges to $W(s)$ in distribution.

We now state the result formally.

**Theorem 1.** *Let $s, Y_n(s)$ and $W(s)$ be as above. Then $Y_n(s)$ converges in distribution to $W(s)$.*

**Proof.** We fill in the details in the preceding summary. The first task is to verify (4) and set up the isometry between $\mathcal{H}_\lambda$ and $\mathcal{H}$. To this end, note that

$$Z_\lambda(s^\phi) = \sum_{j=0}^{N_\lambda} \lambda^{-\frac{i}{2}} S_{N_\lambda}^{(j)}((s^\phi)^{(j)})$$

174

$$= \sum_{j=0}^{N_\lambda} \lambda^{-\frac{i}{2}} \sum_{(N_\lambda,j)} \phi(X_{i_1}) \ldots \phi(X_{i_j})$$

$$= \prod_{i=1}^{N_\lambda} (1 + \lambda^{-\frac{1}{2}} \phi(X_i))$$

and so

$$EZ_\lambda(s^\phi)Z_\lambda(s^\psi) = E \prod_{i=1}^{N_\lambda} (1 + \lambda^{-\frac{1}{2}} \phi(X_i))(1 + \lambda^{-\frac{1}{2}} \psi(X_i))$$

$$= \sum_{n=0}^{\infty} \frac{e^{-\lambda} \lambda^n}{n!} \prod_{i=1}^{n} E(1 + \lambda^{-\frac{1}{2}} \phi(X_i))(1 + \lambda^{-\frac{1}{2}} \psi(X_i))$$

$$= \sum_{n=0}^{\infty} \frac{e^{-\lambda} \lambda^n}{n!} (1 + \lambda^{-1} E\phi(X_1)\psi(X_1))^n$$

$$= e^{\mu(\phi\psi)}$$

$$= E\varepsilon(\phi)\varepsilon(\psi)$$

Next we check that the set of variables $\{Z_\lambda(s) : s \in H\}$ is indeed the Hilbert space generated by $\mathcal{F}_\lambda$. First we show that the correspondence $s \leftrightarrow Z_\lambda(s)$ is in fact an isometry. Then the desired result will follow if we can show that $\{s^\phi : \phi \in \mathcal{E}\}$ is dense in $H$. To carry out this program, consider two sequences $s$ and $t$ in $H$. Then

$$E S_n^{(j)}(s^{(j)}) S_n^{(j')}(t^{(j')}) = \sum_{(n,j)} \sum_{(n,j')} Es^{(j)}(X_{i_1}, \ldots, X_{i_j}) t^{(j')}(X_{i'_1}, \ldots, X_{i'_{j'}})$$

$$= \begin{cases} \binom{n}{j} Es^{(j)}(X_1, \ldots, X_j) t^{(j)}(X_1, \ldots, X_j), & \text{if } j = j', \\ 0, & \text{otherwise,} \end{cases}$$

using property (ii) of the functions of $H$. Thus

$$E \ Z_\lambda(s)Z_\lambda(t) = E \sum_{j=0}^{N_\lambda} \lambda^{-\frac{i}{2}} S_{N_\lambda}^{(j)}(s^{(j)}) \sum_{j'=0}^{N_\lambda} \lambda^{-\frac{i'}{2}} S_{N_\lambda}^{(j')}(t^{(j')})$$

$$= \sum_{n=0}^{\infty} \frac{e^{-\lambda} \lambda^n}{n!} \lambda^{-j} E S_n^{(j)}(s^{(j)}) S_n^{(j)}(t^{(j')})$$

$$= \sum_{n=0}^{\infty} \frac{e^{-\lambda} \lambda^n}{n!} \sum_{j=0}^{n} \lambda^{-j} \binom{n}{j} Es^{(j)}(X_1, \ldots, X_j) t^{(j)}(X_1, \ldots, X_j)$$

$$= \sum_{j=0}^{\infty} e^{-\lambda} j!^{-1} Es^{(j)}(X_1, \ldots, X_j) \, t^{(j)}(X_1, \ldots, X_j) \sum_{n=j}^{\infty} \lambda^{n-j}/(n-j)!$$

$$= \sum_{j=0}^{\infty} (j!)^{-1} Es^{(j)}(X_1, \ldots, X_j) \, t^{(j)}(X_1, \ldots, X_j)$$

$$= \langle s, t \rangle_H \tag{6}$$

and so the map $s \to Z_\lambda(s)$ is an isometry and $\{Z_\lambda(s) : s \in H\}$ is a Hilbert space. It remains to show that $\{s^\phi\}$ is dense in $H$. It is enough to prove that if

$$\int s^{(j)}(x_1, \ldots, x_j) \phi(x_1) \ldots \phi(x_j) \, \mu(dx_1) \ldots \mu(dx_j) = 0 \tag{7}$$

for all $\phi$ in $\mathcal{E}$, then $s^{(j)} \equiv 0$ a.e. $(\mu \times \ldots \times \mu)$ where $\mu$ is the measure corresponding to $F$.

By Lemma A below, the fact that (7) is true for all $\phi$ in $\mathcal{E}$, implies that (7) is actually true for all $\phi$ in the set of functions $f$ square integrable with respect to $\mu$ and satisfying $\mu(f) = 0$. Denote this set of functions by $\mathcal{L}$, then $\mathcal{E}$ is dense in $\mathcal{L}$. Let $e_1, e_2, \ldots$ be an orthonormal basis for $\mathcal{L}$, then the set of functions $e_{i_1} \ldots e_{i_j}$ is an orthonormal basis for the space $\mathcal{L}^{(j)}$ of all functions $f^{(j)}$ of $j$ arguments satisfying $Ef^{(j)}(X_1, \ldots, X_j) = 0$, $E(f^{(j)}(X_1, \ldots, X_j))^2 < \infty$. Then by Parseval's relation, (7) implies for any $\phi$ in $\mathcal{L}$,

$$\sum_{i_1=1}^{\infty} \cdots \sum_{i_j=1}^{\infty} \hat{s}(i_1, \ldots, i_j) \hat{\phi}(i_1) \ldots \hat{\phi}(i_1) = 0 \tag{8}$$

where $\hat{s}(i_1, \ldots, i_j) = E\{s^{(j)}(X_1, \ldots, X_j) e_{i_1}(X_1) \ldots e_{i_j}(X_j)\}$ and $\hat{\phi}(i) = E\phi(X_1) e_i(X_1)$. For fixed integers $i_1, \ldots, i_j$, set $\phi = e_{i_1} + \cdots + e_{i_j}$, then $\hat{\phi}(i) = 1$ if $i$ is in the set $\{i_1, \ldots, i_j\}$ and zero otherwise, so that (8) and the fact that $\hat{s}$ is symmetric imply that $\hat{s}(i_1, \ldots, i_j) = 0$. Hence, once again by Parseval's equation we must have $s^{(j)} \equiv 0$.

Thus it follows from (7) that the functions $\{s^\phi\}$ are dense in $H$, and so $\mathcal{H}_\lambda = \{Z_\lambda(s) : s \in H\}$. In view of this we can extend our basic isometry $I_\lambda : \mathcal{F}_\lambda \to \mathcal{F}$ to an isometry $I_\lambda : \mathcal{H}_\lambda \to \mathcal{H}$, and so we may define a random variable $W(s) = I_\lambda(Z_\lambda(s))$ having the property that

$$E\{W(s))W(t)\} = EZ_\lambda(s)Z_\lambda(t) = \langle s, t \rangle_H$$

176

for all $s, t$ in $H$. In particular, $W$ is an isometry, being a composition of isometries, and $\mathcal{F} = \{W(s) : s \in \mathcal{H}\}$.

Now consider the asymptotic behaviour as $\lambda \to \infty$ of the random vector $(\log Z_\lambda(s^{\phi_1}), \ldots, \log Z_\lambda(s^{\phi_m}))$. Note that

$$\log Z_\lambda(s^\phi) = \sum_{i=1}^{N_\lambda} \log(1 + \lambda^{-\frac{1}{2}} \phi(X_i))$$

$$= \sum_{i=1}^{N_\lambda} \lambda^{-\frac{1}{2}} \phi(X_i) - \frac{1}{2} \sum_{i=1}^{N_\lambda} \phi^2(X_i)/\lambda + o_p(1),$$

and so since $N_\lambda/\lambda \overset{p}{\to} 1$, it follows by the central limit theorem for random numbers of summands (see Anscombe (1952) for a univariate version) that the random vector $\left( \sum_{i=1}^{N_\lambda} \lambda^{-\frac{1}{2}} \phi_1(X_i), \ldots, \sum_{i=1}^{N_\lambda} \lambda^{-\frac{1}{2}} \phi_m(X_i) \right)$ converges in law to $(G(\phi_1), \ldots, G(\phi_m))$. By Lemma B, $\sum_{i=1}^{N_\lambda} \phi_j^2(X_i)/\lambda$ converges to $E\phi_j^2(X_1)$, so that $(Z_\lambda(s^{\phi_1}), \ldots, Z_\lambda(s^{\phi_m}))$ converges in distribution to $(\varepsilon(\phi_1), \ldots, \varepsilon(\phi_m))$ and any linear combination $\sum_{j=1}^m c_j Z_\lambda(s^{\phi_j})$ converges in distribution to $\sum_{j=1}^m c_j \varepsilon(\phi_j)$. The proof is now completed by extending this result to an arbitrary $Z_\lambda(s), s \in H$ and then showing that $Z_\lambda(s)$ and $Y_n(s)$ are asymptotically equal.

For the first part, take $s$ in $H$ and since the functions $s^\phi, \phi \in \mathcal{E}$ generate $H$, we can find $t = \sum_{j=1}^m c_j s^{\phi_j}$ such that $\| s - t \|_H < \varepsilon$, where $\varepsilon > 0$ is arbitrary. Then $Z_\lambda(t)$ converges to

$$\sum_j c_j \varepsilon(\phi_j) = \sum_j c_j I_\lambda(Z_\lambda(s^{\phi_j}))$$

$$= I_\lambda(\sum_j c_j Z_\lambda(s^{\phi_j}))$$

$$= I_\lambda(Z_\lambda(\sum_j c_j s^{\phi_j}))$$

$$= I_\lambda(Z_\lambda(t))$$

$$= W(t)$$

by (3).Then

$$|Ee^{i\xi Z_\lambda(s)} - Ee^{i\xi W(s)}| \leq |Ee^{i\xi Z_\lambda(s)} - Ee^{i\xi Z_\lambda(t)}| + |Ee^{i\xi Z_\lambda(t)} - Ee^{i\xi W(t)}|$$

$$+ |Ee^{i\xi W(t)} - Ee^{i\xi W(s)}| \tag{9}$$

177

Using the inequality $|e^{ix} - 1| < |x|$, we see that he first term of (9) is less than
$E|e^{i\xi(Z_\lambda(s) - Z_\lambda(t))} - 1| \leq E|\xi(Z_\lambda(s) - Z_\lambda(t))| \leq |\xi| \parallel Z_\lambda(s) - Z_\lambda(t) \parallel_{\mathcal{H}_\lambda} = |\xi| \parallel s - t \parallel_H < |\xi|\varepsilon$. The second term of (9) converges to zero since $Z_\lambda(t) \xrightarrow{D} W(t)$, and the third term is less than $|\xi|\varepsilon$ in absolute value by the above argument. Hence $Z_\lambda(s) \xrightarrow{D} W(s)$ for all $s$ in $H$, and in particular $W(s)$ is independent of $\lambda$.

Finally consider

$$E|Y_n(s) - Z_n(s)|^2 = EY_n^2(s) - 2EY_n(s)Z_n(s) + EZ_n^2(s). \qquad (10)$$

To prove that the left hand side of (10) converges to zero, it is enough to show that $EY_n^2(s) \to \parallel s \parallel_H^2$ and $EY_n(s)Z_n(s) \to \parallel s \parallel_H^2$ . An argument similar to that used to prove (6) shows that

$$EY_n^2(s) = \sum_{j=0} a_{n,j} E(s^{(j)}(X_1, \ldots, X_j))^2 / j!$$

where $a_{n,j} = j! n^{-j} \binom{n}{j}$. Since $a_{n,j} \leq 1$ and $a_{n,j} \to 1$ as $n \to \infty$ for each $j$, it follows by dominated convergence that $EY_n^2(s)$ converges to $\sum_{j=0}^\infty E(s^{(j)}(X_1, \ldots, X_j))^2 / j! = \parallel s \parallel_H^2$ . Similarly,

$$ES_m^{(j)}(s^{(j)})S_n^{(k)}(s^{(k)}) = \begin{cases} \binom{\min(m,n)}{j} E(s^{(j)}(X_1, \ldots, X_j))^2 & \text{if } k = j, \\ 0 & \text{otherwise}, \end{cases}$$

so that

$$ES_{N_n}^{(j)}(s^{(j)})S_n^{(j)}(s^{(j)}) = \sum_{m=j}^\infty \frac{e^{-n}n^m}{m!} \binom{\min(m,n)}{j} E(s^{(j)}(X_1, \ldots, X_j))^2$$

and

$$EZ_n(s)Y_n(s) = \sum_{j=0}^\infty n^{-j} \sum_{m=j}^\infty \frac{e^{-n}n^m}{m!} \binom{\min(m,n)}{j} E(s^{(j)})^2$$

$$= \sum_{j=0}^\infty b_{n,j} E(s^{(j)})^2 / j!$$

where

$$b_{n,j} = e^{-n} n^{-j} \sum_{m=j}^\infty \frac{n^m}{m!} \binom{\min(n,m)}{j} j!.$$

178

Then (10) will converge to zero, and the theorem will be proved, if we can show that $b_{n,j}$ satisfies the hypotheses of the dominated convergence theorem. We have

$$b_{n,j} = e^{-n}n^{-j}\sum_{m=j}^{n}\frac{n^m}{m!}\binom{m}{j}j! + e^{-n}n^{-j}\sum_{m=n+1}^{\infty}\frac{n^m}{m!}\binom{n}{j}j!$$

$$= \sum_{m=0}^{n-j}\frac{e^{-n}n^m}{m!} + n^{-j}\binom{n}{j}j!\sum_{m=n+1}^{\infty}e^{-n}n^m/m!$$

$$= Pr(N_n \le n - j) + a_{n,j}Pr(N_n > n).$$

Hence $b_{n,j}$ is bounded by 2 and $\lim b_{n,j}$ converges to 1 for all $j$ since $\lim_n Pr(N_n \le n - j) = \lim_n Pr(N_n > n) = 1/2$. This follows from the fact that $(N_n - n)/\sqrt{n}$ converges to a $N(0,1)$ variate by the central limit theorem.

The theorem is proved, once Lemmas A and B are disposed of.

**Lemma A.** *Let $\mu(\phi) = 0$, $\mu(\phi^2) < \infty$. Then there exists a sequence of functions $\psi_n$ in $\mathcal{E}$ with*

$$\lim_n \int |\phi(x_1)\ldots\phi(x_j) - \psi_n(x_1)\ldots\psi_n(x_j)|^2 dF(x_1)\ldots dF(x_j) = 0.$$

**Proof.** Since $\mathcal{E}$ is dense in the set of all functions $\phi$ with $\mu(\phi) = 0$ and $\mu(\phi^2) < \infty$, we can find a sequence of functions $\psi_n$ in $\mathcal{E}$ with $\| \psi_n - \phi \|^2 / \| \phi \|^2 = \epsilon_n^2$, say, where $\| \phi \| = \{\mu(\phi^2)\}^{\frac{1}{2}}$ and $\epsilon_n = o(1)$. Moreover, since $| \| \psi_n \| - \| \phi \| | \le \| \psi_n - \phi \| = \| \phi \| \epsilon_n$ we deduce that

$$1 - \epsilon_n \le \| \psi_n \| / \| \phi \| \le 1 + \epsilon_n. \tag{11}$$

From the identity

$$\int |\phi(x_1)\ldots\phi(x_j) - \psi_n(x_1)\ldots\psi_n(x_j)|^2 dF(x_1)\ldots dF(x_j)$$

$$= \| \phi \|^{2j}\left(1 - 2\left\{\frac{\mu(\phi\psi_n)}{\| \phi \|^2}\right\}^j + \left\{\frac{\| \psi_n \|}{\| \phi \|}\right\}^{2j}\right) \tag{12}$$

with $j = 1$ we get

$$\epsilon_n^2 = 1 - 2\mu(\phi\psi_n)/ \| \phi \|^2 + \| \psi_n \|^2/\| \phi \|^2$$

179

and so by (11)

$$2\frac{\mu(\phi\psi_n)}{\|\phi\|^2} = 1 - \epsilon_n^2 + \|\psi_n\|^2 / \|\phi\|^2$$

$$\geq 2(1 - \epsilon_n)$$

and

$$\left\{\frac{\mu(\phi\psi_n)}{\|\phi\|^2}\right\}^j \geq (1 - \epsilon_n)^j.$$

Hence (12) is less than $\|\phi\|^{2j} (1 - 2(1 - \epsilon_n)^j + (1 + \epsilon_n)^{2j})$ which converges to zero as $n \to \infty$.

**Lemma B.** Let $Y_1, \ldots, Y_n$ be a sequence of i.i.d. random variables with mean $\mu$ and let $\overline{Y}_0 = \mu$, $\overline{Y}_n = n^{-1}(Y_1 + \cdots + Y_n)$ for $n > 0$. Let $N_\lambda$ be Poisson with mean $\lambda$, and independent of the $Y's$. Then $\overline{Y}_{N_\lambda} \xrightarrow{p} \mu$ as $\lambda \to \infty$.

**Proof.** Using Chebyshev's inequality, we get

$$Pr\big(|\overline{Y}_{N_\lambda} - \mu| > \varepsilon\big) = \sum_{n=0}^{\infty} \frac{e^{-\lambda}\lambda^n}{n!} Pr(|\bar{Y}_n - \mu| > \varepsilon)$$

$$\leq \sum_{n=1}^{\infty} \frac{e^{-\lambda}\lambda^n}{n!} \frac{\sigma^2}{\varepsilon^2 n}$$

$$= \frac{\sigma^2 e^{-\lambda}}{\varepsilon^2} \int_0^\lambda \frac{(e^x - 1)}{x} dx$$

which converges to zero as $\lambda \to \infty$ by L'Hôpital's Rule.

The limit $W(s)$ can be described in terms of the so-called *Wiener integral* and we now discuss how this can be done.

For real $t$, consider the random variable

$$\varepsilon(t\phi) = \exp\{tG(\phi) - \tfrac{1}{2}t^2\mu(\phi^2)\}.$$

Regarded as a function of $t$, $\varepsilon(t\phi)$ is infinitely differentiable in mean square. To compute the derivative, we use the generating function for Hermite polynomials. From e.g. Kendall and Stuart (1963) p155, we have

$$\exp\{tx - \tfrac{1}{2}t^2\} = \sum_{k=0}^{\infty} \frac{t^k}{k!} H_k(x)$$

180

where $H_k(x)$ is the $k$th Hermite polynomial. Thus for any $\phi$ with $\mu(\phi^2) = 1$,

$$\varepsilon(t\phi) = \sum_{k=0}^{\infty} \frac{t^k}{k!} H_k(G(\phi)),$$

and differentiating term by term we get

$$\frac{d^k}{dt^k}\varepsilon(t\phi)\bigg|_{t=0} = H_k(G(\phi)). \tag{13}$$

Consider also the random variable

$$Z_\lambda(s^{t\phi}) = \sum_{k=0}^{\infty} \lambda^{-k/2} S_{N_\lambda}^{(k)}((s^{t\phi})^{(k)})$$

$$= \sum_{k=0}^{\infty} \lambda^{-k/2} t^k S_{N_\lambda}^{(k)}((s^\phi)^{(k)});$$

differentiating again term by term in mean square we obtain

$$\frac{d^k}{dt^k}Z_\lambda(s^{t\phi})\bigg|_{t=0} = \lambda^{-k/2} k! S_{N_\lambda}^{(k)}((s^\phi)^{(k)}). \tag{14}$$

Since mean-square derivatives are preserved under isometries, it follows from (13) and (14) that $S_{N_\lambda}^{(k)}((s^\phi)^{(k)})$ is in $\mathcal{H}_\lambda$ and that

$$I_\lambda(\lambda^{-k/2} k! S_{N_\lambda}^{(k)}((s^\phi)^{(k)})) = H_k(G(\phi)) \tag{15}$$

whenever $\mu(\phi^2) = 1$.

We can now define the $k$-dimensional Wiener integral. Let $s^{(k)}$ be a symmetric function of $k$ variables satisfying $E^{(k)}(x_1, \ldots, x_{k-1}, X_k) = 0$. The $k$-dimensional Wiener integral $I_k(s^{(k)})$ of $s^{(k)}$ is defined by the equation

$$I_k(s^{(k)}) = I_\lambda(\lambda^{-k/2} k! S_{N_\lambda}^{(k)}(s^{(k)});$$

for another approach to the definition see Ito (1951). Note that (15) entails $I_1(\phi) = G(\phi)$ for every $\phi$ with $\mu(\phi) = 0$ and $\mu(\phi^2) = 1$. The integral has the properties

(i) $E I_k(s^{(k)}) = 0$;

(ii) $Cov(I_k(s^{(k)}), I_k(t^{(k)}))$ equals $k! E\{s^{(k)}(X_1, \ldots, X_k) t^{(k)}(X_1, \ldots, X_k)\}$ if $k = l$ and zero otherwise;

(iii) $I_k((s^\phi)^{(k)}) = H_k(I_1(\phi))$.

The properties (i) and (ii) are established by noting that $I_k(s^{(k)}) = \frac{1}{k!}I_\lambda(Z\lambda(s))$, where $s$ is the sequence $(0, 0, \ldots, s^{(k)}, \ldots)$. For example, to prove (i) consider $t = (1, 0, 0, \ldots)$. Then $Z_\lambda(t) = 1$, and $E\,I_k(s^{(k)}) = E\,I_\lambda(Z_\lambda(s))I_\lambda(Z_\lambda(t)) = \langle s, t\rangle_H = 0$. Finally, the limit $W(s)$ ocurring in Theorem 1 can now be expressed in terms of Wiener integrals. From (3) and (4) and using the convention that $s_n^{(j)} = 0$ for $j > n$ we can write

$$W(s) = I_\lambda(Z_\lambda(s))$$
$$= \sum_{k=0}^{\infty} \lambda^{-k/2} I_\lambda(S_{N_\lambda}^{(k)}(s^{(k)}))$$
$$= \sum_{k=0}^{\infty} (k!)^{-1} I_k(s^{(k)}).$$

## Example 2. Asymptotics for first-order degeneracy.

As an example of how Theorem 1 implies the asymptotic results of Chapter 3, consider a $U$-statistic with zero mean whose kernel $h$ is of degree two and is degenerate of order one. Then $s = (0, h, 0, \ldots)$ is in $H$ and using (1) we get $Y_n(s) = n^{-1}\binom{n}{2}U_n = (n-1)U_n/2$. Thus by Theorem 1, $nU_n$ converges in distribution to $2W(s)$, which for $s$ defined above, takes the form $W(s) = \frac{1}{2}I_2(h)$.

The function $h$ can be expanded as in (1) of Section 3.2.2, and so, using the linearity of the Wiener integral we get

$$I_2(h) = \sum_\nu \lambda_\nu I_2((s^{f_\nu})^{(2)})$$

where, as in Section 3.2.2, $f_\nu$ and $\lambda_\nu$ are the eigenfuctions and eigenvalues of the integral equation with kernel $h$.

By property (iii) of the Wiener integral, $I_2((s^{f_\nu})^{(2)}) = H_2(I_1(f_\nu))$ and the r.v.s $I_1(f_\nu)$ are independent standard normals since the eigenfunctions are orthonormal. Thus $nU_n$ is asymptotically distributed as $\sum_\nu \lambda_\nu(Z_\nu^2 - 1)$ as in Chapter 3.

A Berry-Esseen theorem for symmetric statistics has been proved by van Zwet (1984), see also Friedrich (1989). An invariance principle for symmetric statistics is presented in Mandelbaum and Taqqu (1984).

## 4.2 $V$-statistics

Consider a regular statistical functional of the type introduced in Section 1.1:

$$T(F) = \int_{\mathbb{R}_k} \psi(x_1, \ldots, x_k) dF(x_1) \ldots dF(x_k). \tag{1}$$

According to the discussion in Section 2.7, a reasonable estimate of $T(F)$ is $T(F_n)$ where $F_n$ is the empirical distribution function of the sample $X_1, \ldots, X_n$ :

$$F_n(x) = n^{-1} \sum_{i=1}^{n} I\{X_i \leq x\}.$$

We have

$$T(F_n) = \int_{\mathbb{R}_k} \psi(x_1, \ldots, x_k) dF_n(x_1) \ldots dF_n(x_k)$$

$$= n^{-k} \sum_{i_1=1}^{n} \cdots \sum_{i_k=1}^{n} \psi(X_{i_1}, \ldots, X_{i_k}),$$

and such statistics are called $V$-statistics, after von Mises, who introduced them in a fundamental paper (von Mises (1947)). $V$-statistics have obvious connections with $U$-statistics, which we explore in this section.

The asymptotic theory of such statistics is usually handled by means of Taylor series approximations (the so-called method of statistical differentials) alluded to briefly in Section 2.7, and which is described more fully in Serfling (1980) Chapter 6, the paper of von Mises cited above, and Filippova (1962). We do not discuss this theory here, but merely discuss some examples and describe the connection between $U$-statistics and $V$-statistics. Our first result makes this connection clear.

**Theorem 1.** *Let $V_n$ be a $V$-statistic based on a symmetric kernel $\psi$ of degree $k$:*

$$V_n = n^{-k} \sum_{i_1=1}^{n} \cdots \sum_{i_k=1}^{n} \psi(X_{i_1}, \ldots, X_{i_k}).$$

*Then we may write*

$$V_n = n^{-k} \sum_{j=1}^{k} j! S_k^{(j)} \binom{n}{j} U_n^{(j)}$$

where $U_n^{(j)}$ is a $U$-statistic of degree $j$. The kernel $\phi_{(j)}$ of $U_n^{(j)}$ is given by

$$\phi_{(j)}(x_1,\ldots,x_j) = (j!\mathcal{S}_k^{(j)})^{-1} \sum_{(j)}^* \psi(x_{i_1},\ldots,x_{i_k})$$

where the sum $\sum_{(j)}^*$ is taken over all $k$-tuples $(i_1,\ldots,i_k)$ formed from $\{1,2,\ldots,j\}$ having exactly $j$ indices distinct, and where the quantities $\mathcal{S}_n^{(j)}$ are Stirling numbers of the second kind (see e.g. Abramowitz and Stegun (1965), and Section 4.3.1.)

**Proof.** Of the $n^k$ possible $k$-tuples of indices chosen from $\{1,2,\ldots,n\}$, the numbers of $k$-tuples having $1,2,\ldots,n$ indices distinct are determined by the expansion

$$n^k = \sum_{j=1}^{k} \mathcal{S}_k^{(j)} n(n-1)\ldots(n-j+1)$$

$$= \sum_{j=1}^{k} j!\mathcal{S}_k^{(j)} \binom{n}{j}$$

and so the number of terms in the sum $\sum_{(j)}^*$ is $j!\mathcal{S}_k^{(j)}$. Hence

$$n^k V_n = \sum_{i_1=1}^{n} \cdots \sum_{i_k=1}^{n} \psi(X_{i_1},\ldots,X_{i_k})$$

$$= \sum_{j=1}^{k} \sum_{(i_1,\ldots,i_k):j\,\text{among}\,i_1\ldots i_k\,\text{distinct}} \psi(X_{i_1},\ldots,X_{i_k})$$

$$= \sum_{j=1}^{k} \sum_{1 \le i_1 < \ldots < i_j \le n} j!\mathcal{S}_k^{(j)} \phi_{(j)}(X_{i_1},\ldots,X_{i_j})$$

$$= \sum_{j=1}^{k} j!\mathcal{S}_k^{(j)} \binom{n}{j} U_n^{(j)}$$

proving the theorem.

**Example 1. Kernels of degree 2.**

We have

$$n^2 V_n = 2\binom{n}{2} U_n^{(2)} + \binom{n}{1} U_n^{(1)}$$

where $U_n^{(2)}$ has kernel $\phi_{(2)}(x_1, x_2) = \frac{1}{2}(\psi(x_1, x_2) + \psi(x_2, x_1)) = \psi(x_1, x_2)$ and $U_n^{(1)}$ has kernel $\phi_{(1)}(x_1) = \psi(x_1, x_1)$.

**Example 2. Kernels of degree 3.**

For $k = 3$,

$$n^3 V_n = 6\binom{n}{3} U_n^{(3)} + 6\binom{n}{2} U_n^{(2)} + \binom{n}{1} U_n^{(1)}.$$

The kernels are

$$\phi_{(3)}(x_1, x_2, x_3) = \frac{1}{6} \sum_{(3)} \psi(x_{i_1}, x_{i_2}, x_{i_3}) = \psi(x_1, x_2, x_3),$$

$$\begin{aligned}
\phi_{(2)}(x_1, x_2) &= \frac{1}{6}(\psi(x_1, x_1, x_2) + \psi(x_1, x_2, x_1) + \psi(x_2, x_1, x_1) \\
&\quad + \psi(x_1, x_2, x_2) + \psi(x_2, x_1, x_2) + \psi(x_2, x_2, x_1)) \\
&= \frac{1}{2}(\psi(x_1, x_1, x_2) + \psi(x_1, x_2, x_2))
\end{aligned}$$

and

$$\phi_{(1)}(x_1) = \psi(x_1, x_1, x_1).$$

Theorem 1 can be used to derive the asymptotic behaviour of the $V$-statistics from that of the corresponding $U$-statistics. We illustrate with three examples.

**Example 3. Asymptotic normality, k = 3.**

For the case $k = 3$ we can write

$$\begin{aligned}
V_n &= n^{-3} 6\binom{n}{3} U_n^{(3)} + n^{-3} 6\binom{n}{2} U_n^{(2)} + n^{-2} U_n^{(1)} \\
&= \frac{(n-1)(n-2)}{n^2} U_n^{(3)} + \frac{3(n-1)}{n^2} U_n^{(2)} + n^{-2} U_n^{(1)}
\end{aligned}$$

and so

$$\frac{\sqrt{n}(V_n - E(V_n))}{3\sigma_1} = \frac{\sqrt{n}(U_n^{(3)} - \theta_3)}{3\sigma_1} + o_p(1)$$

writing $\theta_3 = EU_n^{(3)}$ and $\sigma_1^2 = Var\, E\{\phi_{(3)}(X_1, X_2, X_3)|X_1\}$. Thus, provided $\sigma_1 > 0$, the asymptotic behaviour of $V_n$ is that of $U_n$, and $\sqrt{n}(V_n - EV_n)$ is asymptotically $N(0, 9\sigma_1^2)$. An exactly similar result holds for general $k$.

**Example 4. The case k=2, first order degeneracy.**

Suppose that $k = 2$ and that $U_n^{(2)}$ has a degeneracy of order 1, so that $n(U_n - \theta_2)$ has a non-degenerate limit distribution, where $\theta_2 = E\psi(X_1, X_2)$. Also write $\theta_1 = E\psi(X_1, X_1)$. Then

$$n(V_n - EV_n) = n(U_n^{(2)} - \theta_2) + (U_n^{(1)} - \theta_1) - (U_n^{(2)} - \theta_2)$$
$$= n(U_n^{(2)} - \theta_2) + o_p(1)$$

so that once again the asymptotic behaviours are identical.

Also note that

$$n(V_n - \theta_2) = n(U_n^{(2)} - \theta_2) + U_n^{(1)} - U_n^{(2)}$$
$$= n(U_n^{(2)} - \theta_2) + \theta_1 - \theta_2 + o_p(1)$$

and so if $U_n^{(2)}$ has a first order degeneracy, the asymptotic distribution of $n(V_n - \theta_2)$ is of the form $\sum \lambda_\nu (Z_\nu^2 - 1) + \theta_1 - \theta_2$.

**Example 5. The Berry-Esseen Theorem for $V$-statistics.**

We treat the case $k = 2$ only, for a general treatment see Janssen (1981). Suppose that the corresponding $U$-statistic $U_n^{(2)}$ is non-degenerate, with $\sigma_1^2 > 0$. Then as in Example 4,

$$n^{\frac{1}{2}}(V_n - E(V_n))/2\sigma_1 = n^{\frac{1}{2}}(U_n^{(2)} - \theta_2)/2\sigma_1 + R_n$$

where $R_n = n^{-\frac{1}{2}}\{(U_n^{(1)} - \theta_1) - (U_n^{(2)} - \theta_2)\}/2\sigma_1$. Denote by $G_n(x)$ and $\Phi_n(x)$ the d.fs of $n^{\frac{1}{2}}(V_n - E(V_n))/2\sigma_1$ and $n^{\frac{1}{2}}(U_n - \theta_2)/2\sigma_1$. Then by the discussion after Theorem 1 of Section 3.3.1, there is a constant $C$ such that

$$\sup_x |\Phi_n(x) - \Phi(x)| \leq C\rho n^{-\frac{1}{2}} \tag{2}$$

where $\rho = E|\psi(X_1, X_2) - \theta_2|^3/2\sigma_1$ and $\Phi$ is the standard normal distribution function.

By a standard argument, for any $\varepsilon > 0$

$$G_n(x) \leq \Phi_n(x + \varepsilon) + Pr(|R_n| \geq \varepsilon)$$

and so

$$|G_n(x) - \Phi(x)| \leq |\Phi_n(x + \varepsilon) - \Phi(x + \varepsilon)| + Pr(|R_n| \geq \varepsilon) + |\Phi(x + \varepsilon) - \Phi(x)|$$

186

which implies that

$$\sup_x |G_n(x) - \Phi(x)| \leq \sup_x |\Phi_n(x) - \Phi(x)| + Pr(|R_n| \geq \varepsilon) + (2\pi)^{-\frac{1}{2}}\varepsilon. \quad (3)$$

We need to bound the second term on the right. Write

$$\begin{aligned}
Pr\left(|R_n| \geq \varepsilon\right) &= Pr\left(|(U_n^{(1)} - \theta_1) - (U_n^{(2)} - \theta_2)| \geq 2n^{\frac{1}{2}}\varepsilon\sigma_1\right) \\
&\leq Pr\left(|U_n^{(1)} - \theta_1| + |U_n^{(2)} - \theta_2| \geq 2n^{\frac{1}{2}}\varepsilon\sigma_1\right) \\
&\leq Pr\left(|U_n^{(1)} - \theta_1| \geq n^{\frac{1}{2}}\varepsilon\sigma_1\right) + Pr\left(|U_n^{(2)} - \theta_2| \geq n^{\frac{1}{2}}\varepsilon\sigma_1\right) \\
&\leq \left(E|U_n^{(1)} - \theta_1|^3 + E|U_n^{(2)} - \theta_2|^3\right) / n^{\frac{3}{2}}\varepsilon^3\sigma_1^3
\end{aligned}$$

by the Markov inequality. By Theorem 3 of Section 3.4.2 we have

$$\max\left(E|U_n^{(1)} - \theta_1|^3, E|U_n^{(2)} - \theta_2|^3\right) \leq C\gamma n^{-\frac{3}{2}}$$

where $\gamma = \max\left(E|\psi(X_1, X_2) - \theta_2|^3, E|\psi(X_1, X_1) - \theta_1|^3\right) / \sigma_1^3$ and so we obtain

$$Pr\left(|R_n| \geq \varepsilon\right) \leq C\gamma n^{-3}\varepsilon^3.$$

Setting $\varepsilon = n^{-\frac{5}{6}}$ and using (2) and (3) shows that we can find a constant $C$ independent of $V_n$ such that

$$\sup_x \left|Pr\left(n^{\frac{1}{2}}(V_n - EV_n) \leq x\right) - \Phi(x)\right| \leq C\gamma n^{-\frac{1}{2}}.$$

## 4.3 Incomplete $U$-statistics

### 4.3.1 Basics

If the sample size $n$ or the degree $k$ of a $U$-statistic are large, the calculation of $U_n$ can be quite onerous, since it involves averaging $\binom{n}{k}$ terms. However, it turns out that in many cases a large proportion of the summands of $U_n$ may be omitted without unduly inflating the variance, because of the dependence between the terms. We are thus led to consider "incomplete" $U$-statistics of the form

$$U_n^{(0)} = m^{-1} \sum_{S \in D} \psi(S) \quad (1)$$

187

where now the sum is taken over the $m$ subsets in some suitably chosen subset $\mathcal{D}$ of $\mathcal{S}_{n,k}$. We call the set $\mathcal{D}$ the *design* of the incomplete $U$-statistic, and is usually chosen to minimize the variance of (1), subject to $m$ remaining fixed.

The variance of $U_n^{(0)}$ (provided $\mathcal{D} \subseteq \mathcal{S}_{n,k}$) is always greater than that of $U_n$, but we hope that $U_n^{(0)}$ will be asymptotically reasonably efficient even though $m$ is of much smaller magnitude than $\binom{n}{k}$.

### Example 1. An incomplete Kendall's tau.

As in Example 5 of Section 2.1.2, let $P_1, \ldots, P_n$ be random points on the plane distributed independently as $P = (X, Y)$. The complete version of Kendall's tau is

$$t_n = \binom{n}{2}^{-1} \{\text{number of concordant pairs} - \text{number of discordant pairs}\}.$$

If we take a design

$$\mathcal{D} = \{\{1, 2\}, \{2, 3\}, \ldots, \{n - 1, n\}, \{n, 1\}\}$$

the incomplete version of this statistic based on $\mathcal{D}$ is

$$t_n^{(0)} = n^{-1}(C_n - D_n)$$

where $C_n$ is the number of indices $i$ for which $P_i$ and $P_{i+1}$ are concordant for $i = 1, 2, \ldots, n$ and $D_n$ is the number of indices $i$ for which $P_i$ and $P_{i+1}$ are discordant, identifying $P_{n+1}$ with $P_1$.

### Example 2. An incomplete version of the sample variance.

Let $X_1, \ldots, X_n$ be a sample from a distribution with variance $\sigma^2$. Then

$$s_0^2 = \frac{1}{2n} \sum_{i=1}^{n} (X_i - X_{i+1})^2$$

is an incomplete version of the usual sample variance if we identify $X_{n+1}$ with $X_1$. Here the design is the same as that of Example 1.

### Example 3. The sample variance (continued).

Instead of the design considered in the previous examples, we could use, for even $n(= 2m)$

$$\mathcal{D} = \{(1, 2), (3, 4), \ldots (2m - 1, 2m)\}$$

and estimate $\sigma^2$ by

$$s_0^2 = \frac{1}{m} \sum_{i=1}^{m} (X_{2i-1} - X_{2i})^2.$$

**Example 4.**

Enlarging on the theme of the last example, consider a $U$-statistic with kernel $\psi$ of degree $k$. If $n = km$ for some integer $m$, take as the design

$$\mathcal{D} = \{\{1, 2, \ldots, k\}, \{k+1, \ldots, 2k\}, \ldots, \{(m-1)k+1, \ldots, mk\}\}$$

and then the corresponding incomplete statistic is

$$U_n^{(0)} = m^{-1} \sum_{j=1}^{m} \psi(S_j)$$

where $S_j = \{(j-1)k+1, \ldots, jk\}$.

In this case, since the summands are independent, the variance of $U_n^{(0)}$ is $\sigma_k^2/m$. The asymptotic relative efficiency (ARE) of the incomplete *vis à vis* the complete statistic is

$$\begin{aligned}
\text{ARE} &= \lim_{n \to \infty} Var\, U_n / Var\, U_n^{(0)} \\
&= \lim_{n \to \infty} \frac{\{k^2 \sigma_1^2/n + o(n^{-1})\}}{\{\sigma_k^2/m\}} \\
&= k\, \sigma_1^2/\sigma_k^2.
\end{aligned}$$

By Theorem 4 of Section 1.3 we have $\sigma_1^2/\sigma_k^2 < 1/k$, so the ARE depends on the ratio $\sigma_1^2/\sigma_k^2$ and may be close to unity.

**Example 5. Random choice of subsets.**

A common method of choosing the subsets of the design is to choose $m$ sets at random from $\mathcal{S}_{n,k}$ either with or without replacement. While not the most efficient method, it has the advantage of simplicity, especially if the calculation of $U_n^{(0)}$ is to be carried out by a computer program.

It is clear from the definition (1) that incomplete $U$-statistics are unbiased. Our first theorem shows that, provided $\mathcal{D} \subseteq \mathcal{S}_{n,k}$, an incomplete statistic is always less efficient than the corresponding complete statistic.

**Theorem 1.** *Let* $U_n^{(0)}$ *be an incomplete U-statistic based on a fixed design* $\mathcal{D} \subseteq \mathcal{S}_{n,k}$, *and let* $U_n$ *be the corresponding complete statistic. Then*

$$Var\, U_n^{(0)} - Var\, U_n = Var(U_n^{(0)} - U_n) \geq 0$$

**Proof.** Let $S_1, \ldots, S_m$ be the sets of the design. Then

$$Var\, U_n^{(0)} = Cov(U_n^{(0)}, m^{-1} \sum_{j=1}^{m} \psi(S_j))$$

$$= m^{-1} \sum_{j=1}^{m} Cov(U_n^{(0)}, \psi(S_j)).$$

Since the random variables $\psi(S_j)$ are identically distributed, by symmetry the quantities $Cov(U_n^{(0)}, \psi(S_j))$ are all the same and so $Var\, U_n^{(0)} = Cov(U_n^{(0)}, \psi(S_j))$ for each set $S_j$ in the design.

In fact, $Cov(U_n^{(0)}, \psi(S))$ is the same for every set in $\mathcal{S}_{n,k}$, so

$$Cov(U_n^{(0)}, \psi(S)) = \binom{n}{k}^{-1} \sum_{(n,k)} Cov(U_n^{(0)}, \psi(S))$$

$$= Cov(U_n^{(0)}, U_n)$$

and so

$$Var(U_n^{(0)} - U_n) = Var\, U_n^{(0)} - 2Cov(U_n^{(0)}, U_n) + Var\, U_n$$

$$= Var\, U_n^{(0)} - Var\, U_n.$$

Thus the incomplete statistic is always less efficient than the complete. To be more precise, we need to be able to compute variances. The next two theorems show how.

**Theorem 2.** *Let* $f_c$ *be the number of pairs of sets in the design that have* $c$ *elements in common. Then*

$$Var\, U_n^{(0)} = m^{-2} \sum_{c=1}^{k} f_c \sigma_c^2.$$

190

**Proof.** We have

$$Var\ U_n^{(0)} = m^{-2} \sum_{S \epsilon \mathcal{D}} \sum_{T \epsilon \mathcal{D}} Cov(\psi(S), \psi(T))$$

$$= m^{-2} \sum_{c=1} f_c \sigma_c^2$$

by now familiar arguments.

The variance can also be expressed in terms of the quantities $\delta_c^2$ occuring in the $H$-decomposition. This form is useful for identifying minimum variance designs.

**Theorem 3.** *For $1 \le \nu \le k$, let $S$ be a set in $\mathcal{S}_{n,\nu}$ and let $n(S)$ be the number of $k$-subsets in the design $\mathcal{D}$ which contain $S$. Define for $1 \le \nu \le k$*

$$A_\nu = \sum_{(n,\nu)} n(S),$$

$$B_\nu = \sum_{(n,\nu)} n^2(S).$$

*Then*

$$Var\ U_n^{(0)} = m^{-2} \sum_{\nu=1}^{k} B_\nu \delta_\nu^2. \tag{2}$$

*The quantities $A_\nu$ and $B_\nu$ are given by*

$$A_\nu = m \binom{k}{\nu} \tag{3}$$

*and*

$$B_\nu = \sum_{c=\nu}^{k} f_c \binom{c}{\nu}. \tag{4}$$

**Proof.** The theorem is most easily proved by introducing the *incidence matrix* of the design, which is defined by $N = (n_{ij})_{n \times m}$ where

$$n_{ij} = \begin{cases} 1 & \text{if index } i \text{ is in set } j, \\ 0 & \text{otherwise.} \end{cases}$$

Note that the $j, j'$ element of $N^T N$ is $\sum_i n_{ij} n_{ij'}$ which is the number of elements $S_j$ and $S_{j'}$ have in common. Thus $f_c$ is the number of off-diagonal elements of $N^T N$ equal to $c$. We also make use of the Stirling numbers of the first and second kind, denoted by $S_\mu^{(\nu)}$ and $\mathcal{S}_\nu^{(\mu)}$ respectively. These are defined by

$$x^\nu = \sum_{\mu=1}^{\nu} \mathcal{S}_\nu^{(\mu)} x(x-1)(x-2)\ldots(x-\mu+1)$$

and

$$x(x-1)\ldots(x-\mu+1) = \sum_{\nu=1}^{\mu} S_\mu^{(\nu)} x^\nu$$

and satisfy

$$\sum_{\nu=\mu}^{\gamma} \mathcal{S}_\nu^{(\mu)} S_\gamma^{(\nu)} = \begin{cases} 1 & \mu = \gamma, \\ 0 & \mu \neq \gamma. \end{cases} \tag{5}$$

See for example, Abramowitz and Stegun (1965) p824 or Riordan (1958), p34.

In view of the interpretation of $f_c$ in terms of elements of $N^T N$, we can write

$$\sum_{c=1}^{k} c^\nu f_c = \sum_{j=1}^{n} \sum_{j'=1}^{m} \left( \sum_{i=1}^{n} n_{ij} n_{ij'} \right)^\nu$$

$$= \sum_{i_1=1}^{n} \cdots \sum_{i_\nu=1}^{n} \left( \sum_{j=1}^{m} n_{i_1 j} \ldots n_{i_\nu j} \right)^2$$

$$= \sum_{\mu=1}^{\nu} \nu! \mathcal{S}_\nu^{(\mu)} B_\mu \tag{6}$$

since the values of $n_{ij}$ are either zero or one.

Multiplying (6) on both sides by $S_\gamma^{(\nu)}$ and summing over $\nu$ from $\mu$ to $\gamma$ yields (4) after using (5) and definition of the Stirling numbers.

Multiplying (4) on both sides by $(-1)^{\gamma-\nu} \binom{\gamma}{\nu}$, summing over $\gamma$ from $\nu$ to $k$ and using the identity

$$\sum_{\gamma=\nu}^{c} (-1)^{\gamma-\nu} \binom{\gamma}{\nu} \binom{c}{\gamma} = \begin{cases} \binom{c}{\nu} & \text{if } c = \nu, \\ 0 & \text{otherwise,} \end{cases}$$

we get

$$f_c = \sum_{\gamma=c}^{k} (-1)^{\gamma-c} \binom{\gamma}{c} B_\gamma. \tag{7}$$

Finally, using Theorem 2 and (7) we get

$$Var\, U_n^{(0)} = m^{-2} \sum_{c=1}^{k} f_c \sigma_c^2$$

$$= m^{-2} \sum_{\gamma=1}^{k} \left\{ \sum_{c=1}^{\gamma} (-1)^{\gamma-c} \binom{\gamma}{c} \sigma_c^2 \right\} B_\gamma$$

$$= \sum_{\gamma=1}^{k} B_\gamma \delta_\gamma^2$$

by (9) of Section 1.6.

Finally, to prove (3), we have

$$mk^2 = \sum_{j=1}^{m} \left( \sum_{i=1}^{n} n_{ij} \right)^\nu = \sum_{i_1=1}^{n} \cdots \sum_{i_\nu=1}^{n} \left( \sum_{j=1}^{m} n_{ij} \ldots n_{\nu j} \right)$$

$$= \sum_{\mu=1}^{\nu} \mu! S_\nu^{(\mu)} A_\mu$$

This leads to (3) in exactly the same way the (6) led to (7).

Next we develop expressions for the variances of incomplete $U$-statistics based on random subset selection.

**Theorem 4.** (i) Suppose the incomplete statistic is constructed by selecting $m$ subsets from $\mathcal{S}_{n,k}$ at random with replacement. Then

$$Var\, U_n^{(0)} = \sigma_k^2/m + (1 - m^{-1})Var\, U_n.$$

(ii) If the selection is without replacement, then

$$Var\, U_n^{(0)} = (N - m)\sigma_k^2/\{m(N - 1)\} + \{N/(N - 1)\}(1 - m^{-1})Var\, U_n$$

where $N = \binom{n}{k}$.

**Proof.** We prove (i) only, (ii) is similar. Let $S_1, \ldots, S_m$ denote the $m$ sets chosen at random with replacement. Then for $i \neq j$,

$$Cov(\psi(S_i), \psi(S_j)) = N^{-2} \sum_{(n,k)} \sum_{(n,k)} Cov(\psi(S), \psi(T))$$

$$= Var\, U_n$$

193

and

$$Var\,\psi(S_i) = N^{-1} \sum_{(n,k)} Var\,\psi(S)$$

$$= \sigma_k^2.$$

Hence

$$Var\,U_n^{(0)} = m^{-2} \sum_{i=1}^{m} \sum_{j=1}^{m} Cov(\psi(S_i), \psi(S_j))$$

$$= m^{-2}\{m(m-1)Var\,U_n + m\sigma_k^2\}$$

proving the result.

**Example 6. Kendall's tau (continued).**

The incomplete version of Kendall's tau introduced in Example 1 has $n = m, k = 2$. Since $f_k = m$ always,

$$Var\,t_n^{(0)} = n^{-1}\{f_1\sigma_1^2 + n\sigma_2^2\}.$$

Direct consideration of the pairs of sets yields $f_1 = 2n$. Alternatively, $B_1 = 4m, B_2 = m$ so that by (7)

$$f_1 = B_1 - 2B_2 = 2n.$$

The variance is thus $n^{-1}\{2\sigma_1^2 + \sigma_2^2\}$. Under independence, $\sigma_1^2 = \frac{1}{9}$ and $\sigma_2^2 = 1$ so that $Var\,t_n^{(0)} = 11/9n$. The ARE of the incomplete *vis à vis* the complete statistic is 4/11 in the case of independence.

**Example 7. The sample variance (continued).**

In this case the design is the same as in Example 2, so that $Var\,s_0^2 = n^{-1}(2\sigma_1^2 + \sigma_2^2)$ as in Example 6. From Example 2 of Section 2.1.3, $\sigma_1^2 = \frac{1}{4}(\mu_4 - \sigma^4)$ and $\sigma_2^2 = \frac{1}{2}(\mu_4 + \sigma^4)$, so that $Var\,s_0^2 = \mu_4/n$.

**4.3.2 Minimum variance designs**

For a fixed $m$, an obvious question is how to choose the design $\mathcal{D}$ to minimise the variance of the incomplete $U$-statistic. This is analogous to the problem in experimental design of allocating treatments to blocks of experimental plots in order to optimise some design criterion. Indeed, there is an obvious connection between incomplete $U$-statistics and incomplete

194

block designs, with the $n$ indices corresponding to "treatments" and the $m$ sets of the design to "blocks". Each set ("block") contains $k$ indices ("treatments"). We require that no set appears more than once in the design, and that all $k$ indices in a set are distinct, so in experimental design terms we have a proper binary design with equal block sizes.

The question of allocating the "treatments" to "blocks" in order to minimise the variance is a difficult combinatorial problem, but we state two principles that allow us to recognise certain designs as optimal, and give some examples.

In view of the symmetry underlying the r.v.s $X_1, \ldots, X_n$ it is reasonable to assume that every index occurs in the same number, r say, of sets of the design. Thus we confine attention to designs that an *equireplicate* in the experimental design sense. Such designs satisfy

$$mk = nr \qquad (1)$$

and are usually called *balanced* in the $U$-statistics literature.

From Theorem 3 of the previous section, in order to achieve a design with minimum variance, it is necessary to minimise the quantities $B_\nu$ over all possible $k$-subsets of $\mathcal{S}_{n,k}$. This in general is extremely difficult, but under certain circumstances we can recognise a minimum.

Note that $B_\nu$ is the sum of squares of integers $n(S)$ that sum to $m\binom{k}{\nu}$. Now it is easily verified that the quadratic $\sum x_i^2$ subject to $\sum x_i = a$ and $x_i \geq 0$ acheives its minimum value when all the $x_i$s are equal. Hence any design for which the quantities $n(S)$ are equal for all $\nu$-sets $S$ will correspond to a minimum $B_\nu$. Since for $\nu = k$, $n(S)$ is zero or one for all $\nu$-subsets, $B_\nu = A_\nu = m$. Also, for equireplicate designs $n(\{i\}) = r$ for all $i$ so that $A_1 = nr, B_1 = nr^2$, and $B_1$ is a minimum. Thus in this case we only need to check that $B_\nu$ is a minimum for $\nu = 2, 3, \ldots, k - 1$.

We can thus state a general result:

**Theorem 1.** *Suppose that for each $\nu = 1, 2, \ldots, k - 1$ that every $\nu$-subset of $\{1, 2, \ldots, n\}$ is contained in the same number $m\binom{k}{\nu}/\binom{n}{\nu}$ of sets of $\mathcal{D}$. Then $\mathcal{D}$ is a minimum variance design.*

**Proof.** Follows directly from the above considerations.

195

**Example 1. Balanced (equireplicate) designs with $k = 2$.**

These are minimum variance since every index appears in the same number of sets of the design, and we can apply Theorem 1. We have $f_1 = B_1 - 2B_2 = nr^2 - 2m = 2m(r-1)$, using (1), and $f_2 = m$. Thus the variance of any balanced design with $k = 2$ is

$$Var\, U_n^{(0)} = m^{-1}(2(r-1)\sigma_1^2 + \sigma_2^2)$$

Such designs have been considered also by Blom (1976) and Brown and Kildea (1978). The asymptotic efficiency is $2r\rho/\{2(r-1)\rho + 1\}$ when $\rho = \sigma_1^2/\sigma_2^2$. Since $0 \le \rho \le \frac{1}{2}$, the efficiency ranges from 0 to 1 depending on the values of $\rho$, and increases as $r$ increases.

**Example 2. BIBDs with $k = 3$.**

Balanced incomplete block designs (see Raghavarao (1971)) for $k = 3$ are equireplicate designs with each pair of treatments occuring in $\lambda$ blocks, so $B_1 = nr^2, B_2 = \binom{n}{2}\lambda^2$ and $B_3 = m$ are all minimising values. The variance is

$$Var\, U_n^{(0)} = m^{-1}(3(r - 2\lambda + 1)\sigma_1^2 + 3\sigma_2^2(\lambda - 1) + \sigma_3^2).$$

Such designs do not always exist. However, an example of designs of this type that do exist is furnished by the series of designs due to Sprott (1954). For each positive integer $t$ there is a minimum variance design with $n = 3t+1$, $m = t(3t+1)$, $r = 3t$ and $\lambda = 2$. The variance of a $U$-statistic based on the designs is $\{9(t-1)\sigma_1^2 + 3\sigma_2^2 + \sigma_3^2\}/\{t(3t+1)\}$ and the ARE is unity.

A second principle leading to minimum variance designs is the following. Suppose for some $1 \le \nu_0 < k$ the quantities $n(S)$ for all $\nu_0$-subsets $S$ are either zero or one, i.e. all $\nu_0$-subsets are contained in at most one block. If $\nu > \nu_0$, then necessarily all $\nu$-subsets are contained in at least one block, so for $\nu \ge \nu_0$ we have $B_\nu = A_\nu$. Since in general $B_\nu \ge A_\nu$, these are minimum $B_\nu's$. In this situation we need only check that $B_1, B_2, \ldots, B_{\nu_0-1}$ are minimised. We state this as Theorem 2.

**Theorem 2.** *Suppose that the design $\mathcal{D}$ minimises $B_\nu$ for $\nu = 1, 2, \ldots,$ $\nu_0 - 1$ and that every $\nu_0$-subset of $\{1, 2, \ldots, n\}$ occurs in at most one set of the design. Then the design is minimum variance.*

**Corollary 1.** *Let $\mathcal{D}$ be a balanced design for which the off-diagonal elements of $NN^T$ are either zero or one. Then $\mathcal{D}$ is a minimum variance design.*

**Proof.** The $i,j$ element of $NN^T$ is just $n(\{i,j\})$, so the result follows directly from Theorem 2.

**Example 3. Example 4 of Section 4.3.1 revisited.**

Since $n(\{i,j\})$ is the number of blocks containing $i$ and $j$, $n(\{i,j\})$ is at most unity and so the design is minimum variance.

**Example 4. (Blom (1976)).**

If $n$ is a multiple of 9, say $n = 9t$ for some integer $t$, the a design for $m = n = 9t$ blocks of $k = 3$ treatments each may be based on $3 \times 3$ Latin squares. Take $t$ $3 \times 3$ Latin squares and arrange the $9t$ treatments in $t$ squares of 9 treatments each. Take for the blocks treatments having the same row, column or letter in any particular square, for a total of $9t$ blocks. Then the design is equireplicate, with each treatment appearing in $r = 3$ blocks, and no pair of treatments can appear in more than one block. The design is thus minimum variance by Corollary 1. We have $B_1 = 9n = 27t$ since the design is equireplicate, and $B_2 = A_2 = 3m = 9t$ by Theorem 3 of Section 4.3.1. Hence $Var\, U_n^{(0)} = (6\sigma_1^2 + \sigma_3^2)/9t$ and the ARE is $9\rho/(6\rho+1)$ where $\rho = \sigma_1^2/\sigma_3^2$ and $0 \le \rho \le \frac{1}{3}$. The ARE thus ranges from 0 to 1 depending on $\rho$.

**Example 5.**

All balanced incomplete block designs are equireplicate, so since a BIBD with $\lambda = 1$ has each pair of varieties occuring in exactly one block it satisfies Theorem 2 with $\nu_0 = 2$. Hence $B_1 = nr^2$ and $B_2 = \binom{n}{2}$ are minimising values. Note that $\lambda = 1$ is equivalent to the off-diagonal elements of $NN^T$ being unity, since $NN^T = (r - \lambda)I + \lambda J$ where $I$ and $J$ are respectively the $n \times n$ identity matrix and a $n \times n$ matrix of ones.

A series of such designs is the series of so-called *Steiner triple systems* (see, e.g. Raghavarao, (1971), p86). These exist for each integer $t$ and have parameters $n = 6t + 3$, $m = (3t + 1)(2t + 1)$, $r = 3t + 1$, $k = 3$ and $\lambda = 1$. The variance of the incomplete $U$-statistic based on this design is

$(9t\sigma_1^2 + \sigma_3^2)/(3t+1)(2t+1)$ and the ARE is 1, provided $\sigma_1^2 > 0$. There is also a Steiner series for $n = 6t + 1, m = t(6t + 1)$.

In fact, by Corollary 1, any equireplicate design for which the off-diagonal elements of $NN^T$ are zero or one will be a minimum variance design.

## Example 6.

In a partially balanced incomplete block design (PBIBD) for $n$ varieties in $m$ blocks with two *associate classes* any two varieties are either *first associates* or *second associates*. All pairs of first associates appear in the same number of blocks, say $\lambda_1$ blocks, and similarly all pairs of second associates appear in $\lambda_2$ blocks. Since the off-diagonal elements of $NN^T$ are $\lambda_1$ and $\lambda_2$, a PBIBD will be a minimum variance design if $\lambda_1 = 1$ and $\lambda_2 = 0$, or vice versa.

## Example 7.

A series of equireplicate designs having all off-diagonal elements of $NN^T$ equal to zeo or one can be constructed using cyclic permutations.

Let $d_1, \ldots, d_k$ be integers between 1 and $n$ and let $P_1, \ldots, P_k$ be the permutation matrices corresponding to the cyclic permutations

$$\begin{pmatrix} 1 & 2 & & n \\ & & \cdots & \\ d_\nu \oplus 1 & d_\nu \oplus 2 & & d_\nu \oplus n \end{pmatrix}, \quad \nu = 1, 2, \ldots, k$$

where $\oplus$ denotes addition (mod $n$). Thus the $i, j$ element of $P_\nu$ is unity if $d_\nu + i = j$ (mod $n$) and zero otherwise.

Let $N = P_1 + \cdots + P_k$. The matrix $N$ will be the incidence matrix of an equireplicate design with $m = n$ and $r = k$ provided the row and column sums of $N$ equal $k$. This will be the case provided the integers $d_\nu$ are distinct. For $\nu \neq \nu'$, the product $P_\nu P_\nu^T$ has its $i, j$ element unity if $i - j = d_{\nu'} - d_\nu$ (mod $n$) and zero otherwise, and $P_\nu P_\nu^T$ is an identity matrix. It follows from these considerations that $NN^T = \sum_{\nu=1}^k \sum_{\nu'=1}^k P_\nu P_\nu^T$ will have off-diagonal elements that are zero or one if the quantities $(d_\nu - d_{\nu'})$ (mod $n$) are distinct. Values of $d_\nu$ suitable for the construction of such designs are given in Table 1.

198

TABLE 1

Values of $d_1, \ldots, d_k$ for different $k$ values. The range of $n$ yielding minimum variance designs appears in parentheses.

| $k$ | $d_1$ | $d_2$ | $d_3$ | $d_4$ | $d_5$ | |
|---|---|---|---|---|---|---|
| 2 | 0 | 1 | | | | $(n \geq 3)$ |
| 3 | 0 | 1 | 3 | | | $(n \geq 7)$ |
| 4 | 0 | 1 | 4 | 6 | | $(n \geq 13)$ |
| 5 | 0 | 1 | 4 | 9 | 11 | $(n \geq 23)$ |

An example of these designs with $k = 4$ and $n = 13$ is

| 1 | 2 | 5 | 7 | | 5 | 6 | 9 | 11 | | 10 | 11 | 1 | 3 |
|---|---|---|---|---|---|---|---|---|---|---|---|---|---|
| 2 | 3 | 6 | 8 | | 6 | 7 | 10 | 12 | | 11 | 12 | 2 | 4 |
| 3 | 3 | 7 | 9 | | 8 | 9 | 12 | 1 | | 13 | 1 | 4 | 6 |
| | | | | | 9 | 10 | 13 | 2 | | | | | |

If $K$ is a positive integer, minimum variance designs for arbitrary $k$ and $m = Kn$ can be constructed for sufficiently large $n$ by considering incidence matrices $N$ of the form

$$N = [N_1 \vdots N_2 \vdots \cdots \vdots N_K]$$

where each of the matrices $N_1, \ldots, N_K$ is derived from permutation matrices as described above. Then $NN^T = N_1 N_1^T + \cdots + N_K N_K^T$, and the off-diagonal elements of $NN^T$ will be either zero or one if all the sets of $d$'s that generate the matrices have distinct differences (mod $n$). As an example, take $K = 2$ and $k = 3$, and $N_1$ based on 0,1,3 and $N_2$ based on 0,4,9. The resulting design is minimum variance for $n \geq 19$.

The variance of such designs is most easily computed by noting that $B_1 = nr^2 = nk^2 K^2$ and $B_\nu = m\binom{k}{\nu}$ for $\nu = 2, 3, \ldots k$. Using (7) of Section 4.3.1 then gives

$$Var\, U_n^{(0)} = m^{-1}(k(kK - 1)\sigma_1^2 + \sigma_k^2).$$

The ARE is $k^2 K \rho / \{k(kK-1)\rho+1\}$ where $\rho = \sigma_1^2 / \sigma_k^2$, provided $K$ is fixed.

### 4.3.3 Asymptotics for random subset selection

We now return to the case where the $m$ subsets forming the design $\mathcal{D}$ are chosen at random from the $\binom{n}{k}$ $k$-subsets available. The basic asymptotics are covered by the following theorem, due to Janson (1984).

**Theorem 1.** Let $U_n^{(0)}$ be a $U$-statistic constructed by selecting $m$ sets at random with replacement from $\mathcal{S}_{n,k}$, and $U_n$ the corresponding complete statistic, assumed to be degenerate of order $d$. Let $\lim_{n \to \infty} n^{d+1} m^{-1} = \alpha$, and assume all necessary variances exist.

(i) If $\alpha = 0$ then $n^{(d+1)/2}(U_n^{(0)} - \theta)$ has the same limit distribution as $n^{(d+1)/2}(U_n - \theta)$;

(ii) If $0 < \alpha < \infty$ then the limit distribution of $m^{\frac{1}{2}}(U_n^{(0)} - \theta)$ is that of the r.v. $\alpha^{\frac{1}{2}} X + \sigma_k Y$, where $X$ has the same distribution as the limiting distribution of $n^{(d+1)/2}(U_n - \theta)$, $Y$ is $N(0,1)$, and $X$ and $Y$ an independent;

(iii) If $\alpha = \infty$, then the limit distribution of $m^{\frac{1}{2}}(U_n^{(0)} - \theta)$ is $N(0, \sigma_k^2)$.

**Proof.** (i) We need only prove that $n^{(d+1)/2}(U_n^{(0)} - U_n)$ converges in probability to zero. Since the mean is zero, it suffices to prove that $Var(U_n^{(0)} - U_n) = o(n^{-(d+1)/2})$. Now by Theorem 2 of Section 4.3.1. we have

$$Var(U_n^{(0)} - U_n) = Var\, U_n^{(0)} - Var\, U_n$$
$$= (\sigma_k^2 + Var\, U_n)/m$$

Hence

$$\lim\ n^{d+1} Var(U_n^{(0)} - U_n) = \lim n^{d+1} m^{-1}(\sigma_k^2 + Var\, U_n)$$
$$= 0$$

proving the result.

(ii) Suppose that the random vector with elements $\{Z_S : S \epsilon \mathcal{S}_{n,k}\}$ has a multinomial distribution $Mult(m; \frac{1}{N}, \dots \frac{1}{N})$ where $N = \binom{n}{k}$, so that we can write $m^{\frac{1}{2}}(U_n^{(0)} - \theta) = m^{-\frac{1}{2}} \sum_{(n,k)} Z_S(\psi(S) - \theta)$. Let $\phi_n$ be the characteristic function (c.f.) of the r.v. $m^{\frac{1}{2}}(U_n^{(0)} - \theta)$ and $\phi$ the limiting c.f. of

$m^{(d+1)/2}(U_n - \theta)$. Then

$$\phi_n(t) = E[\exp(it \ m^{-\frac{1}{2}} \sum Z_S(\psi(S) - \theta)]$$
$$= E[E\{\exp(it \ m^{-\frac{1}{2}} \sum Z_S(\psi(S) - \theta))|X_1, \dots, X_n\}]$$
$$= E[\exp(it \ m^{\frac{1}{2}}U_n)$$
$$\times E\{\exp(it \ m^{\frac{1}{2}} \sum (Z_S - \frac{m}{N})(\psi(S) - \theta))|X_1, \dots, X_n\}].$$

The conditional expectation is the characteristic function of a r.v. of the type considered in Lemma A, so it converges almost surely to $e^{\sigma_k^2 t^2/2}$, since $\sigma_k^2 = \lim_{N \to \infty} N^{-1} \sum (\psi(S) - \theta)^2$ almost surely. Hence

$$\lim_{n \to \infty} \phi_n(t) = \lim_{n \to \infty} E(\exp(it \ m^{\frac{1}{2}}U_n))e^{-\sigma_k^2 t^2/2}$$
$$= \lim_{n \to \infty} E[\exp\{it(m^{\frac{1}{2}}n^{-(d+1)/2})n^{(d+1)/2}U_n\}]e^{-\sigma_k^2 t^2/2}$$
$$= \phi(\alpha^{-\frac{1}{2}}t)e^{-t^2 \sigma_k^2/2}$$

which is the c.f. of a r.v $\alpha^{-\frac{1}{2}}X + Y$ where $X$ has the limiting distribution of $n^{\frac{d+1}{2}}(U_n - \theta)$, $Y$ is $N(0, \sigma_k^2)$ and $X$ and $Y$ are independent.

(iii) The proof of (iii) is the same as that of (ii) with the obvious changes.

To complete the proof we prove Lemma A.

**Lemma A.** *Let $a_1, a_2, \dots$ be a sequence of constants having the properties $\lim_{N \to \infty} N^{-1} \sum_{i=1}^{N} a_i = 0$ and $\lim_{N \to \infty} N^{-1} \sum_{i=1}^{N} a_i^2 = \sigma^2$ and let the r.v.s $Z_1, \dots, Z_N$ have a multinomial distribution, Mult$(m; N^{-1}, \dots, N^{-1})$. Then as $m$ and $N \to \infty$, the limit distribution of*

$$m^{-\frac{1}{2}} \sum_{i=1}^{N} a_i(Z_i - m/N) \qquad (2)$$

*is $N(0, \sigma^2)$.*

**Proof.** The c.f. of $Z_1, \dots, Z_n$ is $(\frac{1}{N}e^{it_1} + \dots + \frac{1}{N}e^{it_N})^m$ so that the c.f. of (2) is

$$e^{-it\bar{a}_N m^{\frac{1}{2}}} \left( \frac{1}{N} e^{ita_1 m^{-\frac{1}{2}}} + \cdots + \frac{1}{N} e^{ita_N m^{-\frac{1}{2}}} \right)^m$$

$$= e^{it\bar{a}_N m^{\frac{1}{2}}} \left( \frac{1}{N} \left\{ N + it \; m^{-\frac{1}{2}} \sum_{i=1}^{N} a_i + \left( \frac{itm^{-\frac{1}{2}}}{2!} \right)^2 \sum_{i=1}^{N} a_i^2 + \cdots \right\} \right)^m$$

$$= e^{it\bar{a}_N m^{\frac{1}{2}}} \left( 1 + \bar{a}_N(it/\sqrt{m}) + \frac{1}{2}\sigma_N^2(it/\sqrt{m})^2 + o(m^{-1}) \right)^m$$

where $\bar{a}_N = \frac{1}{N}\sum_{i=1}^{N} a_i$, and $\sigma_N^2 = \frac{1}{N}\sum_{i=1}^{N} a_i^2$. The log of the above is

$$- it\bar{a}_N m^{\frac{1}{2}} + m(\log(1 + \bar{a}_N \left( it/\sqrt{m} + \frac{1}{2}\sigma_N^2(it/\sqrt{m})^2 + \cdots \right)$$

$$= -ita_N m^{\frac{1}{2}} + m \left( \bar{a}_N(itm^{-\frac{1}{2}}) + (\frac{1}{2}\sigma_N^2 - \frac{1}{2}\bar{a}_N^2)(\frac{it}{\sqrt{m}})^2 + \cdots \right)$$

$$= - \left( \frac{1}{2}\sigma_N^2 - \frac{1}{2}\bar{a}_N^2 \right) t^2 + o(1).$$

Provided $\bar{a}_N \to 0$ and $\sigma_N^2 \to \sigma^2$ the log converges to $- \sigma^2 t^2/2$ and so (2) is asymptotically normal with mean 0 and variance $\sigma^2$.

A similar theorem can be proved in the case of sampling without re-placement. The asymptotic distribution is identical, except in the case $d + 1 = k$. In this case the asymptotic distribution is that of $\alpha_{\frac{1}{2}} X + Y$, where $Y$ is $N(0, \sigma_k^2(1 - \pi))$ and $\pi = \lim_{n \to \infty} m/N = k!/\alpha$. The reader is referred to Janson (1984) for details.

The efficiency of random designs versus optimal designs can be easily evaluated using Theorem 4 of Section 4.3.1 and the formulae in Section 4.3.2. We present some examples.

### Example 1. Balanced designs for $k = 2$ versus random designs.

For the designs of Example 1 of Section 4.3.2, the ratio of variances for statistics based on random and minimum variance designs is

$$m^{-1}(2(r - 1)\sigma_1^2 + \sigma_2^2)/\{m^{-1}(\sigma^2 + (m - 1)Var\, U_n)\}.$$

Assuming that $r$ is fixed, so that $m = O(n)$, this converges to $(2(r - 1)\sigma_1^2 + \sigma_2^2)/(2r\sigma_1^2 + \sigma_2^2)$ which cannot be less that $r/(r + 1)$. In the case when

$r$ increases, the random choice design is asymptotically efficient *vis à vis* the optimal design for the same number of samples. The efficiency of the random design versus the complete statistic is $2r\rho/(1 + 2r\rho)$ where $\rho = \sigma_1^2/\sigma_2^2$ and hence ranges from zero to $r/(1 + r)$.

**Example 2.**

The variances of $U$-statistics based on the designs of the type in Example 2 of Section 4.3.2 are

$$\{9(t - 1)\sigma_1^2 + 3\sigma_2^2 + \sigma_3^2\}/t(3t + 1)$$

where $m = t(3t + 1)$. The ARE relative to random designs is

$$\lim_{t \to \infty} \frac{(9(t - 1)\sigma_1^2 + 3\sigma_2^2 + \sigma_3^2)}{(3t^2 + t - 1)(9\sigma_1^2(3t + 1)^{-1} + O(t^{-1}))} = 1$$

so that random designs are asymptotically efficient.

**Example 3.**

For the designs based on cyclic permutations as in Example 7 of Section 4.3.2,

$$Var\ U_n^{(0)} = m^{-1}(k(kK - 1)\sigma_1^2 + \sigma_k^2)$$

so that provided $\sigma_1^2$ is positive the ARE compared to random choice designs is $1 - \{k\sigma_1^2/\sigma_k^2\}/\{1 + k^2K(\sigma_1^2/\sigma_k^2)\}$ which, since $0 \le k\sigma_1^2/\sigma_k^2 \le 1$, cannot be less than $k/(1 + k)$.

In view of the fact that it is easy to evaluate $U$-statistics based on a random choice of subsets, these efficiency losses seem a small price to pay, particularly, for the case when $k$ is large (say 3 or 4) and optimal designs are complicated.

## 4.3.4 Asymptotics for balanced designs

Suppose we have a sequence $U_n^{(0)}$ of incomplete $U$-statistics based on a kernel $\psi$ and a sequence of designs $\mathcal{D}_n$. Various types of asymptotics are possible; assuming that the design $\mathcal{D}_n$ contains $m_n$ sets, a convenient classification can be made in terms of the limit of the ratio $m_n/n$.

One type of asymptotic behaviour occurs when the number of sets in $\mathcal{D}_n$ is small compared to $n$, or in other words when $m_n/n \to 0$. Provided that the sets in $\mathcal{D}_n$ are not chosen from a restricted part of the sample, the $m_n$ sets will be largely disjoint and the incomplete $U$-statistic will be essentially a sum of $m$ i.i.d. summands. We would thus expect the random variable $m^{\frac{1}{2}}(U_n^{(0)} - \theta)$ to be asymptotically $N(0, \sigma_k^2)$ and this indeed is the case, as the following theorem shows. For different versions of this result, see Blom (1976) and Janson (1984).

**Theorem 1.** *Suppose that $m/n \to 0$ and that $f_c/m^2$ is $O(n^{-1})$ for $c = 1, 2, \ldots, k$. Then $m^{\frac{1}{2}}(U_n^{(0)} - \theta)$ is asymptotically $N(0, \sigma_k^2)$.*

**Proof.** Consider a set $S_{j'}$ that is disjoint from all other sets in the design. If $N = (n_{ij})$ is the incidence matrix of the design, then $\sum_{j \neq j'} \sum_i n_{ij} n_{ij'} = 0$. Since $\sum_i n_{ij} n_{ij'}$ is the $j, j'$ element of $N^T N$, it follows that the set $S_j$ will be disjoint from the other sets in the design if and only if the $j$th row of $N^T N$ (apart from the element on the diagonal) consists only of zeros. Hence if $D$ denotes the number of sets in the design disjoint from the rest, we must have

$$D + f_1 + \cdots + f_{k-1} \geq m \qquad (1)$$

since $f_1 + \cdots + f_{k-1}$ is the number of non-zero off-diagonal elements of $N^T N$, and hence is greater than the number of rows of $N^T N$ containing at least one non-zero element. Similarly, by considering elements rather than rows, we obtain

$$D(m - 1) + f_1 + \cdots + f_{k-1} \leq m(m - 1). \qquad (2)$$

Now by assumption, $nf_c/m^2$ is bounded for $c = 1, 2, \ldots, k-1$, so $f_c/m$ must converge to zero since $nm^{-1}$ converges to infinity. Hence from (1) and (2), $Dm^{-1}$ must converge to unity. Thus writing the sets disjoint from the rest as $S_1, \ldots S_D$, we have

$$U_n^{(0)} = m^{-1} \sum_{S \in \mathcal{D}_n} \psi(S)$$

$$= Dm^{-1} \left( \frac{1}{D} \sum_{j=1}^{D} \psi(S_j) \right) + m^{-1} \sum_{j=D+1}^{m} \psi(S_j)$$

$$= V_1 + V_2 \quad \text{say.}$$

Now $V_1$ and $V_2$ are independent, so that

$$mVar\, U_n^{(0)} = Dm^{-1}\sigma_k^2 + m\, Var V_2, \tag{3}$$

and by Theorem 2 of Section 4.3.1

$$\lim_{n\to\infty} mVar\, U_n^{(0)} = \lim_{n\to\infty} \left(\sum_{c=1}^{k-1} f_c \sigma_c^2 / m + \sigma_k^2\right)$$
$$= \sigma_k^2 \tag{3}$$

since $f_c/m \to 0$ for $c = 1, \ldots, k-1$. Passing to the limit in (3) and using (4) we see that $\lim_{n\to\infty} mVar V_2 = 0$ and so $m^{\frac{1}{2}}(U_n^{(0)} - \theta)$ and $m^{-1/2}D^{1/2}\sum_{j=1}^{D}(\psi(S_j) - \theta)$ have the same limit distribution. This is seen to be $N(0, \sigma_k^2)$ by applying the central limit theorem to the second of the two random variables above.

The case $m/n \to 0$ considered above is generally not so interesting from a statistical point of view, since it covers only cases where the asymptotic relative efficiency is zero. To see this consider

$$ARE = \lim_{n\to\infty} Var\, U_n / Var\, U_n^{(0)}$$
$$= \lim_{n\to\infty} (k^2\sigma_1^2/n)/(f_1 m^{-1}\sigma_1^2 + \cdots f_{k-1} m^{-1}\sigma_{k-1}^2 + \sigma_k^2)$$
$$= 0.$$

Thus such statistics will in general be too inefficient for practical use. Accordingly, we now confine attention to designs for which $m/n$ either converges to a positive number, or increases to $\infty$. We also require our designs to be balanced, which seems reasonable in view of the fact that the r.v.s $X_1, \ldots, X_n$ are assumed to be i.i.d.

Our next result concerns balanced incomplete $U$-statistics based on non-degenerate $U$-statistics of degree two, and is due to Brown and Kildea (1978).

**Theorem 2.** Let $\psi(x_1, x_2)$ be a kernel of degree two, and suppose that $\sigma_1^2 > 0$. Consider a sequence $\mathcal{D}_n$ of balanced designs such that each index appears in $r_n$ sets of $\mathcal{D}_n$. We thus have $m = nr_n/2$ and $f_1 = 2(r_n - 1)$.

(i) If $r_n$ does not depend on $n$, then the statistic $m^{\frac{1}{2}}(U_n^{(0)} - \theta)$ is asymptotically $N(0, \sigma^2)$ where $\sigma^2 = 2(r-1)\sigma_1^2 + \sigma_2^2$.

(ii) If $r_n \to \infty$, then $n^{\frac{1}{2}}(U_n^{(0)} - \theta)$ is asymptotically $N(0, k^2\sigma_1^2)$ i.e. the same as the corresponding complete statistic.

**Proof.** (i) Let $W_{n,M}^{(0)}$ be an incomplete $U$-statistic based on the design $\mathcal{D}_n$ and the kernel defined by

$$I_{\{\psi < M\}}\psi(x_1, x_2) = \begin{cases} \psi(x_1, x_2) & \text{if } \psi(x_1, x_2) < M, \\ 0 & \text{otherwise.} \end{cases}$$

Then the statistic $U_n^{(0)} - W_{n,M}^{(0)}$ is an incomplete $U$-statistic based on kernel $I_{\{\psi > M\}}\psi(x_1, x_2)$, with variance $m^{-1}\{2(r-1)\sigma_{1,M}^2 + \sigma_{2,M}^2\}$, where, for example,

$$\sigma_{2,M}^2 = \int \int I_{\{\psi > M\}}\psi^2(x_1, x_2)dF(x_1)dF(x_2).$$

Since $I_{\{\psi > M\}}\psi^2$ is dominated by the integrable function $\psi^2$, it follows by the dominated convergence theorem that $\lim_{M\to\infty} \sigma_{2,M}^2 = 0$. Since $\sigma_{1,M}^2 \leq \frac{1}{2}\sigma_{2,M}^2$ it then follows that for fixed $n$, $\lim_{M\to\infty} nVar(U_n^{(0)} - W_{n,M}^{(0)}) = 0$ uniformly in $n$. Hence we may (and from now to the end of the proof do) assume that the kernel $\psi$ is bounded by some $M$. We may also assume that $\theta = 0$ (otherwise consider $\psi^* = \psi - \theta$).

The proof is based on the idea that, if the moments of a sequence of r.v.s converge to those of the normal distribution, then the the sequence is necessarily asymptotically normal.

Consider the $\nu$th moment of $m^{\frac{1}{2}}U_n^{(0)}$ :

$$E(m^{\frac{1}{2}}U_n^{(0)})^\nu = m^{-\nu/2} \sum_{j_1=1}^{m} \cdots \sum_{j_\nu=1}^{m} E\{\psi(S_{j_1})\ldots\psi(S_{j_\nu})\} \tag{5}$$

A term in (5) involves $\nu$ sets $S_{j_1}, \ldots, S_{j_\nu}$ of $\mathcal{D}_n$ which may or may not be distinct. We may associate with these sets a *multigraph* whose vertices are the indices in $S = \bigcup_{c=1}^{\nu} S_{j_c}$, and vertices $i$ and $j$ are joined by one or more edges if the set $\{i, j\}$ equals one or more of the sets $S_{j_1}, \ldots, S_{j_2}$. The multigraph thus has at most $2\nu$ and at least two vertices and $\nu$ edges corresponding to the $\nu$ sets.

206

For example, if $\nu = 3$ and we have sets $S_1 = \{1,2\}$, $S_2 = \{2,3\}$ and $S_3 = \{4,5\}$ then the corresponding multigraph takes the form

$$1 \longleftrightarrow 2 \longleftrightarrow 3 \quad 4 \longleftrightarrow 5.$$

On the other hand, if $S_1$ and $S_2$ are as above, but $S_3 = \{1,2\}$ then the multigraph is

$$1 \Longleftrightarrow 2 \longleftrightarrow 3.$$

The multigraph corresponding to a term in (5) may be decomposed into a number of *connected components*, which are sets of vertices all of which are connected to at least one other vertex by an edge, and have the additional property that no two vertices in different components are connected. Thus the first multigraph above consists of two connected components while the second consists of only one.

The number of ways a connected component can be chosen from the $m$ sets of the design is $O(m)$, since the first edge can be chosen in $m$ ways, but successive edges must be chosen from the $O(r)$ sets having an index in common with those already chosen. Thus the number of terms in (5) corresponding to multigraphs having $c$ connected components is $O(m^c)$.

Any term whose multigraph has a connected component with only one edge must be zero, since such components involve a factor $\psi(S)$ in the term independent of the others and $E\{\psi(S)\} = 0$ for all sets $S$ by assumption. Thus if a non-zero term has $c$ connected components, with $\nu_1, \nu_2, \ldots, \nu_c$ edges respectively, then each $\nu_j \geq 2$ and hence $\nu = \nu_1 + \cdots + \nu_c \geq 2c$. If $\nu$ is odd, then $\nu - 1 \geq 2c$ and so there can be at most $(\nu - 1)/2$ connected components in any non-zero term. It follows that there are $O(m^{(\nu-1)/2})$ non-zero terms in (5) and hence all odd moments converge to zero.

Now consider the even moments. Arguing as above, the non-zero terms of (5) either involve $\nu/2$ components each with two edges, or fewer than $\nu/2$ components. Hence for even $\nu$,

$$E\{(m^{\frac{1}{2}}U_n^{(0)})^\nu\} = m^{-\frac{\nu}{2}} \sum {}^* \prod_{i=1}^{\nu/2} E\{\psi(S_{i1})\psi(S_{i2})\} + o(1) \qquad (6)$$

where $\sum^*$ denotes summation over all terms whose graphs have $\nu/2$ components each with two edges, and $S_{i1}$ and $S_{i2}$ are the two edges (sets) in the $i$th component.

207

The proof is completed by employing the following device: imagine a set $\{Y_S : S \in \mathcal{D}_n\}$ of jointly normal r.vs with zero mean and covariance given by

$$Cov(Y_S, Y_T) = \begin{cases} \sigma_2^2 & \text{if } S = T, \\ \sigma_1^2 & \text{if } |S \cap T| = 1, \\ 0 & \text{if } S \cap T = \emptyset. \end{cases}$$

Define a r.v. $Z_m = m^{-1} \sum_{S \in \mathcal{D}} Y_S$, then by repeating the above analysis we see that for each integer $\nu$

$$E(Z_m^\nu) = m^{-\nu} \sum\nolimits^* \prod E\{Y_{S_{i_1}} Y_{S_{i_2}}\} + o(1) \tag{6}$$

But $Z_m$ is normal with mean 0 and variance $\sigma^2/m = (2(r-1)\sigma_1^2 + \sigma_2^2)/m$, which may be seen by using the proof of Theorem 2 of Section 4.3.1. Hence $E(Z_n^\nu) = 0$ for odd $\nu$, and for even $\nu$

$$E(Z_m^\nu) = m^{-\nu/2}\nu!2^{-\nu/2}\sigma^\nu \big/ (\nu/2)!$$

using the standard formulae for normal moments. Moreover, $E(Y_S Y_T) = E\psi(S)\psi(T)$ for all $S$ and $T$ in the design, so that using (6), (7) and (8) we obtain for even $\nu$

$$E(m^{\frac{1}{2}}U_n^{(0)})^\nu = (\nu!)2^{-\nu/2}\sigma^2/(\nu/2)! + o(1)$$

and the proof of (i) is complete.

(ii) The proof of (ii) is trivial by comparison. We have by Theorem 1 of Section 4.3.1, and using the relation $2m = nr$

$$\begin{aligned} Var(n^{\frac{1}{2}}U_n^{(0)} - n^{\frac{1}{2}}U_n) &= n(Var\, U_n^{(0)} - Var\, U_n) \\ &= (n/m)(2(r-1)\sigma_1^2 + \sigma_2^2) - 4\sigma_1^2 + o(1) \\ &= 2r^{-1}(2(r-1)\sigma_1^2 + \sigma_2^2) - 4\sigma_1^2 + o(1) \\ &= o(1) \end{aligned}$$

and so $n^{\frac{1}{2}}U_n^{(0)}$ and $n^{\frac{1}{2}}U_n$ have identical asymptotic behaviour.

**Example 1. Kendall's tau (continued).**

The design in Example 1 of Section 4.3.1 is balanced, with $r = 2$ and $n = m$ so Theorem 1 is applicable. The statistic $n^{\frac{1}{2}}t_n^{(0)}$ is asymptotically normal with mean $\tau$ and asymptotic variance $(2\sigma_1^2 + \sigma_2^2)/n$, which reduces in the case of independence to $11/9n$.

## Example 2. The Hodges-Lehmann estimator.

Suppose $X_1, \ldots, X_n$ are a random sample from a symmetric distribution with a bounded continuous density and median $\theta$. The Hodges-Lehmann estimator of $\theta$ discussed in Example 6 of Section 2.2.6 is the median $\xi_n$ of the quantities

$$\left\{ \tfrac{1}{2}(X_i + X_j) : 1 \le i < j \le n \right\}.$$

An "incomplete" version of this estimator is

$$\xi_n^0 = \text{median} \left\{ \tfrac{1}{2}(X_i + X_j) : \{i, j\} \in \mathcal{D}_n \right\}$$

for some suitable sequence of balanced designs $\mathcal{D}_n$, in which we suppose that $r$ is held fixed as $n$ increases. The statistic $\xi_n^0$ is not a $U$-statistic, but its distribution may be expressed in terms of the quantity

$$\sum_{\{i,j\} \in \mathcal{D}} I\{X_i + X_j \le 2\theta + 2xn^{-\frac{1}{2}}\}$$

which is. Set $Y_i = X_i - \theta$, and let $G$ and $G * G$ be the distribution functions of $Y_i$ and $Y_i + Y_j$ respectively. Let $g$ be the density of $Y_i$. The function $G * G$ has a bounded continuous derivative since $g$ does, and is symmetric about zero.

We can now write

$$Pr\left(n^{\frac{1}{2}}(\xi_n^{(0)} - \theta) \le x\right) = Pr\left(m^{-1} \sum_{\{i,j\} \in \mathcal{D}_n} I\{Y_i + Y_j \le 2xn^{-1/2}\} \ge \tfrac{1}{2}\right)$$

$$= Pr\left(U_n^{(0)} \ge \tfrac{1}{2}\right)$$

where $U_n^{(0)}$ is the incomplete $U$-statistic based on design $\mathcal{D}_n$ and kernel $\psi_n(y_1, y_2) = I\{y_1 + y_2 \le 2n^{-\frac{1}{2}}x\}$.

From the results of Chapter 2 it follows that

$$E\ U_n^{(0)} = E\{\psi_n(Y_1, Y_2)\} = G * G(2n^{-\frac{1}{2}}x) = p_n \quad \text{say,}$$

and

$$\psi_n^{(1)}(y) = E\psi_n(y, Y_1) = Pr\left(Y_1 + y \geq n^{-\frac{1}{2}}x\right)$$
$$= G(n^{-\frac{1}{2}}x - y).$$

Hence (we write $\theta_{n,c}^2$ for $\theta_c^2$ since $\psi$ depends on $n$)

$$\sigma_{n,1}^2 = Var(\psi_n^{(1)}(Y_1))$$
$$= \int_{-\infty}^{\infty} (G(n^{-\frac{1}{2}}x - y) - p_n)^2 g(y) dy.$$

Further, $\sigma_{n,2}^2 = Var(\psi_n(Y_1, Y_2)) = p_n(1 - p_n)$. Using the facts that $G * G$ has a bounded first derivative and is symmetric about zero, a Taylor series expansion of $G * G$ about 0 yields

$$p_n = \frac{1}{2} + 2n^{-\frac{1}{2}}xg_0 + o(n^{-\frac{1}{2}}) \tag{6}$$

where $g_0 = \int_{-\infty}^{\infty} g^2(u) du$. Hence

$$\lim_{n \to \infty} E(U_n^{(0)}) = \frac{1}{2}, \tag{8}$$

$$\lim_{n \to \infty} \sigma_{n,2}^2 = \frac{1}{4}, \tag{9}$$

and by the dominated covergence theorem

$$\lim_{n \to \infty} \sigma_{n,1}^2 = \int_{-\infty}^{\infty} (G(y) - \frac{1}{2})^2 g(y) dy$$
$$= Var(G(Y))$$
$$= \frac{1}{12} \tag{10}$$

since $G(Y)$ is uniformly distributed on $[0, 1]$.

Since the kernel $\psi_n$ depends on $n$, Theorem 1 cannot be applied directly. However, as $n \to \infty$, $\psi_n$ converges to the kernel $\psi(y_1, y_2) = I\{y_1 + y_2 \geq 0\}$, so consider the incomplete $U$-statistic $W_n^{(0)}$ based on $\psi$ and $\mathcal{D}_n$. We can apply the last theorem to deduce that $m^{\frac{1}{2}}(W_n^{(0)} - \frac{1}{2})$ is asymptotically normal $N(0, (2r + 1)/12)$, since the methods above show that for the kernel $\psi, \sigma_1^2 = \frac{1}{12}$ and $\sigma_2^2 = \frac{1}{4}$. Now

$$m^{\frac{1}{2}}(U_n^{(0)} - p_n) = m^{\frac{1}{2}}(W_n^{(0)} - \frac{1}{2}) + m^{\frac{1}{2}}(U_n^{(0)} - W_n^{(0)} + \frac{1}{2} - p_n) \tag{11}$$

210

and in view of (8), (9) and (10) it easily follows that the second term in (11) is asymptotically negligible, and that the asymptotic distribution of $m^{\frac{1}{2}}(U_n^{(0)} - p_n)$ is $N(0, (2r+1)/12)$. Using this fact and Lemma B, and denoting the d.f. of the standard normal by $\Phi$, we get

$$
\begin{aligned}
\lim_{n\to\infty} \Pr\left(n^{\frac{1}{2}}(\xi_n^{(0)} - \theta) \leq x\right) &= \lim_{n\to\infty} \Pr\left(U_n^{(0)} \geq \tfrac{1}{2}\right) \\
&= \lim_{n\to\infty} \Pr\left(m^{\frac{1}{2}}(U_n^{(0)} - p_n) \geq m^{\frac{1}{2}}(\tfrac{1}{2} - p_n)\right) \\
&= \Phi(2x g_0 (6r)^{\frac{1}{2}}/(2r+1)^{\frac{1}{2}})
\end{aligned}
$$

since $\lim_{n\to\infty} m^{\frac{1}{2}}(\tfrac{1}{2} - p_n) = -2(\tfrac{r}{2})^{\frac{1}{2}} x g_0$ by (7). The incomplete Hodges-Lehmann estimator has thus asymptotically a normal distribution with a mean of $\theta$ and variance $(2r+1)/24 r g_0^2 n$. The standard (i.e. complete) Hodges-Lehmann estimator is asymptotically normal with mean $\theta$ and variance $1/12 g_0^2 n$ so that the ARE of the incomplete statistic is $2r/(2r+1)$, which can be made arbitrarily close to 1 by taking $r$ sufficiently large. A considerable saving in computation can be achieved for only a slight efficiency loss.

Not all designs lead to asymptotically normal statistics. In the degenerate case, Weber (1981) gives examples of designs leading to statistics that have a variety of asymptotic behaviours. However, we may identify two further situations which lead to normal asymptotics. The first occurs when the ratio $m/n$ tends to infinity, and the second when the designs are minimum variance. These behaviours when $k = 2$ have been dealt with in the previous theorem. Our next two results extend these, under somewhat restrictive assumptions, to general $k$, and various degrees of degeneracy:

**Theorem 3.** *Suppose $U_n^{(0)}$ is an incomplete $U$-statistic based on a series of designs $\mathcal{D}_n$, and suppose that $U_n^{(0)}$ is dth-order degenerate, so that*

$$
\sigma_c^2 = 0 \quad c \leq d \quad \text{and} \quad \sigma_{d+1}^2 > 0.
$$

*Let $f_{c,n}$ be the number of pairs of sets in $\mathcal{D}_n$ that have c elements in common, and further suppose that*

$$
\lim_{n\to\infty} n^{d+1} f_c/m_n^2 = \begin{cases} d!\binom{k}{d} & \text{if } c = d, \\ 0 & \text{otherwise.} \end{cases}
$$

211

Then $n^{\frac{d+1}{2}}(U_n^{(0)} - \theta)$ has the same limit distribution as the corresponding complete statistic.

**Proof.** From Theorem 1 of Section 4.3.1 we may write, denoting the corresponding complete statistic by $U_n$,

$$Var\{n^{(d+1)/2}(U_n^{(0)} - U_n)\} = n^{d+1}(Var\, U_n^{(0)} - Var\, U_n)$$
$$= \sum_{c=1}^{k} n^{d+1} f_c \sigma_c^2 / m^2 - n^{d+1}\binom{k}{d+1}(n^{-d-1}) + o(1)$$
$$= o(1),$$

and so the result follows by Slutsky's theorem.

Note that Theorem 2(ii) is a special case of Theorem 3 with $k = 2$ and $d = 0$.

The designs in Theorem 2 are all minimum variance designs. Our next result is a theorem covering the minimum variance designs based on cyclic permutations described in Example 7 of Section 4.3.2. These are designs for any $k$, and $m = Kn$ for some integer $K$, and exist for sufficiently large $n$. The designs may be described by means of the sets of quantities $d_1, \ldots, d_k$ defined in in Example 7 of Section 4.3.2. Suppose $D_1, \ldots, D_k$ are these sets; they have the properties

(i) all differences (mod $n$) of integers in a set are distinct;

(ii) all differences (mod $n$) of integers in distinct sets are distinct.

Suppose now we have a sequence of such designs with $m = nK_n$. Incomplete $U$-statistics based on such a sequence will be asymptotically normal, as the next theorem shows.

**Theorem 4.** Let $U_n^{(0)}$ be a sequence of $U$-statistics based on the sequence of designs $\mathcal{D}_n$ described in Example 7. Then

(i) If $K_n$ is constant (i.e. does not depend on $n$) then $m^{1/2}(U_n^{(0)} - \theta)$ is asymptotically $N(0, k(kK - 1)\sigma_1^2 + \sigma_k^2)$,

(ii) If $K_n \to \infty$ as $n \to \infty$, then $m^{\frac{1}{2}}(U_n^{(0)} - \theta)$ is asymptotically $N(0, k^2\sigma_1^2)$.

212

**Proof.** (i) Let $D_1, \ldots, D_k$ be the $K$ sets of $d's$ generating the designs, so that the blocks are of the form

$$S_{ij} = (i-1) \oplus D_j = \{(i-1) \oplus d_{1j}, \ldots, (i-1) \oplus d_{kj}\}$$

where $D_j = \{d_{ij}, \ldots, d_{kj}\}$ and $\oplus$ denotes addition mod $n$. If $M$ is the largest integer in the sets $D_1, \ldots, D_K$, the blocks have the property that $S_{ij}$ and $S_{i'j'}$ are disjoint provided $i - i'$ exeeds $M$ (mod $n$). (We may imagine the $n$ treatments equally spaced round the circumference of a circle of circumference $n$. Then $S_{ij}$ and $S_{i'j'}$ will be disjoint provided the shortest arc between $i$ and $i'$ exceeds $M$ in length.) Now define a random vector $\mathbf{Y}_i = (\psi(S_{i1}) - \theta, \ldots, \psi(S_{iK}) - \theta)$. Provided the shortest arc between $i$ and $i'$ is longer than $M$, $\mathbf{Y}_i$ and $\mathbf{Y}_{i'}$ will be independent. Moreover, for any integer $\nu$, $\mathbf{Y}_i$ and $\mathbf{Y}_{i\oplus\nu}$ have the same distribution, since the joint distribution of $X_{i\oplus d_{11}}, \ldots, X_{i\oplus d_{kK}}$ does not depend on $i$. Thus the sequence of random vectors is stationary in the sense that the joint distributions of

$$\mathbf{Y}_{i_1}, \ldots, \mathbf{Y}_{i_\nu} \quad \text{and} \quad \mathbf{Y}_{i_1 \oplus \mu}, \ldots, \mathbf{Y}_{i_\nu \oplus \mu}$$

are identical for all $i_1, \ldots, i_\nu$ and $\mu$. A minor adaption of the Hoeffding-Robbins (1948) $M$-dependent central limit theorem shows that the random vector $n^{-\frac{1}{2}} \sum_{i=1}^{n} \mathbf{Y}_i$ is asymptotically normal, and so if $\mathbf{1}$ is a $K$-vector of ones, $m^{-\frac{1}{2}} \sum_{i=1}^{n} \mathbf{Y}_i^T \mathbf{1}$ is also asymptotically normal. But

$$m^{-\frac{1}{2}} \sum_{i=1}^{n} \mathbf{Y}_i^T \mathbf{1} = m^{\frac{1}{2}} \sum_{i=1}^{n} \sum_{j=1}^{K} (\psi(S_{ij}) - \theta) = m^{\frac{1}{2}}(U_n^{(0)} - \theta)$$

so that $m^{\frac{1}{2}}(U_n^{(0)} - \theta)$ as asymptotically normal with mean zero and variance $k(kK-1)\sigma_1^2 + \sigma_k^2$.

(ii) Writing as usual

$$n^{\frac{1}{2}}(U_n^{(0)} - \theta) = n^{\frac{1}{2}}(U_n^{(0)} - \theta) + n^{\frac{1}{2}}(U_n^{(0)} - U_n)$$

we see that the second term on the right has mean zero and variance

$$n(m^{-1}(k(kK_n - 1)\sigma_1^2 + \sigma_k^2)) - k^2\sigma_1^2 + o(1)$$
$$= Kn^{-1}(k(kK_n - 1)\sigma_1^2 + \sigma_k^2) - k^2\sigma_1^2 + o(1)$$
$$= o(1).$$

213

Our final result in this section is an incomplete version of the Poisson convergence result described in Theorem 2 of Theorem 3.2.4. Recall that if we have a sequence $X_1, X_2, \ldots$ of independent and identically distributed random variables, and a sequence $\psi_n(x_1, \ldots, x_k)$ of zero-one kernels of degree $k$ then under certain conditions $T_n = \sum_{(n,k)} \psi_n(X_{i_1}, \ldots, X_{i_k})$ converges to a Poisson law. An "incomplete" version of this result is due to Berman and Eagleson (1983):

**Theorem 5.** *Suppose that $\mathcal{D}_n$ is a balanced design containing $m$ sets and that*

$$\lim_{n \to \infty} m \ E\psi_n(X_i, \ldots, X_k) = \lambda \tag{12}$$

*and*

$$\lim_{n \to \infty} n \ E\{\psi_n(X_1, \ldots, X_k)\psi_n(X_2, \ldots, X_{k+1})\} = 0. \tag{13}$$

*Then the r.v. $T_n^{(0)} = \sum_{S \epsilon \mathcal{D}_n} \psi_n(S)$ converges in distribution to a Poisson distribution with parameter $\lambda$.*

**Proof.** Expanding the $\nu^{th}$ moment of $T_n^{(0)}$, we get

$$E\{(T_n^{(0)})^\nu\} = \sum_{j_1=1}^{m} \cdots \sum_{j_\nu=1}^{m} E\{\psi_n(S_{j_1}) \ldots \psi_n(S_{j_\nu})\}. \tag{14}$$

As in Theorem 2, each term of (14) has assoicated with it a multigraph made up of several connected components. Divide the terms of (14) up into two classes, the first consisting of those terms for which the sets $S_{j_1}, \ldots, S_{j_\nu}$ are disjoint, and the second consisting of all other terms. The second type of term corresponds to a multigraph with a number of connected components, say $s$ in all, where $s < \nu k$. The expectation of (14) is the product of terms like

$$E\{\psi(S_{j_1}) \ldots \psi(S_{j_c})\} \tag{15}$$

corresponding to a single connected component, and where every set in the product (15) has an element in common with one other of the sets $S_{j_1}, \ldots, S_{j_c}$. We need to show that expressions like (15) are $O(n^{-1})$. To this

214

end, note that, since $\psi_n$ is either zero or one, and assuming that $|S_{j_1} \cap S_{j_2}| = c$, we have

$$E\{\psi_n(S_{j_1}) \ldots \psi_n(S_{j_c})\} \leq E\{\psi_n(S_{j_1})\psi_n(S_{j_2})\}$$

$$\leq E\{\psi_n(X_1, \ldots, X_k)\psi_n(X_2, \ldots, X_{k+1})\}$$

arguing as in the proof of Theorem 2 of Section 3.2.4. Hence by (13), $E\{\psi_n(S_{j_1}) \ldots \psi_n(S_{j_c})\}$ is $o(n^{-1})$ and thus any term with $s$ connected components is $o(n^{-s})$. A slight extension of the arguments of Theorem 2 shows that there are $O(n^s)$ such terms, and so the sum of terms of the second type is $o(1)$. To handle terms the first type, introduce random variables $\{Y_S, S \in \mathcal{D}_n\}$ that are independent zero-one r.vs with $Pr(Y_S = 1) = \lambda/m$. Then $\tilde{T}_n = \sum_{S \in \mathcal{D}} Y_S$ has asymptotically the same moments as the sum of the terms of the first type and is distributed binomially with parameters $m$ and $\lambda/m$. Its asymptotic distribution is hence Poisson, and its moments converge to those of the Poisson. Using the argument of Theorem 2, it follows that the moments of $T_n$ converge to those of the Poisson, and the theorem is proved.

## 4.4  Bibliographic details

The paper by Efron and Stein (1981) is a good source of information on symmetric statistics, although it does not treat asymptotics. The nice characterisation of $U$-statistics contained in Theorem 2 of Section 4.1.1 is due to Lenth (1983), as are some of the other results of that section. The asymptotic results are taken from Rubin and Vitale (1980) and Dynkin and Mandelbaum (1983).

The fundemental paper on $V$-statistics is by von Mises (1947); Serfling (1980) has much more detail than we provide. The representation of $V$-statistics in terms of $U$-statistics can be found in Janssen (1981) and Lee (1985).

Basic material on incomplete $U$-statistics is found in Blom (1976). Theorem 3 of Section 4.3.1 is from Lee (1982), as is the material on minimum variance designs in Section 4.3.2. The results on asymptotic distributions are due to Janson (1984) in the random selection case and Brown and Kildea (1978) in the balanced case. The "incomplete" Poisson convergence theorem (Theorem 5 of Section 4.3.4) is from Berman and Eagleson (1983).

# CHAPTER FIVE

# Estimating Standard Errors

This chapter treats the following problem: given a $U$-statistic, how do we measure its standard error? In a related vein, how does the fact we are estimating $Var\, U_n$ (by $\widehat{Var}\, U_n$ say) affect the asymptotic distribution of $n^{\frac{1}{2}}(U_n - \theta)/(\widehat{Var}\, U_n)^{\frac{1}{2}}$? We discuss methods based on jackknifing and bootstrapping, and treat the application of these techniques to both complete and incomplete statistics.

## 5.1 Standard errors via the jackknife

The jackknife is a widely applicable statistical tool used for both bias reduction and the estimation of standard errors. We provide a brief general discussion before treating the application of the technique to $U$-statistics; readers requiring a fuller account are referred to the monograph by Efron (1982).

### 5.1.1 The jackknife estimate of variance

We assume as usual that $X_1, \ldots, X_n$ is a random sample of independent and identically distributed variables, with common distribution function $F$. In addition a functional $T(F)$ is given, and an estimator $T_n$ of $T(F)$ which depends on the sample size $n$. We consider *pseudovalues*

$$\hat{T}_i = nT_n - (n-1)T_{n-1}(-i)$$

where $T_{n-1}(-i)$ is the statistic $T_{n-1}$ computed on the sample of $n-1$ variables formed from the original data set by deleting the $i$th data value. The jackknifed estimate of $T(F)$ based on $T_n$ is simply the average of the pseudovalues:

$$\tilde{T}_n = n^{-1} \sum_{i=1}^{n} \hat{T}_i.$$

The estimator $\tilde{T}_n$ typically is less biased than $T_n$ (if $T_n$ has a bias of order $n^{-1}$ then typically $\tilde{T}_n$ has a bias of order $n^{-2}$) and an estimate of

$Var\, T_n$ (or $Var\, \tilde{T}_n$) is given by the *jackknife estimate of variance*

$$\widehat{Var}(JACK) = \frac{1}{n(n-1)} \sum_{i=1}^{n} (\hat{T}_i - \tilde{T}_n)^2.$$

An alternative expression for $\widehat{Var}(JACK)$ directly in terms of the "leave one out" statistics $\hat{T}_n(-i)$ is

$$\widehat{Var}(JACK) = \frac{n-1}{n} \sum_{i=1}^{n} (T_{n-1}(-i) - \bar{T}_{n-1}(\cdot))^2 \qquad (1)$$

where $\bar{T}_{n-1}(\cdot) = n^{-1} \sum_{i=1}^{n} T_{n-1}(-i)$.

In the case of $U$-statistics, Theorem 2 of Section 4.1 tells us the the jackknifed version $\tilde{U}_n$ coincides exactly with $U_n$, and in any event, $U_n$ is unbiased. Rather, the usefulness of the jackknife in the context of $U$-statistics is that (1) furnishes a consistent estimate of $Var\, U_n$.

The explicit application of the jackknife to the problem of estimating the standard errors of $U$-statistics was first considered by Arvesen (1969) although an equivalent formulation appears in Sen (1960). Our first result, giving an explicit representation for $\widehat{Var}(JACK)$, is due to Arvesen.

**Theorem 1.** *Let $U_n$ be a $U$-statistic of degree $k$ based on a kernel $\psi$. Then*

$$\widehat{Var}(JACK) = n^{-2}(n-1) \binom{n-1}{k}^{-2} \sum_{c=0}^{k} (cn - k^2) S_c$$

*where the quantities $S_c$ are defined by*

$$S_c = \sum_{|S \cap T| = c} \psi(S)\psi(T) \qquad c = 0, 1, \dots, k.$$

*Thus, $S_c$ is the sum of products over all pairs of sets having $c$ indices in common.*

**Proof.** If $\sum_{(n-1,k)}^{(-i)}$ denotes summation over all $k$-subsets of $\{1, 2, \dots, n\}$ that do not contain $i$, we can write

$$U_{n-1}(-i) = \binom{n-1}{k}^{-1} \sum_{(n-1,k)}^{(-i)} \psi(X_{i_1}, \dots, X_{i_k})$$

and

$$\widehat{Var}(JACK) = \frac{n-1}{n} \left( \sum_{i=1}^{n} U_{n-1}^2(-i) - nU_n^2 \right) \qquad (2)$$

since $n^{-1} \sum_{i=1}^{n} U_{n-1}(-i) = U_n$ by Theorem 2 of Section 4.2. Now

$$\sum_{i=1}^{n} U_{n-1}^2(-i) = \sum_{i=1}^{n} \binom{n-1}{k}^{-2} \sum_{(n-1,k)}^{(-i)} \psi(S) \sum_{(n-1,k)}^{(-i)} \psi(T) \qquad (3)$$

$$= \binom{n-1}{k}^{-2} \sum_{c=0}^{k} (n-2k+c) \sum_{|S\cap T|=c} \psi(S)\psi(T) \qquad (4)$$

since for any pair of sets $S$ and $T$ with $c$ elements in common, the term $\psi(S)\psi(T)$ occurs in $n - |S \cap T| = n - 2k + c$ terms of the right of (3).

Also

$$U_n^2 = \binom{n}{k}^{-2} \sum_{c=0}^{k} \sum_{|S\cap T|=c} \psi(S)\psi(T) \qquad (5)$$

so that substituting (4) and (5) into (2) we get

$$\widehat{Var}(JACK) = \frac{n-1}{n} \left\{ \binom{n-1}{k}^{-2} \sum_{c=0}^{k} (n-2k+c)S_c - \binom{n}{k}^{-2} \sum_{c=0}^{k} S_c \right\}$$

$$= n^{-2}(n-1) \binom{n-1}{k}^{-2} \sum_{c=0}^{k} (nc - k^2)S_c.$$

Our next result investigates the bias of the jackknife estimate of variance.

**Theorem 2.** *The bias involved in using (1) as an estimate of $Var\, U_n$ is positive and is given by*

$$BIAS(JACK) = n \sum_{j=2}^{k} (j-1)(n-j)^{-1} \binom{k}{j}^2 \binom{n}{j}^{-1} \delta_j^2$$

*where the quantities $\delta_c^2$ are the usual conditional variances of the $H$-decomposition components.*

**Proof.** Using the $H$-decomposition, write

$$U_n = \theta + \sum_{c=1}^{k} \binom{k}{c} H_n^{(c)}$$

219

and

$$U_n(-i) = \theta + \sum_{c=1}^{k} \binom{k}{c} H_n^{(c)}(-i)$$

so that

$$\sum_{i=1}^{n}(U_n(-i) - U_n)^2$$

$$= \sum_{i=1}^{n}\sum_{j=1}^{k}\sum_{j'=1}^{k} \binom{k}{j}\binom{k}{j'} \left(H_n^{(j)}(-i) - H_n^{(j)}\right)\left(H_n^{(j')}(-i) - H_n^{(j')}\right).\quad (6)$$

If $h^{(j)}$ are the kernels of the component $U$-statistics $H_n^{(j)}$, Theorem 3 of Section 1.6 tells us that $Eh^{(j)}h^{(j')} = 0$ for $j \neq j'$ so that taking expectations of both sides of (6) yields

$$E\frac{n-1}{n}\sum_{i=1}^{n}(U_n(-i) - U_n)^2$$

$$= \frac{n-1}{n}\sum_{i=1}^{n}\sum_{j=1}^{k}\binom{k}{j}^2 E\left(H_n^{(j)}(-i) - H_n^{(j)}\right)^2.\quad (7)$$

Consider now a fixed component $H_n^{(j)}$. Using Theorem 1, we have

$$E\frac{n-1}{n}\sum_{i=1}^{n}\left(H_n^{(j)}(-i) - H_n^{(j)}\right)^2 = n^{-2}(n-1)\binom{n-1}{j}^{-2}\sum_{c=0}^{j}(nc - j^2)E(S_c)$$

and $ES_c = 0$ for $c = 0, 1, 2, \ldots, j-1$ and $ES_j = \binom{n}{j}\delta_j^2$ again by Theorem 3 of Section 1.6.

Thus

$$E\frac{n-1}{n}\sum_{i=1}^{n}\left(H_n^{(j)}(-i) - H_n^{(j)}\right)^2 = n^{-2}(n-1)\binom{n-1}{j}^{-2}j(n-j)\binom{n}{j}\delta_j^2$$

$$= n^{-1}(n-1)\binom{n-1}{j}^{-1}j\delta_j^2$$

and so from (7)

$$E\ \widehat{Var}(JACK) = n^{-1}(n-1)\sum_{j=1}^{k}\binom{k}{j}^2\binom{n-1}{j}^{-1}j\delta_j^2.\quad (8)$$

220

Hence

$$BIAS(JACK) = E\widehat{Var}(JACK) - Var\, U_n$$

$$= \sum_{j=2}^{k} n(j-1)(n-j)^{-1} \binom{k}{j}^2 \binom{n}{j}^{-1} \delta_j^2$$

using (8) and Theorem 4 of Section 1.6.

Note that the estimate $\widehat{Var}(JACK)$ is unbiased when $k = 1$ and its bias is of order $n^{-2}$ when $k \geq 2$.

**Example 1. The sample variance.**

From earlier examples, $\delta_2^2 = \sigma^4$ and so the bias in the jackknife estimate of $Var\, s^2$ is $2\sigma^4/(n-1)(n-2)$.

The representation of Theorem 1 provides a convienient tool for investigating the consistency of the estimator $\widehat{Var}(JACK)$. The quantities $S_c$ of Theorem 1 are closely related to $U$-statistics: define a kernel $\Psi^{(c)}$ of degree $2k - c$ by

$$\Psi^{(c)}(x_1,\ldots,x_{2k-c}) = \{(2k-c)!\}^{-1} \sum_{(2k-c)} \psi(x_{i_1},\ldots,x_{i_k})(x_{i_{k-c+1}},\ldots,x_{i_{2k-c}})$$

so that $\Psi^{(c)}$ is a symmetrised version of $\psi(x_1,\ldots,x_k)\psi(x_{k-c+1},\ldots,x_{2k-c})$. Let $U_n^{(c)}$ be the $U$-statistic based on $\Psi^{(c)}$, then $U_n^{(c)}$ is just a normalised version of $S_c$, in fact $U_n^{(c)} = N_c^{-1} S_c$ where $N_c = \binom{n}{k}\binom{k}{c}\binom{n-k}{k-c}$ is the number of ways of choosing pairs of $k$-sets $S$ and $T$ with $|S \cap T| = c$. The $U$-statistic $U_n^{(c)}$ is an unbiased estimator of $E\psi(X_1,\ldots,X_k)\psi(X_{k-c+1},\ldots,X_{2k-c}) = \sigma_c^2 + \theta^2$ so that $U_n^{(c)}$ converges almost surely to $\sigma_c^2 + \theta^2$ by Theorem 3 of Section 3.4.2. Thus

$$\widehat{Var}(JACK) = \frac{k^2}{n}\sigma_1^2 + R_n \tag{9}$$

where $R_n$ is $o(n^{-1})$ with probability one, and hence $n(\widehat{Var}(JACK) - Var\, U_n)$ converges to zero with probability one.

A similar argument leads to an asymptotic expression for the variance. Define

$$\sigma_1(c,d) = Cov\big(\Psi^{(c)}(X_1,\ldots,X_{2k-c}), \Psi^{(d)}(X_{2k-c},\ldots,X_{4k-c-d-1})\big)$$

so that by Theorem 2 of Section 1.4,

$$Cov(U_n^{(c)}, U_n^{(d)}) = (2k-c)(2k-d)\sigma_1(c,d)n^{-1} + o(n^{-1}).$$

Hence

$$Var\{\widehat{Var}(JACK)\} = Var\left\{ n^{-2}(n-1)\binom{n-1}{k}^{-2} \sum_{c=0}^{k}(cn-k)N_cU_n^{(c)} \right\}$$

$$= k^4 \left\{ 4k^2\sigma_1(0,0) - 4k(2k-1)\sigma_1(0,1) + (2k-1)^2\sigma_1(1,1) \right\} n^{-3}$$

$$+ o(n^{-3}). \tag{10}$$

Thus the mean squared error of $\widehat{Var}(JACK)$ is determined largely by its variance, and its bias is insignificant in large samples.

## Example 2. The sample variance (continued).

The derived kernels introduced above are

$$\Psi^{(0)}(x_1, x_2, x_3, x_4) = \tfrac{1}{12}\{(x_1-x_2)^2(x_3-x_4)^2 + (x_1-x_3)^2(x_2-x_4)^2$$

$$+ (x_1-x_4)^2(x_2-x_3)^2\}$$

$$\Psi^{(1)}(x_1, x_2, x_3) = \tfrac{1}{12}\{(x_1-x_2)^2(x_1-x_3)^2 + (x_1-x_2)^2(x_2-x_3)^2$$

$$+ (x_1-x_3)^2(x_3-x_2)^2\}$$

and

$$\Psi^{(2)}(x_1, x_2) = \tfrac{1}{4}(x_1-x_2)^4.$$

Elementary calculations yield

$$E\Psi^{(0)}(x_1, X_2, X_3, X_4) = \tfrac{1}{2}E(x_1-X_2)^2\sigma^2$$

$$= \tfrac{1}{2}((x_1-\mu)^2 + \sigma^2)\sigma^2,$$

$$E\Psi^{(1)}(x_1, X_2, X_3) = \tfrac{1}{12}\left( 2\left\{ 2(x_1-\mu)^2\sigma^2 - 2(x-\mu)\mu_3 + \mu_4 + \sigma^4 \right\} \right.$$

$$\left. + \{(x-\mu)^2 + \sigma^2\}^2 \right)$$

and

$$E\Psi^{(2)}(x_1, X_2) = \tfrac{1}{4}\left\{ x_1-\mu)^4 + 6(x_1-\mu)^2\sigma^2 - 4(x_1-\mu)\mu_3 + \mu_4 \right\}$$

which are of the form $\sum_{i=0}^{4} a_{ij}(x_i-\mu)^i$ for $j = 0, 1, 2$. Thus using Theorem 2 of Section 2.1.4 again we get

$$\sigma_1(c,d) = Cov\left( \sum_{i=0}^{4} a_{ic}(X_1-\mu)^i, \sum_{i=0}^{4} a_{id}(X_1-\mu)^i \right)$$

$$= \sum_{i=0}^{4}\sum_{j=0}^{4} a_{ic}a_{jd}(\mu_{i+j} - \mu_i\mu_j) \tag{11}$$

222

so

$$Var(\widehat{Var}(JACK)) = \{256\sigma_1(0,0) - 384\sigma_1(0,1) + 144\sigma_1(1,1)\}\, n^{-3} + o(n^{-3})$$

where the quantities $\sigma_1(0,0)$, $\sigma_1(0,1)$ and $\sigma_1(1,1))$ are given by (11).

**Example 3. The sample mean.**

In this case $\Psi^{(0)}(x_1, x_2) = x_1 x_2$ and $\Psi^{(1)}(x_1) = x_1^2$, so that

$$\sigma_1(0,0) = Cov(X_1 X_2, X_2 X_3) = \mu^2 \sigma^2$$

$$\sigma_1(0,1) = Cov(X_1 X_2, X_2) = \mu \mu_3 + 2\sigma^2 \mu^2$$

and

$$\sigma_1(1,1) = Cov(X_1^2, X_1^2) = \mu_4 + 4\mu_3 \mu + 4\sigma^2 \mu^2 - \sigma^4.$$

Hence

$$Var(\widehat{Var}(JACK)) = (4\sigma_1(0,0) - 4\sigma_1(0,1) + \sigma_1(1,1))n^{-3} + o(n^{-3})$$
$$= (\mu_4 - \sigma^4)n^{-3} + o(n^{-3}).$$

Note in this case that $\widehat{Var}(JACK) = \{n(n-1)\}^{-1} \sum_{i=1}^{n}(X_i - \bar{X})^2 = s^2/n$.

The estimator $\widehat{Var}$ (JACK) was first introduced in an equivalent form by Sen (1960) although not identified as the jackknife estimator. Sen defines "components" of the $U$-statistic $U_n$ by

$$V_i = \binom{n-1}{k-1}^{-1} \sum_{(n-1,k-1)}^{(-i)} \psi(X_i, X_{i_1}, \ldots, X_{i_{k-1}})$$

where the sum is taken over all $(k-1)$-subsets of $\{1, 2, \ldots, i-1, i+1, \ldots, n\}$. Then $U_n = n^{-1} \sum_{i=1}^{n} V_i$ and Sen's estimator of $Var\, U_n$ is $k^2 S^2/n$ where

$$S^2 = \frac{1}{(n-1)} \sum_{i=1}^{n} (V_i - U_n)^2.$$

The estimator $k^2 S^2/n$ is just $((n-k)/n)^2\, \widehat{Var}(JACK)$. To see this, write

$$U_n = \binom{n}{k}^{-1} \sum_{(n,k)} \psi(X_{i_1}, \ldots, X_{i_k})$$

$$= \binom{n}{k}^{-1} \left\{ \sum_{(n-1,k-1)}^{(-i)} \psi(X_i, X_{i_1}, \ldots X_{i_{k-1}}) + \binom{n-1}{k} U_n(-i) \right\}$$

$$= \binom{n}{k}^{-1} \binom{n-1}{k-1} V_i + \binom{n}{k}^{-1} \binom{n-1}{k} U_n(-i)$$

$$= \frac{k}{n} V_i + \left(1 - \frac{k}{n}\right) U_n(-i).$$

Thus $(n - k)(U_n - U_n(-i)) = k(V_i - U_n)$ and hence

$$\widehat{Var}(JACK) = \frac{(n-1)^2}{(n-k)^2} k^2 S^2/n.$$

Since $\widehat{Var}(JACK)$ is a consistent estimator of $Var\, U_n$, in the sense that

$$n(\widehat{Var}(JACK) - Var\, U_n) \xrightarrow{wp1} 0,$$

it follows that $n^{\frac{1}{2}}(U_n - \theta)/(\widehat{Var}(JACK))^{\frac{1}{2}}$ converges to a standard normal distribution for non-degenerate $U_n$.

### 5.1.2 Jackknifing functions of $U$-statistics

Consider using a function $g(U_n)$ of $U_n$ to estimate $g(\theta)$. In general, $g(U_n)$ will not be unbiased, and we might expect that jackknifing $g(U_n)$ will reduce its bias in addition to providing a standard error. Introduce the notations $g_n$ for $g(U_n)$ and $g_n(-i)$ for $g(U_{n-1}(-i))$. Then the jackknifed estimate of $g(\theta)$ is $\hat{g} = \sum_{i=1}^{n} \hat{g}_i$ where the $\hat{g}_i$ are the psuedovalues $\hat{g}_i = ng_n - (n-1)g_n(-i)$. The estimate of $Var\, g(U_n)$ is $\widehat{Var}(JACK) = (n-1)\sum_{i=1}^{n}(\hat{g}_i - \hat{g})^2$. Now assume that the function $g$ has a bounded second derivative in a neighbourhood of $\theta$, so that

$$g(u) = g(\theta) + (u - \theta)g'(\theta) + \tfrac{1}{2}(u - \theta)^2 g''(\xi)$$

for some $\xi$ between $u$ and $\theta$. Similarly we may expand $g$ about $U_n$ and obtain

$$g(U_n(-i)) = g(U_n) + (U_n(-i) - U_n)g'(U_n) + \tfrac{1}{2}(U_n(-i) - U_n)^2 g''(\xi_i) \quad (1)$$

where $\xi_i$ lies between $U_n(-i)$ and $U_n$, provided that $g$ has a bounded second derivative in this region. Summing both side of (1) and dividing by $n$ yields

$$\frac{1}{n}\sum_{i=1}^{n} g_n(-i) = g_n + \tfrac{1}{2}\sum_{i=1}^{n}(U_n(-i) - U_n)^2 g''(\xi_i) \quad (2)$$

and using the relationship

$$\frac{1}{n}\sum_{i=1}^{n} g_n(-i) = (ng_n - \hat{g})/(n-1)$$

224

we obtain

$$\hat{g} = g - \frac{n-1}{2n} \sum_{i=1}^{n} (U_n(-i) - U_n)^2 g''(\xi_i)$$

$$= g + R_n$$

say, and so

$$n^{\frac{1}{2}}(\hat{g} - g(\theta)) = n^{\frac{1}{2}}(g(U_n) - g(\theta)) - n^{\frac{1}{2}} R_n. \tag{3}$$

By Lemma A below, the first term on the right of equation (3) converges in distribution to $N(0, k^2 \sigma_1^2 / g'(\theta)^2)$. Since the second derivative of $g$ is assumed to be bounded in a neighbourhood of $\theta$, and $U_n(-i)$ and $U_n$ are consistent estimators of $\theta$, it suffices to show that $n^{\frac{1}{2}} \sum_{i=1}^{n} (U_n(-i) - U_n)^2 \overset{D}{\longrightarrow} 0$ in order to show that $n^{\frac{1}{2}} R_n \overset{P}{\longrightarrow} 0$. But this is equivalent to $n^{\frac{1}{2}} \hat{V}ar(JACK) \overset{D}{\longrightarrow} 0$ which follows from (9) of Theorem 5.1.1. We have in fact proved the following theorem:

**Theorem 1.** *Let $g$ have a bounded second derivative in a neighbourhood of $\theta$. Then $n^{\frac{1}{2}}(\hat{g} - g(\theta))$ converges in distribution to a $N(0, k^2 \sigma_1^2 / g'(\theta)^2)$ distribution provided $U_n$ is non-degenerate and $g'(\theta) \neq 0$.*

We need also to prove Lemma A:

**Lemma A.** *Let $X_n \overset{P}{\longrightarrow} \theta$ and $n^{\frac{1}{2}}(X_n - \theta) \overset{D}{\longrightarrow} N(0, \sigma^2)$. Suppose that $g$ has a continuous derivative in $(\theta - \varepsilon, \theta + \varepsilon)$, with $g'(\theta) \neq 0$. Then $n^{\frac{1}{2}}(g(X_n) - g(\theta))g'(\theta) \overset{D}{\longrightarrow} N(0, \sigma^2)$.*

**Proof.** Denote by $A_n$ the event $\{|X_n - \theta| < \varepsilon\}$. Then if $A_n$ occurs

$$g(X_n) = g(\theta) + (X_n - \theta)g(\xi_n)$$

where $\xi_n$ is between $X_n$ and $\theta$, so that $\xi_n \overset{P}{\longrightarrow} \theta$. Hence

$$n^{\frac{1}{2}}(g(X_n) - g(\theta)) = n^{\frac{1}{2}}(X_n - \theta)g'(\xi_n)I\{A_n\} + YI\{A_n'\} \tag{4}$$

for some $Y$. Moreover, $Pr(|Y|I\{A_n'\} > \varepsilon) \leq Pr(A_n')$ which converges to zero, so that $YI\{A_n'\} \overset{P}{\longrightarrow} 0$ and the asymptotic distributions of the left hand side and the first term on the right side of (4) coincide. Moreover,

$$n^{\frac{1}{2}}(X_n - \theta)g'(\xi_n)I\{A_n\} = n^{\frac{1}{2}}(X_n - \theta)g'(\theta)\frac{g'(\xi_n)}{g'(\theta)}I\{A_n\}. \tag{5}$$

225

Since $\xi_n \xrightarrow{p} \theta$ and $I\{A_n\} \xrightarrow{p} 1$ the term on the left of (5) has the same asymptotic distribution as $n^{\frac{1}{2}}(X_n - \theta)g'(\theta)$ and the lemma is proved.

Next we show that the jackknife estimate $\widehat{Var}(JACK)$ of $Var\,g(U_n)$ is strongly consistent, provided $g$ has a continuous first derivative at $\theta$.

**Theorem 2.** *The jackknife estimate of variance satisfies $n(\widehat{Var}(JACK) - Var\,g(U_n)) \xrightarrow{wp1} 0$ provided $g'$ is continuous in a neighbourhood of $\theta$.*

**Proof.** Write

$$
\begin{aligned}
g_n(-i) &= g + (U_n(-i) - U_n)g'(\xi_{i,n}) \\
&= g + (U_n(-i) - U_n)g'(\theta) + (U_n(-i) - U_n)(g'(\xi_{i,n}) - g'(\theta)) \\
&= g + A_i + B_i
\end{aligned}
$$

say, where $\xi_{i,n}$ lies between $U_n(-i)$ and $U_n$, and so converges to $\theta$ with probability one. Writing $\bar{g}_n = n^{-1}\sum_{i=1}^{n} g_n(-i)$ and averaging the above over $i$ yields

$$
g_n(-i) - \bar{g}_n = A_i - (B_i - \bar{B})
$$

since $\bar{A} = 0$, and hence

$$
n\widehat{Var}(JACK) = (n-1)\sum_{i=1}^{n}(g_n(-i) - \bar{g}_n)^2
$$

$$
= (n-1)\left\{\sum_{i=1}^{n} A_i^2 - 2\sum_{i=1}^{n} A_i(B_i - \bar{B}) + \sum_{i=1}^{n}(B_i - \bar{B})^2\right\}.
$$

Now

$$
(n-1)\sum_{i=1}^{n}(B_i - \bar{B})^2 \leq (n-1)\sum_{i=1}^{n} B_i^2
$$

$$
\leq (n-1)\sum_{i=1}^{n}(U_n(-i) - U_n)^2\varepsilon^2
$$

for sufficiently large $n$ with probability one, since $\xi_{i,n} \xrightarrow{p} \theta$ for each $i$. Because the random variable $(n-1)\sum_{i=1}^{n}(U_n(-i) - U_n)^2$ is $n$ times the jackknife estimate of $Var\,U_n$ which converges with probability one to $k^2\sigma_1^2$

226

by (9), and since $\varepsilon^2$ is arbitrarily small, it follows that $(n-1)\sum_{i=1}^{n}(B_i-\bar{B})^2$ converges to zero with probability one, and hence by the Cauchy-Schwartz inequality, so does $2(n-1)\sum_{i=1}^{n}A_i(B_i-\bar{B})$. Finally, $(n-1)\sum A_i^2$ is just $(g'(\theta))^2 n$ times the jackknifed estimate of $Var\,U_n$. Since this converges to $k^2\sigma_1^2$ with probability one by (9), the theorem is proved, since $Var\,g(U_n) = Var\,U_n(g'(\theta))^2 + o(n^{-1})$.

### 5.1.3 Extension to functions of several $U$-statistics

The results above carry over with the obvious modifications to functions of several $U$-statistics. If $U_{1,n},\ldots,U_{m,n}$ are $m$ $U$-statistics of order $k_1,\ldots,k_m$ based on kernels $\psi^{(1)},\ldots,\psi^{(m)}$ and estimating parameters $\theta_1,\ldots,\theta_m$, what can we say about the random variable $g(U_{1,n},\ldots,U_{m,n})$ where $g$ is a function of $m$ variables? Assume that the $U$-statistics are based on the same sample $X_1,\ldots,X_n$. Then we can estimate $g(\theta_1,\ldots,\theta_m)$ by jackknifing the statistic $g(U_{1,n},\ldots,U_{m,n})$ and estimate the variance of the jackknifed estimator $\hat{g}$ by the usual estimator $\widehat{Var}(JACK)$ where now $g_n(-i)$ denotes the vector $g(U_{1,n}(-i),\ldots,U_{m,n}(-i))$.

Provided $g$ has continuous first order partial derivatives in a neighbourhood of $\theta = (\theta_1,\ldots,\theta_m)$, the methods of the last section can be used to prove that

$$n\widehat{Var}(JACK) \xrightarrow{wp1} \sum_{i=1}^{k}\sum_{j=1}^{k} k_i k_j g_i' g_j' \xi_1(i,j)$$

where $\xi_1(i,j) = Cov(\psi^{(i)}(X_1,\ldots,X_{k_i})\psi^{(j)}(X_{k_i},X_{k_i+1},\ldots,X_{k_i+k_j-1}))$ and $g_i'$ denotes the partial derivative $\partial g/\partial u_i|_{\mathbf{u}=\theta}$. Under the additional assumption of continuity of the second order partial derivatives, we can also show that

$$n^{\frac{1}{2}}(\hat{g} - g(\theta_1,\ldots,\theta_m)) \xrightarrow{D} N\left(0, \sum_{i=1}^{m}\sum_{j=1}^{m} k_i k_j g_i' g_j' \xi_1(i,j)\right).$$

### Example 1. The sample correlation coefficient.

Let $(X_1,Y_1),\ldots,(X_n,Y_n)$ be a random sample from a bivariate distribution, and define

$$U_{1,n} = \tfrac{1}{2}\sum_{1\le i<j\le n}(X_i-X_j)(Y_i-Y_j),$$

227

$$U_{2,n} = \tfrac{1}{2} \sum_{1 \le i < j \le n} (X_i - X_j)^2,$$

and

$$U_{3,n} = \tfrac{1}{2} \sum_{1 \le i < j \le n} (Y_i - Y_j)^2.$$

If $Var\, X_1 > 0$ and $Var\, Y_1 > 0$, the function $g(u_1, u_2, u_3) = u_1/(u_2 u_3)^{\frac{1}{2}}$ has continuous second order partial derivatives at the point $(\sigma_{XY}, \sigma_X^2, \sigma_Y^2)$, where $\sigma_{XY}$ denotes $Cov(X_1, Y_1) = EU_{1,n}$, $\sigma_X^2$ denotes $Var\, X_1 = EU_{2,n}$ and $\sigma_Y^2$ denotes $Var\, Y_1 = EU_{3,n}$.

Clearly the r.v. $g(U_{1,n}, U_{2,n}, U_{3,n})$ is just the sample correlation coefficient, and $\rho = g(\sigma_{XY}, \sigma_X^2, \sigma_Y^2)$ is the population correlation. We have

$$n^{\frac{1}{2}}(\hat{g} - \rho) \xrightarrow{\;\mathcal{D}\;} N(0, \sigma^2)$$

where the asymptotic variance $\sigma^2$ is given by

$$\sigma^2 = \frac{\rho^2}{4} \left\{ \frac{\mu_{4,0}}{\sigma_X^2} + \frac{\mu_{0,4}}{\sigma_Y^2} + \frac{2\mu_{2,2}}{\sigma_X^2 \sigma_Y^2} + \frac{4\mu_{2,2}}{\sigma_{XY}^2} - \frac{4\mu_{3,1}}{\sigma_{XY} \sigma_x^2} - \frac{4\mu_{1,3}}{\sigma_{XY} \sigma_Y^2} \right\}$$

where $\mu_{i,j} = E(X_1 - \mu_X)^i (Y_1 - \mu_Y)^j$.

Due to the consistency of $\widehat{Var}(JACK)$, we also have

$$\frac{n^{\frac{1}{2}}(\hat{g} - \rho)}{(\widehat{Var}(JACK))^{\frac{1}{2}}} \xrightarrow{\;\mathcal{D}\;} N(0, 1).$$

For a simulation experiment on jackknifing the correlation coefficient, see Efron (1981).

**Example 2. The one-way random effects model.**

The one way random effects model is

$$Y_{ij} = \mu + a_i + \xi_{ij} \qquad i = 1, 2, \ldots, I; \quad j = 1, 2, \ldots, J;$$

where $a_1, \ldots, a_I$ are independent identically distributed random variables with variance $\sigma_a^2$, and the $\xi_{ij}$ are also i.i.d., independent of the $a's$ and have variance $\sigma^2$. The quantity $\sigma_a^2$ is often of interest.

228

Define $V_i = \overline{Y}_i$, the mean of the $i$th sample (so that $V_1, \ldots, V_I$ are i.i.d. with mean $\mu$, variance $\sigma_a^2 + \sigma^2/J$), and $W_i = (J-1)^{-1} \sum_{j=1}^{J} (Y_{ij} - \overline{Y}_{i.})^2$, (so that the $W_i$ are the (i.i.d.) sample variances of the $I$ groups). Finally let $X_i = (V_i, W_i)$ and define two $U$-statistics by $U_{1,I} = \frac{1}{2} \binom{I}{2}^{-1} \sum_{1 \le i < j \le I} (V_i - V_j)^2$ and $U_{2,I} = I^{-1} \sum_{i=1}^{I} W_i$. Let $g(u_1, u_2)$ be the function $g(u_1, u_2) = u_1 - u_2/J$. The statistic $U_{1,I}$ estimates $\theta_1 = \sigma_a^2 + \sigma^2/J$ and $U_{2,I}$ estimates $\theta_2 = \sigma^2$, so that the jackknifed version $\hat{g}$ of $g(U_{1,I}, U_{2,I})$ estimates $\sigma_a^2$. As $I \to \infty$ with $J$ held fixed,

$$I^{\frac{1}{2}}(\hat{g} - \sigma_a^2)/(\widehat{Var}(JACK))^{\frac{1}{2}} \xrightarrow{\mathcal{D}} N(0,1) \tag{1}$$

and

$$\lim I\widehat{Var}(JACK) = 4\xi_1(1,1) - 4\xi_1(1,2)/J + \xi_1(2,2)/J^2$$

where

$$\xi_1(1,1) = \tfrac{1}{4}(\mu_4 - (\sigma_a + \sigma^2/J)^2), \quad \mu_4 = E(V_1 - \mu)^4,$$
$$\xi_1(1,2) = \tfrac{1}{2}(E\{(V_1, -V_2)^2 W_2\} - (\sigma_a + \sigma^2/J)\sigma^2)$$

and

$$\xi_1(2,2) = Var\, W_1.$$

In a similar manner, we can estimate the ratio $\lambda = \sigma_a^2/\sigma^2$ which in this case is of the form $g(\theta_1, \theta_2) = (\theta_1/\theta_2) - J^{-1}$. Provided $\sigma^2 > 0$, (1) once again holds true and

$$\lim_{I \to \infty} I\widehat{Var}(JACK) = \left(4\xi_1(1,1) - 4(\lambda + J^{-1})\xi_1(1,2) + (\lambda + J^{-1})^2 \xi_1(2,2)\right)/\sigma^4.$$

### 5.1.4 Additional results

Sen (1977b) proves the strong consistency of the variance estimator for both functions of $U$-statistics and linear combinations of $U$-statistics that arise in the case of $V$-statistics (see Section 4.2). Also proved in this reference are weak and strong invariance principles for the jackknifed statistic.

Several authors have considered the modifications that must be made to the standard asymptotic results when the $U$-statistic is "Studentised" i.e. when it is standardised by the jackknife estimate of variance, rather than a known asymptotic variance, as in Chapter 3. Cheng (1981) and Helmers (1985) consider Berry-Esseen rates. Vandemaele and Veraverbeke (1985) deal with large deviations.

Krewski (1978) extends Arvesen's results to the case where the basic r.v.s are derived from sampling without replacement from a finite population. Majumdar and Sen (1978) extend the results of Sen (1977b) to the finite population sampling case.

## 5.2 Bootstrapping $U$-statistics

As an alternative to the jackknife, we may use the *bootstrap* technique for the estimation of various characteristics such as standard errors, bias or percentage points of the sampling distributions of $U$-statistics. We sketch below the basic ideas behind this interesting technique, which is due to Efron. For more detail, we refer the reader to Efron (1982).

Consider a functional $\theta(F)$ and an estimator $\hat{\theta}_n$ of $\theta(F)$ based on an i.i.d. sample $X_1, \ldots, X_n$ distributed as $F$. We may use a computer to draw a *bootstrap sample* $X_1^*, \ldots, X_m^*$ which is independently and identically distributed as $F_n$, the empirical distribution function of the original sample:

$$F_n(x) = n^{-1} \sum_{i=1}^{n} I\{X_i \leq x\}.$$

Suppose interest centres on some numerical characteristic of the sampling distribution of $\hat{\theta}_n$ such as its mean, bias or standard deviation, which we denote by $T(n, F)$. By simulating bootstrap samples or direct calculation we can evaluate this characteristic in the case of samples of size $m$ drawn from the distribution $F_n$; that is, we can calculate $T(m, F_n)$. The basic idea of the bootstrap is that $T(n, F_n)$ should approximate $T(n, F)$.

### Example 1. The sample mean.

We begin by considering bootstrap estimates of functionals relating to the sample mean. For the variance of the sample mean, we have $T(n, F) = Var(\bar{X}_n) = \sigma^2/n$ so that $T(n, F_n) = Var(X_1^*)/n = n^{-2} \sum_{i=1}^{n} (X_i - \bar{X}_n)^2$. Alternatively, we may be interested in the distribution of $(\bar{X}_n - \mu)/\sigma$ so that $T(n, F)$ might be $Pr\left(\sqrt{n}(\bar{X}_n - \mu)/\sigma \leq x\right)$. In this case, the conventional estimate of $T(n, F)$ is $\Psi(x)$, while the bootstrapped estimate is $T(n, F_n) = Pr\left(\sqrt{n}(\bar{X}_n^* - \bar{X}_n)/s_n \leq x\right)$ where $s_n^2 = VarX_n^*$. There is theoretical evidence to suggest that the latter estimate is better than the former.

One way of examining the comparative merits of these approximations is via Edgeworth expansions. In the case of the sample mean, we have (see e.g. Feller (1966 p.512))

$$Pr\left(\frac{\sqrt{n}(\bar{X}_n - \mu)}{\sigma} \leq x\right) = \Phi(x) + \frac{\mu_3(1 - x^2)}{\sigma^3 n^{\frac{1}{2}}}\phi(x) + o(n^{-\frac{1}{2}})$$

where $\Phi$ and $\phi$ are respectively the distribution function and the density of the standard normal distribution and $\mu_3 = E(X_1 - \mu)^3$. But Singh (1981) shows that

$$Pr\left(\frac{\sqrt{n}(\bar{X}_n^* - \bar{X}_n)}{s_n} \leq x\right) = \Phi(x) + \frac{\mu_3(1 - x^2)}{\sigma^3 n^{\frac{1}{2}}}\phi(x) + o(n^{-\frac{1}{2}})$$

and hence the difference

$$Pr\left(\sqrt{n}\frac{(\bar{X}_n - \mu)}{\sigma} \leq x\right) - Pr\left(\sqrt{n}\frac{(\bar{X}_n^* - \bar{X}_n)}{S_n} \leq x\right)$$

is $o(n^{-\frac{1}{2}})$, while the difference between $Pr\left(\sqrt{n}(\bar{X}_n - \mu)/\sigma \leq x\right)$ and $\Phi(x)$ is $O(n^{-\frac{1}{2}})$.

A similar question arises for $U$-statistics, if we consider the problem of estimating

$$Pr\left(\sqrt{n}(U_n - \theta)/k\sigma_1^2 \leq x\right) = G_n(x, F)$$

say. A reasonable conjecture is that the bootstrapped version of this probability, say $G_n(x, F_n)$, is a better estimate of $G_n(x, F)$ than $\Phi(x)$. A similar theorem has been proved by Beran (1984). Athreya et.al. (1984) consider laws of large numbers for $U$-statistics calculated from bootstrapped samples.

The situation is a little simpler if our aim is merely to estimate the standard error of $U_n$ via bootstrapping. For $c = 1, 2, \ldots, k$, denote the quantities $E\psi_c^2(X_1, \ldots, X_c)$ by $\xi_c$, and let $\xi_0 = \theta^2$. Then from Chapter 2 we have

$$Var\, U_n = \sum_{c=1}^{k}\binom{n}{k}^{-1}\binom{k}{c}\binom{n-k}{k-c}\sigma_c^2$$

231

$$= \sum_{c=1}^{k} \binom{n}{k}^{-1} \binom{k}{c} \binom{n-k}{k-c} (\xi_c - \xi_0)$$

$$= \sum_{c=0}^{k} \binom{n}{k}^{-1} \binom{k}{c} \binom{n-k}{k-c} \xi_c - \xi_0. \tag{1}$$

The quantities $\xi_c$ are regular statistical functionals of degree $2k - c$ and are given by

$$\xi_c(F) = \int \Psi^{(c)}(x_1, \ldots, x_{2k-c}) \prod_{i=1}^{2k-c} dF(x_i)$$

where the kernels $\Psi^{(c)}$ introduced in the last section are symmetrised versions of the functions $\psi(x_1, \ldots x_k)\psi(x_{k-c+1}, \ldots, x_{2k-c})$. Thus we can write

$$Var\, U_n = \sum_{c=0}^{k} \binom{n}{k}^{-1} \binom{k}{c} \binom{n-k}{k-c} \xi_c(F) - \xi_0(F)$$

$$= \sigma^2(n, F)$$

say. Thus in the case of estimating the variance, the bootstrap estimate is $\sigma^2(n, F_n)$ and so requires no simulations for its calculation, since we can compute the statistics $\xi_c(F_n)$ directly from the data. In fact the statistics $\xi_c(F_n)$ are just $V$-statistics:

$$\xi_c(F_n) = n^{-(2k-c)} \sum_{i_1=1}^{n} \cdots \sum_{i_{2k-c}=1}^{n} \Psi^{(c)}(X_{i_1}, \ldots, X_{i_{2k-c}})$$

$$= V_n^{(c)}$$

say. The bootstrap estimate of variance is just a linear combination of $V$-statistics:

$$\widehat{Var}(BOOT) = \sum_{c=0}^{k} \binom{n}{k}^{-1} \binom{k}{c} \binom{n-k}{k-c} V_n^{(c)} - V_n^{(0)}. \tag{2}$$

The sampling distribution of $\widehat{Var}(BOOT)$ can be simply studied via the representation (2). It is convenient to represent the $V$-statistics in terms of $U$-statistics in the manner of Theorem 1 of Section 4.2. Let $U_n^{(c,j)}, j = 1, 2, \ldots, 2k - c$ be $U$-statistics based on kernels $\Psi^{(c,j)}$ derived from the symmetric kernels $\Psi^{(c)}$ as in Theorem 1 of Section 4.2. Then

$$V_n^{(c)} = U_n^{(c,2k-c)} + \binom{2k-c}{2} \left\{ U_n^{(c,2k-c-1)} - U_n^{(c,2k-c)} \right\} n^{-1} + o(n^{-1})$$

almost surely, and

$$EV_n^{(c)} = \theta^{(c,2k-c)} + \binom{2k-c}{2}\{\theta^{(c,2k-c-1)} - \theta^{(c,2k-c)}\}n^{-1} + o(n^{-1})$$

where $\theta^{(c,j)} = EU_n^{(c,j)} = E\Psi^{(c,j)}$. From Section 4.2 we have

$$\Psi^{(c,2k-c)}(x_1,\ldots,x_{2k-c}) = \Psi^{(c)}(x_1,\ldots,x_{2k-c})$$

and

$$\Psi^{(c,2k-c-1)}(x_1,\ldots,x_{2k-c-1}) = (2k-c-1)^{-1}\{\Psi^{(c)}(x_1,x_1,\ldots,x_{2k-c-1})$$
$$+ \ldots + \Psi^{(c)}(x_1,x_2,\ldots,x_{2k-c-1},x_{2k-c-1}\}$$

so that $\theta^{(c,2k-c)} = \xi_c$ and $\theta^{(c,2k-c-1)} = E\Psi^{(c)}(X_1,X_1,\ldots X_{2k-c-1})$. Hence

$$EV_n^{(c)} - \xi_c = \binom{2k-c}{2}\{\theta^{(c,2k-c-1)} - \theta^{(c,2k-c)}\}n^{-1} + o(n^{-1}) \qquad (3)$$

and the bias incurred in estimating $\xi_c$ with $V_n^{(c)}$ is thus $O(n^{-1})$. Picking out the significant terms of (2), and noting that $EV_n^{(c)}$ is bounded, we get

$$E\widehat{Var}(BOOT) - Var\, U_n = \left\{\binom{n}{k}^{-1}\binom{n-k}{k} - 1\right\}(EV_n^{(0)} - \xi_0)$$
$$+ k\binom{n}{k}^{-1}\binom{n-k}{k-1}(EV_n^{(1)} - \xi_1) + o(n^{-2})$$

which, using (3) is

$$\left(k^2\binom{2k-1}{2}\{\theta^{(1,2k-2)} - \xi_1\} - k^2\binom{2k}{2}\{\theta^{(0,2k-1)} - \xi_0\}\right)n^{-2} + o(n^{-2})$$

and so finally we get

$$BIAS(\widehat{Var}(BOOT)) = \frac{k^2}{2n^2}\left\{(2k-1)(2k-2)(\theta^{(1,2k-2)} - \xi_1)\right.$$
$$\left. - 2k(2k-1)(\theta^{(0,2k-1)} - \xi_0)\right\} + o(n^{-2}).$$

In terms of the original kernel $\psi$,

$$\theta^{(0,2k-1)} = \frac{1}{2k-1}\{\xi_1 + 2(k-1)\theta E\psi(X_1,X_1,X_2,\ldots,X_{k-1})\}$$

and

$$\theta^{(1,2k-2)} = \frac{1}{2k-1}\{(k-2)E(\psi(X_1,X_1,\ldots,X_{k-1})\psi(X_{k-1},X_k,\ldots,X_{2k-2}))$$
$$+E(\psi(X_1,X_1,X_2,\ldots,X_{k-1})\psi(X_1,X_k,\ldots,X_{2k-2})) + \tfrac{1}{2}(k-1)\xi_1\}.$$

## Example 1. Kernels of degree 2.

For kernels of degree 2, the formulae above take the form

$$\theta^{(0,3)} = \tfrac{1}{3}\{2\xi_1 + \theta E\psi(X_1,X_1)\},$$

$$\theta^{(1,2)} = \tfrac{1}{3}\{2E\psi(X_1,X_1)\psi(X_1,X_2) + \xi_2\}$$

and

$$BIAS(\widehat{Var}(BOOT)) = 12\{(\theta^{(1,2)} - \xi_1) - 2(\theta^{(0,3)} - \theta^2)\}n^{-2} + o(n^{-2}).$$

## Example 2. Sample variance.

In this case $\psi(x,x) = 0$ and so $\theta^{(0,3)} = \tfrac{2}{3}\xi_1$, $\theta^{(1,2)} = \tfrac{1}{3}\xi_2$ and

$$BIAS(\widehat{Var}(BOOT)) = 4(\sigma_2^2 - 7\sigma_1^2)n^{-2} + o(n^{-2})$$
$$= (9\sigma^4 - 5\mu_4)n^{-2} + o(n^{-2}).$$

A similar approach enables as to compute the asymptotic expansion of the quantity $Var(\widehat{Var}(BOOT))$: we can write (2) as

$$\frac{k^2}{n}(V_n^{(1)} - V_n^{(0)}) + R_n$$

where $Var\,R_n = o(n^{-3})$ and also write $Var\,V_n^{(c)} = Var\,U_n^{(c,2k-c)} + o(n^{-1})$. Combining these, we can write the variance of (2) as

$$Var(\widehat{Var}(BOOT)) = Var\{k^2 n^{-1}(U_n^{(0,2k-1)} - U_n^{(0,2k)})\} + o(n^{-3})$$
$$= k^4\{4k^2\sigma_1(0,0) - 4k(2k-1)\sigma_1(0,1)$$
$$+ (2k-1)^2\sigma_1(1,1)\}n^{-3} + o(n^{-3})$$

234

where the quantities $\sigma_1(i, j)$ are those appearing in (10) of Section 5.1.1. Inspection of the last equation and (10) of Section 5.1.1 reveals that up to order $n^{-3}$, the variances of the bootstrap and jackknife estimates of $Var\, U_n$ are identical. Since the bias terms, although different for the two estimators, are of order $n^{-2}$, the mean squared errors of the two estimates are also equal, up to order $n^{-3}$.

Another possible estimate suggested by equation (2) is to replace the $V$-statistics with the corresponding $U$-statistics $U_n^{(c, 2k-c)}$ having kernels $\Psi^{(c)}$. The result is

$$\widehat{Var}(UBOOT) = \sum_{c=0}^{k} \binom{n}{k}^{-1} \binom{k}{c} \binom{n-k}{k-c} U_n^{(c, 2k-c)} - U_n^{(0, 2k)} \quad (4)$$

which is unbiased since $EU_n^{(c, 2k-c)} = \xi_c$ for $c = 0, 1, 2, \ldots k$. A more convienient expression for $\widehat{Var}(UBOOT)$ is derived from (4):

$$\widehat{Var}(UBOOT) = \binom{n}{k}^{-1} \sum_{c=0}^{k} \binom{k}{c} \binom{n-k}{k-c} N_c^{-1} S_c - U_n^{(0, 2k)}$$

$$= \binom{n}{k}^{-2} \sum_{c=0}^{k} S_c - U_n^{(0, 2k)}$$

$$= U_n^2 - U_n^{(0, 2k)}$$

using the notation of Section 5.1.1. Since $U_n^{(c, 2k-c)}$ and $V_n^{(c)}$ are asymptotically equivalent, the variance of $\widehat{Var}(UBOOT)$ coincides with that of $\widehat{Var}(BOOT)$ up to order $n^{-3}$.

Since the square of the bias of both $\widehat{Var}(JACK)$ and $\widehat{Var}(BOOT)$ are $O(n^{-4})$ and the variances of these estimators are equal up to $O(n^{-3})$, a comparison of the two estimators on the basis of their mean squared errors will involve computing the $n^{-4}$ term in the asymptotic expansion of the variances of the estimators. This is feasible if $k = 2$ and we assume that $\psi(x, x) = 0$:

Using Theorem 1 of Section 4.2 we get

$$n^4 V_n^{(0)} = nU_n^{(0,1)} + 14 \binom{n}{2} U_n^{(0,2)} + 36 \binom{n}{2} U_n^{(0,2)} + 24 \binom{n}{4} U_n^{(0,4)}$$

$$n^3 V_n^{(1)} = nU_n^{(1,1)} + 6 \binom{n}{2} U_n^{(1,2)} + 6 \binom{n}{3} U_n^{(1,3)} \quad (5)$$

and

$$n^2 V_n^{(2)} = n U_n^{(2,1)} + 2 \binom{n}{2} U_n^{(2)}.$$

Denoting $U_n^{(c,2k-c)}$ by $U_n^{(c)}$ and using the formulae for the kernels $\psi^{(c,j)}$ given in Chapter 4 and the fact that $\psi(x, x) = 0$, we obtain from (5)

$$n^4 V_n^{(0)} = 24 \binom{n}{4} U_n^{(0)} + 24 \binom{n}{3} U_n^{(1)} + 4 \binom{n}{2} U_n^{(2)}$$

$$n^3 V_n^{(1)} = 6 \binom{n}{3} U_n^{(1)} + 2 \binom{n}{2} U_n^{(2)}$$

and

$$n^2 V_n^{(2)} = 2 \binom{n}{2} U_n^{(2)}.$$

The $U$-statistics $U_n^{(0)}, U_n^{(1)}$ and $U_n^{(2)}$ have kernels $\psi^{(0)}, \psi^{(1)}, \psi^{(2)}$ given by

$$\psi^{(0)}(x_1, x_2, x_3, x_4) = \tfrac{1}{3} \{ \psi(x_1, x_2)\psi(x_3, x_4) + \psi(x_1, x_3)\psi(x_2, x_4)$$
$$+ \psi(x_1, x_4)\psi(x_2, x_3) \},$$

$$\psi^{(1)}(x_1, x_2, x_3) = \tfrac{1}{3} \{ \psi(x_1, x_2)\psi(x_2, x_3) + \psi(x_2, x_3)\psi(x_3, x_4)$$
$$+ \psi(x_3, x_1)\psi(x_1, x_2) \}$$

and

$$\psi^{(2)}(x_1, x_2) = \psi(x_1, x_2).$$

Substituting these in the formula for $\widehat{Var}(BOOT)$, we get

$$\widehat{Var}(BOOT) = 4(U_n^{(1)} - U_n^{(0)})n^{-1} + (26U_n^{(0)} - 32U_n^{(1)} + 6U_n^{(2)})n^{-2} + o(n^{-2}).$$

From Theorem 1 of Section 5.1.1 and the relationship $U_n^{(c)} = N_c^{-1} S_c$ we get a similar expression for $\widehat{Var}(JACK)$:

$$\widehat{Var}(JACK) = 4(U_n^{(1)} - U_n^{(0)})n^{-1} + 4(U_n^{(0)} - 2U_n^{(1)} + U_n^{(2)})n^{-2} + o(n^{-2}).$$

Now define

$$J_1 = 4(U_n^{(1)} - U_n^{(0)}),$$
$$J_2 = 4(U_n^{(0)} - 2U_n^{(1)} + U_n^{(2)}),$$

and

236

$$B_2 = (26U_n^{(0)} - 32U_n^{(1)} + 6U_n^{(2)}).$$

Then the mean squared errors of these two estimators of $Var\,U_n$ are

$$MSE(\widehat{Var}(JACK)) = n^{-2}Var\,J_1 + 2n^{-3}Cov(J_1, J_2)$$
$$+ (BIAS(JACK))^2 + o(n^{-4})\ (6)$$

and

$$MSE(\widehat{Var}(BOOT)) = n^{-2}Var\,J_1 + 2n^{-3}Cov(J_1, B_2)$$
$$+ (BIAS(BOOT))^2 + o(n^{-4})\ (7)$$

A simple but tedious calculation yields

$$Var\,J_1 = \{256\sigma_1(0,0) - 384\sigma_1(0,1) + 144\sigma_1(1,1)\}n^{-1}$$
$$+ \{1152\delta_2(0,0) - 1152\delta_2(0,1) + 288\delta_2(1,1)\}n^{-2} + o(n^{-2}),$$
$$Cov(J_1, J_2) = \{576\sigma_1(0,1) - 288\sigma_1(1,1) + 96\sigma_1(1,2)$$
$$- 256\sigma_1(0,0) - 128\sigma_1(0,2)\}n^{-1} + o(n^{-1})$$

and

$$Cov(J_1, B_2) = \{2784\sigma_1(0,1) - 1152\sigma_1(1,1) + 144\sigma_1(1,2)$$
$$- 1664\sigma_1(0,0) + 192\sigma_1(0,2)\}n^{-1} + o(n^{-1})$$

where $\sigma_c(a,b) = Cov(\psi^{(a)}(S), \psi^{(b)}(T))$, $|S \cap T| = c$ and $\delta_2(a,b) = \sigma_2(a,b) - 2\sigma_1(a,b)$. Substituting these results into (6) and (7), and using the formulae for the biases developed earlier, yields

$$MSE(\widehat{Var}(JACK)) = \{256\sigma_1(0,0) - 384\sigma_1(0,1) + 144\sigma_1(1,1)\}n^{-3}$$
$$+ \{1152\delta_2(0,0) - 1152\delta_2(0,1) + 288\delta_2(1,1)$$
$$+ 1152\sigma_1(0,1) - 576\sigma_1(1,1) + 192\sigma_1(1,2)$$
$$- 512\sigma_1(0,0) - 256\sigma_1(0,2) + 4(\sigma_2^2 - 2\sigma_1^2)^2\}n^{-4}$$
$$+ o(n^{-4})$$

and

$$MSE(\widehat{Var}(BOOT)) = \{256\sigma_1(0,0) - 384\sigma_1(0,1) + 144\sigma_1(1,1)\}n^{-3}$$
$$+ \{1152\delta_2(0,0) - 1152\delta_2(0,1) + 288\delta_2(1,1)$$
$$+ 5568\sigma_1(0,1) - 2304\sigma_1(1,1) + 288\sigma_1(1,2)$$
$$- 3328\sigma_1(0,0) + 384\sigma_1(0,2)\} + 16(\sigma_2^2 - 7\sigma_1^2)\}n^{-4}$$
$$+ o(n^{-4}).$$

A similar calculation yields

$$MSE(\widehat{Var}(UBOOT)) = \{256\sigma_1(0,0) - 384\sigma_1(0,1) + 144\sigma_1(1,1)\}n^{-3}$$
$$+ \{11526_2(0,0) - 11526_2(0,1) - 288\delta_2(1,1)$$
$$+ 576\sigma_1(0,1) - 288\sigma_1(1,1) + 96\sigma_1(1,2)$$
$$- 256\sigma_1(0,0) - 128\sigma_1(0,2)\}n^{-4} + o(n^{-4}).$$

## 5.3 Variance estimation for incomplete $U$-statistics

In the final section of this chapter we give a brief treatment of variance estimation for incomplete $U$-statistics. We begin by considering balanced incomplete statistics. Note that, for the remainder of this section only, we denote an incomplete $U$-statistic by $U_N^*$ rather than $U_n^{(0)}$ as is done in Chapter 4. This is done to avoid a clash of notation with the statistics $U_n^{(0)}, \ldots, U_n^{(k)}$ introduced in Section 5.1.

### 5.3.1 The balanced case

Let $U_n^*$ be an incomplete $U$-statistic based on some balanced design $\mathcal{D}$ not necessarily having minimum variance, and let $\psi(S_j) = \psi(X_{i_1}, \ldots, X_{i_k})$ where $S_j = \{i_1, \ldots, i_k\}$ is the $j$th set in the design. Suppose each index is contained in $r$ sets of $\mathcal{D}$, and as in Chapter 4, denote the incidence matrix of the design by $N$.

### The jackknife estimator

Define $U_n^*(-i) = (m-r)^{-1} \sum^{(-i)} \psi(S_j)$ where the sum $\sum^{(-i)}$ is taken over all $(m-r)$ sets in the design which do not contain $i$. Then $U_n(-i) = (m-r)^{-1} \sum_{j=1}^m (1 - n_{i,j})\psi(S_j)$ and $n^{-1} \sum_{i=0} U_n(-i) = U_n^*$, and so

$$U_n^*(-i) - U_n^* = m^{-1}(m-r)^{-1} \sum_{j=1}^m (r - mn_{ij})\psi(S_j). \tag{1}$$

Let $S_c^* = \sum \psi(S_j)\psi(S_{j'})$, where the sum extends over all $f_c$ pairs of sets in the design having $c$ elements in common. $S_c^*$ is an unbiased estimator of $f_c(\sigma_c^2 + \theta^2)$.

238

The jackknife estimate of $Var\,U_n^*$ is

$$\widehat{Var}(JACK) = n^{-1}(n-1)\sum_{i=0}^{n-1}(U_n^*(-i) - U_n^*)^2$$

which by using (1) and the fact that $mk = nr$, can be written

$$m^{-2}(n-k)^{-2}(n-1)\sum_{c=0}^{k}(cn-k^2)S_c^*.$$

Thus

$$E\widehat{Var}(JACK) = m^{-2}(n-k)^{-2}(n-1)\sum_{c=0}^{k}(cn-k^2)f_c(\sigma_c^2+\theta^2)$$

$$= m^{-2}(n-k)^{-2}(n-1)\sum_{c=1}^{k}(cn-k^2)f_c\sigma_c^2 \qquad (4)$$

and the bias of the jackknife estimate of variance is, from (4) and Theorem 2 of Section 4.3.1

$$BIAS = m^{-2}\sum_{c=1}^{k}\left\{(n-1)(cn-k^2)(n-k)^{-2}-1\right\}f_c\sigma_c^2$$

$$= nm^{-2}(n-k)^{-2}\sum_{c=1}^{k}\left\{(n-1)(c-1)-(k-1)^2\right\}f_c\sigma_c^2.$$

Using the identities (8) of Section 2.1.6 and (7) of Section 4.3.1 we can express the bias as

$$BIAS = nm^{-2}(n-k)^{-2}$$

$$\sum_{c=1}^{k}\left[(n-1)(c+1)B_{c+1} + \{(n-1)(c-1)-(k-1)^2\}B_c\right]\delta_c^2$$

where the quantities $B_c, c = 1, \ldots, k$ are those defined in Theorem 3 of Section 4.3.1, and $B_{k+1} = 0$.

A crude condition sufficient for the bias to be positive that is adequate for most practical applications is $n \geq (k-1)^2 - 1$. Since the $B_c$'s are positive, this guarantees that all terms in the above sum are positive, except for the first.

The assumption of balance implies that $B_1 = nr^2$ and since $B_2$ is the sum of squares of $\binom{n}{2}$ positive terms whose sum is $m\binom{k}{2}$ then necessarily $B_2 \geq m^2\binom{k}{2}^2/\binom{n}{2}$. Thus we must have $2(n-1)B_2 - (k-1)^2 B_1 \geq 0$ and the first term is also positive.

Further, if the design now has the zero-one property of the Corollary to Theorem 2 of Section 4.3.2 (the off diagonal elements of $NN^T$ zero or one) then $B_c = m\binom{k}{c}$ for $c = 2, 3, \ldots, k$, and the parameters satisfy $n \geq 1 + r(k-1)$. Then for $c = 2, 3, \ldots, k$

$$(n-1)(c+1)B_{c+1} + \{(n-1)(c-1) - (k-1)^2\}B_c = m\binom{k}{c}\{(k-1)(n-k)\} \geq 0,$$

and

$$2(n-1)B_2 - (k-1)^2 B_1 = (n-1) - r(k-1) \geq 0.$$

Thus in the special case of these zero-one designs, the bias is positive for all values of $n \geq k+1$.

For the zero-one designs, the only non-zero $f_c's$ in the expression (4) are $f_1 = mk(r-1)$ and $f_k = m$. The bias is given in terms of the $\sigma_c^2$ by

$$BIAS = nm^{-2}(n-k)^{-2}\left[-(k-1)^2 mk(r-1)\sigma_1^2 + m(k-1)(n-k)\sigma_k^2\right]$$

which is asymptotically like

$$-k^2(k-1)n^{-2}[(r-1)/r]\sigma_1^2 + k(k-1)r^{-1}n^{-1}\sigma_k^2 + o(n^{-1}).$$

For the case where $\lim mn^{-1}$ is finite, i.e. $r$ is bounded as $m, n$ increase, the bias is $O(n^{-1}) = O(m^{-1})$ as is $Var\, U_n^*$, so the jackknife will be unsatisfactory in this case. On the other hand, if $\lim mn^{-1}$ is infinite, and $r$ increases without bound as $n, m$ increase then the bias is $O(m^{-1})$, compared to the variance which is $O(n^{-1})$, and the jackknife will be a more attractive estimator. Thus the jackknife should be used only when the ratio $mn^{-1}$ is large.

**Example 1.**

(c.f. Example 1 of Section 4.3.2). Take $k = 2, m = n$ and any balanced design. Then $r = 2$ and $Var\, U_n^* = n^{-1}(2\sigma_1^2 + \sigma_2^2)$. The bias of the jackknife estimate of $Var\, U_n^*$ is $(n-2)^{-2}\{(n-2)\sigma_2^2 - \sigma_1^2\}$.

**Example 2.**

(c.f. Example 5 of Section 4.3.2). Take $k = 3, n = (6t+1), m = t(6t+1)$ for a positive integer $t$. A minimum variance design is one based on Steiner triple systems. We have $r = 3t, mn^{-1} = t$ and $Var U_n^* = m^{-1}(9t\sigma_1^2 + \sigma_3^2) \sim 3t^{-1}\sigma_1^2/2$. The bias of the jackknife is $t^{-2}(\sigma_3^2/3 - \sigma_1^2)$.

**A bootstrap estimate**

Consider once again balanced designs not necessarily of minimum variance. The conventional bootstrap estimate of $Var U_n^*$ would be

$$\widehat{Var}(BOOT) = m^{-2} \sum_{c=1}^{k} f_c \tilde{\sigma}_c^2,$$

where the quantities $\tilde{\sigma}_c^2$ are the $\sigma_c^2$ evaluated as though the empirical d.f. $F_n$ of the sample was the true d.f. of $X_1, \ldots, X_n$: i.e.

$$\tilde{\sigma}_c^2 = n^{-(2k-c)} \sum_{i_1=0}^{n-1} \cdots \sum_{i_{2k-c}=0}^{n-1} \psi(X_{i_1}, \ldots, X_{i_k}) \psi(X_{i_{k-c+1}} \cdots X_{i_{2k-c}})$$

$$- \left( n^{-k} \sum_{i_1=0}^{n-1} \cdots \sum_{i_k=0}^{n-1} \psi(X_{i_1}, \ldots, X_{i_k}) \right)^2.$$

Since incomplete $U$-statistics are used on the grounds of computational parsimony, presumably expressions of the form $\tilde{\sigma}_c^2$ will be deemed too computationally expensive. Accordingly, we may seek estimates of $\sigma_c^2$ based solely on the $m$ quantities $\psi(S_j)$ already calculated. An obvious suggestion is

$$\hat{\sigma}_c^2 = f_c^{-1} S_c^* - f_0^{-1} S_0^*$$

which is an unbiased estimate of $\sigma_c^2$. We are then led to consider an "unbiased bootstrap" of the form

$$\widehat{Var}(UBOOT) = m^{-2} \sum_{c=1}^{k} f_c \hat{\sigma}_c^2$$

$$= m^{-2} \left( \sum_{c=1}^{k} S_c^* - f_0^{-1} S_0^* (m^2 - f_0) \right)$$

$$= (U_n^*)^2 - f_0^{-1} S_0^*.$$

241

Consider a kernel $\psi^*(x_1, \ldots, x_k)$ defined by

$$\psi^*(x_1, \ldots, x_k) = \psi(x_1, \ldots, x_k) - \theta.$$

Then the unbiased bootstrap estimate $\widehat{Var}(UBOOT)$ is the same calculated for $\psi^*$ and for $\psi$, so without loss of generality we may assume that $\theta = 0$.

To investigate the second order properties of the estimate, suppose for ease of calculation that $k = 2$ and $n = m$.

Let the $n$ sets of the design be $\{n, 1\}$, $\{1, 2\}, \ldots, \{n - 1, n\}$, and let $T_j = \psi(X_{j-1}, X_j)$, $j = 1, \ldots, n$. Define $S^* = S_0^* + S_1^* + S_2^*$; then $S^* = \left( \sum_{j=1}^n T_j \right)^2$, $S_1^* = 2 \sum_{j=1}^n T_j T_{j+1}$ and $S_2^* = \sum_{j=1}^n T_j^2$. The random variables $T_j$ have zero means, and second order moments given by

$$Cov(T_i, T_j) = \begin{cases} \sigma_2^2 & \text{if } i = j, \\ \sigma_1^2 & \text{if } |i - j| = 1 \mod n, \\ 0 & \text{otherwise.} \end{cases}$$

The expected values of $S^*, S_1^*$ and $S_2^*$ are given by

$$E(S^*) = n(2\sigma_1^2 + \sigma_2^2),$$
$$E(S_1^*) = 2n\sigma_1^2,$$
$$E(S_2^*) = n\sigma_2^2,$$

and lengthy but straightforward calculations yield

$$Var(S^*) = O(n^2),$$
$$Var(S_1^*) = 4(C_1 - 5\sigma_1^4)n + O(1),$$
$$Var(S_2^*) = (C_1 - 3\sigma_2^4)n + O(1),$$
$$Cov(S_1^*, S_2^*) = 2(C_{12} - 4\sigma_1^2\sigma_2^2)n + O(1),$$
$$Cov(S^* S_1^*) = 2\sigma_1^4 n^2 + O(n)$$

and

$$Cov(S^*, S_2^*) = (\sigma_1^2\sigma_2^2 + \sigma_2^4)n^2 + O(n),$$

where

$$C_1 = E\{T_1^2 T_2^2\} + 2E\{T_1 T_2^2 T_3\} + E\{T_1 T_2 T_3 T_4\},$$
$$C_2 = E\{T_1^4\} + 2E\{T_1^2 T_2^2\},$$

and

$$C_{12} = E\{T_1 T_2 (T_1^2 + T_2^2)\} + E\{T_1 T_2 T_3 (T_1 + T_2 + T_3)\}.$$

In terms of $S^*, S_1^*, S_2^*$ we can write for the present case

$$\widehat{Var}(UBOOT) = (-3S^* + nS_1^* + nS_2^*)/n^2(n-3)$$

so that from the above results we obtain

$$Var\{\widehat{Var}(UBOOT)\} = (4C_1 + C_2 + 4C_{12} - 32\sigma_1^4 - 22\sigma_1^2\sigma_2^2 - 9\sigma_2^2)n^{-3} + o(n^{-3}).$$

Thus the (mean squared error)$^{\frac{1}{2}}$ of the unbiased bootstrap is $O(n^{-3/2})$, compared to $O(n^{-1})$ for the jackknife, and is thus the bootstrap estimate is to be preferred to the jackknife when the ratio $mn^{-1}$ is small compared to $n$ and $m$.

### 5.3.2 Incomplete $U$-statistics based on random choice

We now turn to the estimator of the standard errors of incomplete $U$-statistics of the type considered in Example 5 of Section 4.3.1 and in Section 4.3.2, those based on random choice of subsets. Specifically, we confine attention to the case where the $m$-sets of the design are chosen with replacement from the $\binom{n}{k}$ available. We write $N = \binom{n}{k}$ and denote by $S_1, \ldots, S_N$ the $N$ subsets of $\{1, 2, \ldots, N\}$. Define $U_n^*(-i)$ now to be the statistic

$$U_n^*(-i) = n(n-k)^{-1}m^{-1}\sum^{(-i)} \psi(S_j)$$

when the sum ranges over all sets chosen that do not contain $i$. Let $\bar{U}_n^* = n^{-1}\sum_{i=1}^n U_n(-i)$. The proposed jackknife estimate is $\widehat{Var}(IJACK)$ (for "incomplete jackknife") given by

$$\widehat{Var}(IJACK) = n^{-1}(n-1)\sum_{i=1}^n (U_n^*(-i) - \bar{U}_n^*)^2.$$

There is a close connection between this estimator and the estimator of the variance of the corresponding complete $U$-statistic discussed in Section 5.1.1. A calculation similar to that of Theorem 1 of Section 5.1.1 shows that

$$\widehat{Var}(IJACK) = (n-1)(n-k)^{-2}m^{-2}\sum_{c=0}^k (nc - k^2)S_c^* \qquad (1)$$

243

where $S_c^*$ is now given by

$$S_c^* = \sum_{|S \cap T| = c} Y(S)Y(T)\psi(S)\psi(T) \qquad (2)$$

the sum being taken over all pairs of subsets $S$ and $T$ of $\{1, 2, \ldots, n\}$ with $c$ elements in common. The quantities $Y(S)$ are the elements of a random $N \left(= \binom{n}{k}\right)$-vector having a multinomial $Mult\left(m, \frac{1}{N}, \frac{1}{N}, \ldots, \frac{1}{N}\right)$ distribution.

From (2) we can write

$$E(S_c^* | X_1, \ldots, X_n) = \sum_{|S \cap T| = c} E(Y(S)Y(T))\psi(S)\psi(T)$$

$$= \begin{cases} \sum_{|S \cap T| = c} \frac{m(m-1)}{N^2}\psi(S)\psi(T) & c = 0, \ldots k-1; \\ \sum_{|S \cap T| = k} \frac{m^2 + m(N-1)}{N^2}\psi(S)\psi(T) & c = k, \end{cases}$$

$$= \begin{cases} \frac{m(m-1)}{N^2} S_c & c = 0, 1, \ldots, k-1; \\ \frac{m(m-1)+mN}{N^2} S_k & c = k; \end{cases}$$

where $S_c = \sum_{|S \cap T| = c} \phi(S)\phi(T)$. Hence

$$E(\widehat{Var}(IJACK)|X_1, \ldots X_n)$$

$$= n-1)(n-k)^{-2}m^{-2}\sum_{c=0}^{k-1}(nc - k^2)\frac{m(m-1)}{N^2}S_c$$

$$+ (n-1)(n-k)^{-2}m^{-2}k(n-k)mN^{-1}S_k$$

$$= m^{-1}(m-1)\widehat{Var}(JACK)$$

$$+ k(n-1)(n-k)^{-1}m^{-1}N^{-1}S_k,$$

and so

$$[E\{\widehat{Var}(IJACK)\}$$

$$= m^{-1}(m-1)E\widehat{Var}(JACK) + k(n-1)(n-k)^{-1}m^{-1}(\sigma_k^2 + \theta^2)$$

where $\widehat{Var}(JACK)$ is the estimate of $Var\, U_n$ discussed in Section 5.1.1. From the formula

$$Var\, U_n^* = m^{-1}(m-1)Var\, U_n + m^{-1}\sigma_k^2$$

244

we obtain

$$BIAS(IJACK) = m^{-1}(m-1)BIAS(JACK)$$
$$+ m^{-1}(n-k)^{-1}\{k(n-1)\theta^2 + n(k-1)\sigma_k^2\}$$

where $BIAS(JACK)$ is the bias of the estimator $\widehat{Var}(JACK)$ of the variance of the complete statistic $U_n$, and is given by formula (8) of Section 5.1.1. We see that the bias of $\widehat{Var}(IJACK)$ is always positive and, up to $O(m^{-1})$, is the same as that of the corresponding complete estimator.

Turning now to the variance of the estimator, we need to calculate the variances and covariances of the quantities $S_c^*$ given by (2). We have for $c = 0, 1, \ldots, k-1$, and writing $\eta_c = \sigma_c^2 + \theta^2$ and $m_{(c)} = m(m-1)\ldots(m-c+1)$

$$Var\, S_c^* =$$
$$\sum_{|S \cap T|=c} \sum_{|S' \cap T'|=c} [\, E\{Y(S)Y(T)Y(S')Y(T')\}E\{\psi(S)\psi(T)\psi(S')\psi(T')\}$$
$$- E\{Y(S)Y(T)\}E\{Y(S')Y(T')\}E\{\psi(S)\psi(T)\}E\{\psi(S')\psi(T')\}\,] \quad (3)$$

Using standard formulae for multinomial joint moments, (3) becomes

$$Var\, S_c^* = m^{(4)}N^{-4} \sum_{|S \cap T|=c} \sum_{|S' \cap T'|=c} \{Cov(\psi(S)\psi(T), \psi(S')\psi(T')) + \eta_c^2\}$$
$$- N_c^2 m^2 (m-1)^2 N^{-4} \eta_c^2 + O(m^3 n^{-2c})$$
$$= m^{(4)}N^{-4}Var\, S_c + O(m^3 n^{-2c}).$$

The order of the remainder is derived by noting that for terms for which $S, T, S'$ and $T'$ are all distinct, $E\{Y(S)Y(T)Y(S')Y(T)\} = m^{(4)}N^{-4}$. The number of terms for which this is not the case is of order $N_c$, and these terms differ from $m^{(4)}N^{-4}$ by $O(m^3 N^{-3})$, so that the remainder is $O(m^3 n^{-2c})$.

Similar arguments yield for $0 \le c, d < k$

$$Cov(S_c^*, S_d^*) = m^{(4)}N^{-4}Cov(S_c, S_d) + O(m^3 n^{-c-d})$$

for $c < k$

$$Cov(S_c^*, S_k^*) = (m^{(4)}N^{-4} + m^{(3)}N^{-3})Cov(S_c, S_k) + O(m^2 n^{-c})$$

and

$$Var\, S_k^* = (m^{(4)}N^{-4} + 2m^{(3)}N^{-3} + m^{(2)}N^{-2})Var\, S_k + O(m).$$

Using these results and (1), we get

$$Var(\widehat{Var}(IJACK)) = Var(\widehat{Var}(JACK)) + O(m^{-1}n^{-2}) \qquad (4)$$

Typically we have $m^{-1}n$ tending to zero, so that in this case the variance of $\widehat{Var}(IJACK)$ is $O(n^{-3})$ as in the complete case, and in fact coincides with $Var(\widehat{Var}(JACK))$ up to $o(n^{-3})$. In the case when $m^{-1}n$ converges to a positive number, the variance differs from that of the complete case by $O(n^{-3})$ rather than $o(n^{-3})$.

We can also consider another variant of the "unbiased bootstrap". From Theorem 4 of Section 4.3.1 we have

$$Var\, U_n^* = m^{-1}\{(m-1)Var\, U_n + \sigma_k^2\}$$

$$= m^{-1}\left\{ (m-1)\left[\sum_{c=0}^{k} \binom{n}{k}^{-1}\binom{k}{c}\binom{n-k}{k-c}(\eta_c - \eta_0)\right] + \eta_k - \eta_0 \right\} \quad (5)$$

and a full bootstrap estimate of $Var\, U_n^*$ would replace $\eta_0, \ldots, \eta_k$ in (5) by the appropriate $V$-statistics. In the present context, this will be computationally too expensive, so the $\eta_c$ need to be replaced by suitable quantities already calculated. Obvious contenders are the quantities $S_c^*$ : their expectations are given by (using (2))

$$ES_c^* = m^{(2)}N^{-2}ES_c$$
$$= m^{(2)}N^{-2}N_c\eta_c$$

for $c = 0, 1, \ldots, k-1$, and

$$ES_k^* = (m^{(2)}N^{-2} + mN^{-1})ES_k$$
$$= m((m-1)N^{-1} + 1)\eta_k.$$

Replacing $\eta_c$ by $(m^{(2)}N^{(2)}N_c)^{-1}S_c^*$ for $c = 0, 1, \ldots, k-1$ and $\eta_k$ by the quantity $m^{-1}\{N^{-1}(m-1)+1\}^{-1}S_k^*$ in (5) yields an "incomplete unbiased bootstrap" given by

$$Var(IUBOOT) = \{m^{-2} - m^{-1}(m-1)^{-1}N_0^{-1}N^2\}S_0^* + m^{-2}\sum_{c=1}^{k} S_c^* \quad (6)$$

$$= (U_n^*)^2 - m^{-1}(m-1)^{-1}N_0^{-1}n^2 S_0^*.$$

246

The variance of this estimator is from (6)

$$
\begin{aligned}
Var(\widehat{Var}(IUBOOT)) &= Var\big[\{m^{-2} - m^{-1}(m-1)^{-1}N_0^{-1}N^2\}S_0^* \\
&\qquad + m^{-2}S_c^*\big] + o(n^{-3}) \\
&= k^4 n^{-2} m^{-4} Var\, S_0^* - 2k^2 n^{-1} m^{-4} Cov_1(S_0^*, S_1^*) \\
&\qquad + m^{-4} Var\, S_1^* + o(n^{-3}) \\
&= k^4 n^{-2} m^{-4}(m^{(4)}N^{-4}Var\, S_0 + O(m^3)) \\
&\qquad - 2k^2 n^{-1} m^{-4}(m^{(4)}N^{-4}Cov(S_0, S_1) + O(m^3 n^{-1})) \\
&\qquad + m^{-4}(m^{(4)}N^{-4}Var\, S_1 + O(m^3 n^{-2})) \\
&= N^{-4}[k^4 n^{-2}Var\, S_0 - 2k^2 n^{-1}Cov(S_0, S_1) + Var\, S_1] \\
&\qquad + O(m^{-1}n^{-2}). \tag{7}
\end{aligned}
$$

Moreover, from Section 5.1.1 we have

$$
Cov(N_c^{-1}S_c, N_d^{-1}S_d) = (2k-c)(2k-d)\sigma_1(c,d)n^{-1} + o(n^{-1}) \tag{8}
$$

so finally substituting (8) into (7) we get

$$
\begin{aligned}
Var(\widehat{Var}(IUBOOT)) &= k^4\big\{4k^2\sigma_1(0,0) - 4k(2k-1)\sigma_1(0,1) \\
&\qquad + (2k-1)^2\sigma_1(1,1)\big\}n^{-3} + O(m^{-1}n^{-2}).
\end{aligned}
$$

Thus, in the case when $m^{-1}n \to 0$, the variance of the "incomplete unbiased bootstrap" is equal to that of the "complete unbiased bootstrap" up to $o(n^{-3})$.

## 5.4 Bibliographic details

The monograph by Efron (1979) is an excellent source of information on bootstrapping and jackknifing, as is his paper with Stein (Efron and Stein (1981)). The details of jackknifing $U$-statistics were first worked out by Arvesen (1969), and most of Section 5.5.1–5.1.3 is derived from his paper. An earlier formulation not couched in jackknife terms is in Sen (1960). The formula for the bias of the jackknife estimate of variance is in Lee (1985), as is some of the material on bootstrapping.

# CHAPTER SIX

# Applications

## 6.1   Introduction

In this final chapter we present several applications of the theory developed so far, focussing on three broad areas. In Section 6.2 we investigate tests and confidence intervals for statistical parameters such as correlation coefficients and symmetry indices, concentrating particularly on correlation coefficients for circular and spherical data. The general theme of this section is to define a parameter measuring some desired property, express it as a regular statistical functional, and investigate the sampling properties of the corresponding $U$-statistic that estimates it.

Our second theme is that of applications of the Poisson convergence theorem (Theorem 2 of Section 3.2.4) and this is the subject of Section 6.3. We cover the multiple comparison of correlation coefficients and applications to testing randomness in spatial patterns.

Section 6.4 deals briefly with sequential estimation which has been the subject of much recent research and we close the chapter with a very brief survey of some applications not covered in earlier sections.

## 6.2 Applications to the estimation of statistical parameters

Suppose $\theta$ is a parameter of interest that can be expressed as a regular statistical functional:

$$\theta = E\psi(X_1, \ldots, X_k) = \int \psi(x_1, \ldots, x_k)\, dF(x_1) \ldots dF(x_k)$$

The previous chapters have indicated how to estimate $\theta$ on the basis of a sample $X_1, \ldots, X_n$ assumed i.i.d. $F$, and how to derive the large sample properties of the estimate. In this section we deal with a range of applications of this principle, concentrating particularly on measures of association and tests of independence. We also discuss tests of symmetry and normality, various tests arising in the comparison of populations and a test for "new better than used".

### 6.2.1 Circular and spherical correlation

Let $\Theta$ and $\Phi$ be two random angles. (For a general discussion of angular random variables, see Mardia (1972).) The problem of defining a "circular correlation" between $\Theta$ and $\Phi$ has received attention from several writers: see e.g. Johnson and Wehrly (1977), Jupp and Mardia (1980), Rivest (1982), Fisher and Lee (1983). Most of these references treat a random angle as a random unit vector in the plane, and make use of standard multivariate ideas. The first two references cited use canonical correlations, while the second two employ the singular values of the matrix $E(XY^T)$, where $X = (\cos\Theta, \sin\Theta)$ and $Y = (\cos\Phi, \sin\Phi)$. For example, Fisher and Lee (1983) consider a correlation $\rho$ based on the product of the singular values which can be written

$$\rho = \frac{E\{\sin(\Theta_1 - \Theta_2)\sin(\Phi_1 - \Phi_2)\}}{[E\{\sin^2(\Theta_1 - \Theta_2)\}E\{\sin^2(\Phi_1 - \Phi_2)\}]^{\frac{1}{2}}} \tag{1}$$

where $(\Theta_1, \Phi_1)$ and $(\Theta_2, \Phi_2)$ are independent and have the same distribution as $(\Theta, \Phi)$.

Jupp and Mardia (1980) propose a list of desirable properties that should be possessed by angular correlations. These are

(i) $|\rho| \leq 1$;

(ii) $\rho = \pm 1$ if and only if $\phi = f(\theta)$ for some specified type of function $f$;

(iii) $\rho = 0$ if $\Phi$ and $\Theta$ are independent;

(iv) $\rho$ is invariant under rotations, and changes sign under reflection.

The correlation $\rho$ defined in (1) has all these properties. In particular, $\Phi = \theta + \alpha(\mathrm{mod}\,2\pi)$ for some angle $\alpha$ if and only if $\rho = 1$ and $\Phi = -\theta + \alpha(\mathrm{mod}\,2\pi)$ if and only if $\rho = -1$.

As in Example 1 of Section 5.1.3, the correlation $\rho$ is a function of regular functionals of degree 2, and the statistic

$$\hat{\rho}_n = U_n^{(1)} / \left(U_n^{(2)}U_n^{(3)}\right)^{\frac{1}{2}}$$

based on the same function of the corresponding $U$-statistics (where for example $U_n^{(1)}$ is the $U$-statistic with kernel $\sin(\theta_1 - \theta_2)\sin(\phi_1 - \phi_2)$) is a consistent estimate of $\rho$. The sample correlation $\hat{\rho}_n$ has the same properties (i)-(iv) as does $\rho$. We can apply the results of Section 5.1.3 to prove the asymptotic normality of the jackknifed version of $\hat{\rho}_n$ :

$$n^{\frac{1}{2}}(\hat{\rho}_n(JACK) - \rho)/(\widehat{Var}(JACK))^{\frac{1}{2}} \xrightarrow{D} N(0,1). \qquad (2)$$

This result holds true for non-degenerate $U_n^{(1)}$ provided that neither of the quantities $E\sin^2(\Theta_1 - \Theta_2)$ and $E\sin^2(\Phi_1 - \Phi_2)$ is zero, for in this case the necessary derivatives of the function $x/(yz)^{\frac{1}{2}}$ are not continuous. This situation however only arises in the trivial case when $\Theta$ or $\Phi$ is constant.

For an arbitrary circular variate $\Psi$, the trignometric moments $\alpha_p$ and $\beta_p$ are defined by

$$\alpha_p = E\cos p\Psi, \qquad \beta_p = E\sin p\Psi.$$

If the function $A(\Psi)$ is defined by

$$A(\Psi) = \alpha_1^2 + \beta_1^2 + \alpha_2\beta_1^2 - \alpha_1^2\alpha_2 - 2\alpha_1\beta_1\beta_2$$

then assuming the independence of $\Theta$ and $\Phi$, it can be shown that the first conditional variance corresponding to $U_n^{(1)}$ is given by $\frac{1}{4}A(\Theta)A(\Phi)$, which will be zero if either $A(\Theta)$ or $A(\Phi)$ is zero. This can happen when the resultant length of $\Theta$ or $\Phi$ is zero (the resultant length of a circular variate $\Psi$ is the quantity $\sqrt{\alpha_1^2 + \beta_1^2}$; for details see Mardia (1972) p.45). For example, if the marginal distributions of $\Theta$ and $\Phi$ are uniform, the resultant lengths of $\Theta$ and $\Phi$ are both zero and so the $U$-statistic $U_n^{(1)}$ has a first-order degeneracy. By a slight modification of the proof of Theorem 1 of Section 5.1.2 we see that $n\hat{\rho}(JACK)$ and $n\hat{\rho}_n = nU_n^{(1)}/(U_n^{(2)}U_n^{(3)})^{\frac{1}{2}}$ have identical limit distributions, which in view of the fact that $U_n^{(2)}$ and $U_n^{(3)}$ converge in probability to $\mu_2 = E\sin^2(\Theta_1 - \Theta_2)$ and $\mu_3 = E\sin^2(\Phi_1 - \Phi_2)$ respectively must equal the limit distribution of $nU_n^{(1)}/(\mu_2\mu_3)^{\frac{1}{2}}$.

By Theorem 1 of Section 3.2.2, the limit distribution of $nU_n^{(1)}$ in the case of uniform independent $\Theta$ and $\Phi$ is that of

$$\sum_{\nu=1}^{\infty} \lambda_\nu(Z_\nu^2 - 1)$$

251

where the $Z_\nu$ are i.i.d. $N(0,1)$ variates and the $\lambda_\nu$ are the eigenvalues of the integral equation

$$\left(\frac{1}{2\pi}\right)^2 \int_0^{2\pi} \int_0^{2\pi} \sin(\theta_1 - \theta_2) \sin(\phi_1 - \phi_2) f(\theta_2, \phi_2) d\theta_2 d\phi_2 = \lambda f(\theta_1, \phi_1). \quad (3)$$

The eigenvalues of (3) are $\pm\frac{1}{4}$, each of multiplicity two, corresponding to the eigenfunctions $\cos(\theta - \phi) \pm \sin(\theta - \phi)$ and $\cos(\theta + \phi) \pm \sin(\theta + \phi)$, so that the limit distribution of $nU_n^{(1)}$ is that of $\frac{1}{4}(X_1 - X_2)$ where $X_1$ and $X_2$ are independent $\chi_2^2$ variates. The limit distribution of $n\hat{\rho}_n$ and $n\hat{\rho}_n(JACK)$ is that of $\frac{1}{2}(X_1 - X_2)$ since $\mu_2 = \mu_3 = \frac{1}{2}$ for uniform $\Theta$ and $\Phi$. These results may be used to construct tests of independence for circular variates, as described in Fisher and Lee (1983).

A generalisation of the above (see Fisher and Lee (1986)) leads to tests of independence for random variables defined on spheres or hyperspheres. A random point on a $k$-dimensional hypersphere can be thought of as a random $k$-vector of length unity, and if $X$ and $Y$ are two such vectors, a measure of association between them may be defined by

$$\rho_k = \rho_k(X, Y) = C_k \det E(XY^T)$$

where $C_k$ is a normalising constant designed to ensure that $|\rho_k| \leq 1$. If the bivariate observations $(X_1, Y_1), \ldots, (X_k, Y_k)$ are independently distributed as $(X, Y)$, then an alternative expression for $\rho_k$ is obtained by using the relationship

$$k! \det E(XY^T) = E \det(X_1, \ldots, X_k) \det(Y_1, \ldots, Y_k)$$

which may be deduced from the Cauchy-Schwartz inequality. It follows that

$$C_k = (k!)^{-1} \{ \det(X_1, \ldots, X_k)^2 \det(Y_1, \ldots, Y_k)^2 \}^{-\frac{1}{2}}$$
$$= \{ \det E(XX^T) \det E(YY') \}^{-\frac{1}{2}}$$

and so

$$\rho_k = \frac{\det E(XY^T)}{\{ \det E(XX^T) \det E(YY^T) \}^{\frac{1}{2}}},$$

where $\det(X_1, \ldots, X_k)$ denotes the determinant of the matrix with columns $X_1, \ldots, X_k$.

Note that $\det(X_1, \ldots, X_k)$ is proportional to the volume of the random parallelopiped formed using the unit vectors $X_1, \ldots, X_k$ so $\rho_k$ is the correlation between the volumes formed by the random matrices $(X_1, \ldots, X_k)$ and $(Y_1, \ldots, Y_k)$. The properties of $\rho_k$ include:

(i) If $H$ is an orthogonal matrix; then $\rho_k(HX, Y) = \text{sgn}\{\det(H)\}\rho_k(X, Y)$;

(ii) If $X$ and $Y$ are independent then $\rho_k(X, Y) = 0$;

(iii) $|\rho_k| \leq 1$;

(iv) $\rho_k(X, Y) = \pm 1$ if and only if $Y = HX$ for some orthogonal matrix $H$ with $\det(H) = \pm 1$.

Thus $\rho_k$ is invariant under orthogonal rotations, changes sign under reflections, behaves properly under independence and has value $+1$ or $-1$ if and only if we have a perfect relationship between $X$ and $Y$. In the case $k = 2$ we have the circular correlation introduced above. In view of the alternative representation

$$\rho_k = \frac{E \det(X_1, \ldots, X_k) \det(Y_1, \ldots, Y_k)}{[E \det^2(X_1, \ldots, X_k) E \det^2(Y_1, \ldots, Y_k)]^{\frac{1}{2}}}$$

an obvious estimator for $\rho_k$ is

$$\hat{\rho}_k = \frac{\sum_{(n,k)} \det(X_{i_1}, \ldots, X_{i_k}) \det(Y_{i_1}, \ldots, Y_{i_k})}{\{\sum_{(n,k)} \det^2(X_{i_1}, \ldots, X_{i_k}) \sum_{(n,k)} \det^2(Y_{i_1}, \ldots, Y_{i_k})\}^{\frac{1}{2}}}.$$

Thus, $\hat{\rho}$ is once again a function of $U$-statistics, $\hat{\rho}_k = U_n^{(1)}/(U_n^{(2)} U_n^{(3)})^{\frac{1}{2}}$, where, for example, $U_n^{(1)}$ is the $U$-statistic based on the kernel

$$K^{(1)}\{(x_1, y_1), \ldots, (x_k, y_k)\} = \det(x_1, \ldots, x_k) \det(y_1, \ldots, y_k).$$

As in the previous case the asymptotic distribution of $\hat{\rho}_k$ is dependent on the order of the degeneracy of $U_n^{(1)}$. In general, $U_n^{(1)}$ will be non-degenerate, and the usual normal theory will apply to both $\hat{\rho}_k$ and its jackknifed version. However, if $X$ and $Y$ are independent, this will not be the case and $U_n^{(1)}$ has a degeneracy of order $k - 1$. To see this, recall that the $c$th conditional variance $\sigma_c^2$ is given by

$$\sigma_c^2 = var[K_c^{(1)}\{(X_1, Y_1), \ldots, (X_c Y_c)\}]$$

where

$$K_c^{(1)}\{(x_1, y_1), \ldots, (x_c, y_c)\}$$
$$= E[K^{(1)}\{(x_1, y_1), \ldots, (x_c, y_c), (X_{c+1}, Y_{c+1}), \ldots, (X_k, Y_k)\}].$$

When the $X_i$ are independent of the $Y_i$, it follows that

$$
\begin{aligned}
K_c^{(1)}\{(x_1, y_1), \ldots, (x_c, y_c)\} &= E\{\det(x_1, \ldots, x_c, X_{c+1}, \ldots, X_k)\} \\
&\quad \times E\{\det(y_1, \ldots, y_c, Y_{c+1}, \ldots, Y_k)\} \\
&= \det\{x_1, \ldots, x_c, E(X_{c+1}), \ldots, E(X_k)\} \\
&\quad \times \det\{y_1, \ldots, y_c, E(Y_{c+1}), \ldots, E(Y_k)\}, \quad (4)
\end{aligned}
$$

and so $\sigma_2^2 = 0$ for $c = 1, \ldots k - 2$ and $\sigma_{k-1}^2 > 0$ provided neither $E(X)$ nor $E(Y)$ is zero, since determinants of matrices with identical columns are zero.

In the case of circular variates (i.e. $k = 2$) the limit distribution is in general normal as described above, and for spherical variates ($k = 3$) the limit distribution is that of a linear combination of $\chi^2$ variates. If the marginal distributions of $X$ and $Y$ possess rotational symmetry about a polar axis, then we can be a bit more explicit in the case $k = 3$:

An alternative representation of a point on the sphere is one in terms of spherical polar coordinates, i.e. one of the form $(\theta, \phi)$, where $0 \le \theta \le \pi$ and $0 < \phi < 2\pi$. Then the unit vector $X$ is

$$X^T = (\sin \Theta \cos \Phi, \sin \Theta \sin \Phi, \cos \Theta)$$

and assuming rotational symmetry about the polar axis $(0, 0, 1)$ (which we also take to be the mean direction) the density of $X$ can be written

$$g_X(x) = g_X(\theta, \phi) = \frac{1}{2\pi} h_X(\theta) \qquad (5)$$

where $h_X$ is some density on $(0, \pi)$, and similarly for $Y$. The integral equation determining the coefficients of the $\chi_1^2$ variates in the limit distribution of $nU_n^{(1)}$ takes the form

$$\int_{S_3} \int_{S_3} K_2^{(1)}((x_1, y_1), (x_2, y_2)) f(x_2, y_2) g_X(x_2) g_Y(y_2) \, dx_2 dy_2 = \lambda f(x_1, y_1)$$

254

The function $K_2^{(1)}$ can be expressed in spherical polar co-ordinates by writing

$$x_i^T = (\sin\theta_i \cos\phi_i, \sin\theta_i \sin\phi_i, \cos\theta_i),$$
$$y_i^T = (\sin\theta_i^* \cos\phi_i^*, \sin\theta_i^* \sin\phi_i^*, \cos_i^*).$$

Then if $\lambda_1(X)$ denotes the resultant length of $X$, we get from (4)

$$K_2^{(1)}(\theta_1,\phi_1,\theta_1^*,\phi_1^*;\ \theta_2,\phi_2,\theta_2^*,\phi_2^*) =$$
$$\lambda_1(X)\lambda_1(Y)\sin\theta_1 \sin\theta_1^* \sin\theta_2 \sin\theta_2^* \sin(\phi_2-\phi_1)\sin(\phi_2^*-\phi_1^*)$$

and using (5), the integral equation becomes

$$\frac{1}{(2\pi)^2}\int_0^{2\pi}\int_0^{2\pi}\int_0^{\pi}\int_0^{\pi} K_2^{(1)}(\theta_1,\ldots,\phi_2)\,f(\theta_2,\theta_2^*,\phi_2,\phi_2^*)$$
$$h_X(\theta_2)\,h_Y(\theta_2^*)\,d\theta_2\,d\theta_2^*\,d\phi_2\,d\phi_2^* = \lambda f(\theta_1,\theta_1^*,\phi_1,\phi_1^*). \quad (6)$$

The eigenvalues of (6) are

$$\pm\lambda_1(X)\lambda_1(Y)\{1-\lambda_2(X)\}\{1-\lambda_2(Y)\}/16$$

each of multiplicity two, where $\lambda_2(X) = E\cos 2\Theta$. Hence the limit distribution of $nU_n^{(1)}$ is that of $(3/16)\lambda_1(X)\lambda_1(Y)\{1-\lambda_2(X)\}\{1-\lambda_2(Y)\}W$, where $W$ has the double exponential density $\frac{1}{4}\exp(-1/2|w|)$.

For distributions with rotational symmetry,

$$\det\{E(XX^T)\} = \{1-\lambda_2(X)\}^2\{1+\lambda_2(X)\}/32,$$

so that the asymptotic distribution of the statistic

$$n\hat{\rho}[\{1+\hat{\lambda}_2(X)\}\{1+\hat{\lambda}_2(Y)\}]^{\frac{1}{2}}/\{6\hat{\lambda}_1(X)\hat{\lambda}_1(Y)\},$$

where $\hat{\lambda}_1$ and $\hat{\lambda}_2$ are consistent estimates of $\lambda_1$ and $\lambda_2$, is double exponential when $X$ and $Y$ are independent.

$U$-statistics can also be used to derive a nonparametric measure of association between circular variates $\Theta$ and $\Phi$ analogous to Kendall's tau. Real-valued random variables are perfectly associated (in the sense of having a Kendall's tau of $\pm 1$) if and only if there is a monotone relationship between $X$ and $Y$, i.e. if there is a monotone function $g$ with $Y = g(X)$. A relationship for circular variates analogous to a monotone relationship between linear (i.e. real valued) variates is a relationship $\Phi = g(\Theta)$ having the property that if $\Theta$ moves through a complete revolution in a particular sense (either clockwise or anticlockwise) then $\Phi = g(\Theta)$ also moves through a complete revolution in the same or opposite sense. If the sense is the same, we have the analogue of a monotone increasing relationship; if different, the relationship is monotone decreasing.

Now consider three points on a torus, say $p_i = (\theta_i, \phi_i), i = 1, 2, 3$. Call the three points (toroidally!) concordant if there is an "increasing" (in the above sense) relationship $g$ such that $\phi_i = g(\theta_i)$, $i = 1, 2, 3$, and discordant if the relationship is "decreasing". Assuming all the $\theta_i$ and $\phi_i$ are distinct, any three points on a torus will be either concordant or discordant in this sense.

Now let $P = (\Theta, \Phi)$ be a random point on the torus, and let $P_1, P_2, P_3$ be independently distributed as $P$. We can define a correlation $\Delta$ analogous to Kendall's tau by

$$\Delta = Pr\left((P_1, P_2, P_3) \text{ are concordant}\right) - Pr\left((P_1, P_2, P_3) \text{ are discordant}\right).$$

If the kernel $\delta$ is defined by

$$\delta(P_1, P_2, P_3) = \begin{cases} 1 & \text{if } p_1, p_2, p_3 \text{ are concordant;} \\ -1 & \text{if } p_1, p_2, p_3 \text{ are discordant;} \end{cases}$$

then $\Delta = E\psi(P_1, P_2, P_3)$ and is estimated by the $U$-statistic $\hat{\Delta}_n$ say of degree 3 with kernel $\delta$.

The statistic $\hat{\Delta}_n$ has the properties

(i) $-1 \le \hat{\Delta}_n \le 1$;

(ii) $\hat{\Delta}_n = +1$ if there is a "monotone increasing" $g$ with $\phi_i = g(\theta_i)$ for $i = 1, \ldots, n$, and $-1$ if there is a "monotone decreasing" $g$ with the same property;

(iii) if $\Theta$ and $\Phi$ are independent, and $P_1, \ldots, P_n$ are distributed independently as $P$, then $\hat{\Delta}_n(P_1, \ldots, P_n)$ is distributed symmetrically about 0.

256

The distribution theory for $\hat{\Delta}_n$ under independence of $\Theta$ and $\Phi$ is easily worked out due to an alternative representaion for $\delta$. Let $p_i = (\theta_i, \phi_i), i = 1, 2, 3$. Then

$$\delta(p_1, p_2, p_3) = \mathrm{sgn}(\theta_1 - \theta_2)\mathrm{sgn}(\theta_2 - \theta_3)\mathrm{sgn}(\theta_3 - \theta_1)$$
$$\times \mathrm{sgn}(\phi_1 - \phi_2)\mathrm{sgn}(\phi_2 - \phi_3)\mathrm{sgn}(\phi_3 - \phi_1). \qquad (7)$$

and since $\delta$ depends only on differences, it does not depend on the choice of origin for the angular measurements. Moreover, since $\delta$ depends only on the sign of these differences, it is invariant under any monotone transformation of the data. Accordingly, to work out the asymptotic properties of our statistic under independence, we may suppose that the $\Theta$'s and $\Phi$'s are uniformly distributed on $[0,1]$.

Define

$$\delta_2(p_1, p_2) = E\{\delta(p_1, p_2, P_3)\}, \quad \sigma_2^2 = Var\{\delta_2(P_1, P_2)\},$$
$$\delta_1(p_1) = E\{\delta_2(p_1, P_2)\}, \quad \sigma_1^2 = Var\{\delta_1(P_1)\},$$

so that $E\{\delta_1(P_1)\} = E\{\delta_2(P_1, P_2)\} = 0$. Routine algebra yields

$$\delta_2(p_1, p_2) = g(\theta_1 - \theta_2)g(\phi_1 - \phi_2),$$

where $g(x) = \mathrm{sgn}(x)(1 - 2|x|)$. Note that

$$\int_0^1 \int_0^1 g(x - y)dx\,dy = 0$$

and that

$$\delta_1(p_1) = E\{\delta_2(p_1, P_2)\} = \int_0^1 g(\theta_1 - \theta_2)d\theta_2 \int_0^1 g(\phi_1 - \phi_2)d\phi_2 = 0.$$

Also, $\sigma_1^2 = Var\{\delta_1(P_1)\} = 0$, $\sigma_2^2 = Var\{\delta_2(P_1, P_2)\} = \frac{1}{9}$ and $\sigma_3^2 = Var\{\delta(P_1, P_2, P_3)\} = 1$. Thus $\hat{\Delta}_n$ is degenerate of order one, and

$$Var\,\hat{\Delta}_n = \binom{n}{3}^{-1} \sum_{c=1}^3 \binom{3}{c}\binom{n-3}{3-c}\sigma_c^2 = \frac{2}{(n-1)(n-2)}.$$

257

To find the asymptotic distribution of $n\hat{\Delta}_n$, the appropriate integral equation is

$$\int_0^1 \int_0^1 g(\theta_1 - \theta_2)g(\phi_1 - \phi_2)f(\theta_2,\phi_2)d\theta_2 d\phi_2 = \lambda f(\theta_1,\phi_1) \qquad (8)$$

which has eigenvalues of the form $(\pm\mu_k)(\pm\mu_l)$, where $\pm\mu_1, \pm\mu_2, \ldots$ are the eigenvalues of the differential equation

$$\int_0^1 g(\theta_1 - \theta_2)f(\theta_2)d\theta_2 = \mu f(\theta_1).$$

These are of the form $\mu_k = i/\pi k$, so that the eigenvalues of (8) are $\pm 1/\pi^2 kl$; $k,l = 1,2,\ldots$, each of multiplicity two. The form of the limit distribution of $n\Delta_n$ is thus

$$\frac{3}{\pi^2}\sum_{l=1}^{\infty}\sum_{k=1}^{\infty} W_{lm}/lm$$

where the r.v.s $W_{lm}$ are independently distributed as the difference of two $\chi_2^2$ variates. Information on the small sample distribution of $n\hat{\Delta}_n$ under independence is given in Fisher and Lee (1982). When $\Theta$ and $\Phi$ are not independent, the limiting distribution is normal and the usual jackknifing methods are appropriate.

A correlation similar to $\Delta$ to measure the association between a linear variate $X$ and a circular variate $\Theta$ is described in Fisher and Lee (1981) and we refer the reader to that paper for details. An extension of the nonparametric angular-angular correlation coefficient to the sphere may be found in Jupp (1987).

### 6.2.2 Testing for symmetry

Let $\phi(x_1, x_2, x_3)$ be a kernel of three variables with the property

$$\phi(a + bx_1, a + bx_2, a + bx_3) = \phi(x_1, x_2, x_3)\text{sgn}(b).$$

If $X_1, X_2, X_3$ are independent r.v.s distributed as $X$, then the parameter

$$\omega = E\phi(X_1, X_2, X_3)$$

258

is a measure of symmetry, in that symmetry of the distribution of $X$ implies $\omega = 0$, although the converse is not true.

Two particular kernels have been suggested in a paper by Davis and Quade (1978): let $x_{(1)} \leq x_{(2)} \leq x_{(3)}$ denote the quantities $x_1, x_2$ and $x_3$ arranged in ascending order. Then define

$$\phi^{(1)}(x_1, x_2, x_3) = \begin{cases} 1 & \text{if } x_{(3)} - x_{(2)} > x_{(2)} - x_{(1)}, \\ 0 & \text{if } x_{(3)} - x_{(2)} = x_{(2)} - x(1), \\ -1 & \text{if } x_{(3)} - x_{(2)} < x_{(2)} - x(1), \end{cases}$$

so that $\phi^{(1)}$ measures the propensity of the mean of $x_1, x_2$ and $x_3$ to be greater than the median of $x_1, x_2, x_3$, or equivalently, for the median of $x_1, x_2, x_3$ to be closer to the maximum value than the minimum.

Also define

$$\phi^{(2)}(x_1, x_2, x_3) = \frac{(x_{(3)} - x_{(2)}) - (x_{(2)} - x_{(1)})}{x_{(3)} - x_{(1)}}.$$

If $\bar{x}$, $\tilde{x}$ and $R$ denote the mean, median and range of $x_1, x_2, x_3$ we can write

$$\phi^{(1)}(x_1, x_2, x_3) = \text{sgn}(\bar{x} - \tilde{x}), \quad \text{and} \quad \phi^{(2)}(x_1, x_2, x_3) = 3(\bar{x} - \tilde{x})/R.$$

$U$-statistics $U_n^{(1)}$ and $U_n^{(2)}$ based on these kernels have asymptotically normal distributions, so that an asymptotic test for symmetry is performed by referring $n^{\frac{1}{2}} U_n^{(1)} / (\widehat{Var}(JACK))^{\frac{1}{2}}$ to the percentage points of the standard normal distribution, and similarly for $U_n^{(2)}$. The two statistics seem about equal in small sample power, according to simulations reported in the above reference. More extensive simulations aere reported in Randles, Fligner, Policello and Wolfe (1980). See also Gupta (1967).

### 6.2.3 Testing for normality

Locke and Spurrier (1976) study the statistic $U_n/s^p$ where $U_n$ is a location-invariant $U$-statistic designed to check departures from symmetry, $s^2$ is the usual sample variance, and $p$ is chosen to make the ratio scale invariant (except for sign). A possible candidate for $U_n$ is the sample skewness, with kernel

$$\phi(x_1, x_2, x_3) = \sum_{i=1}^{3} (x_i - \tfrac{1}{2}(x_1 + x_2 + x_3))^3.$$

Also proposed by Locke and Spurrier are $U$-statistics based on kernels of the form

$$\phi^{(p)}(x_1, x_2, x_3) = (x_{(3)} - x_{(2)})^p - (x_{(2)} - x_{(1)})^p$$

which we denote by $U_n^{(p)}$. The statistics $T_{n,p} = U_n^{(p)}/s^p$ are then used to test the normality of the underlying sample. Since $s^2$ is a consistent estimate of $\sigma^2$, it is clear from Chapter 5 that the studentised statistic $n^{\frac{1}{2}}(T_{np} - \theta_p/\sigma^2)/(\widehat{Var}(JACK))^{\frac{1}{2}}$ is asymptotically normal, where $\theta_p = E\phi^{(p)}(X_1, X_2, X_3)$. Since $\theta_p = 0$ for symmetric distributions, $T_{n,p}$ may be used as a test for symmetry, or as a test for normality against non-symmetric alternatives.

If $X_1, \ldots, X_n$ are normally distributed, then Locke and Spurrier show that

$$Var\, T_{n,1} = \begin{cases} a_1 & n = 3, \\ \frac{1}{4}(a_1 + 3a_2) & n = 4, \\ 6(n-3)!(n!)^{-1}\{a_1 + 3(n-3)a_2 \\ \qquad + 1.5(n-3)(n-4)a_3\}, & n \geq 5, \end{cases}$$

where $a_1, a_2, a_3$ are constants derived from the expectations of normal order statistics, and are given by $a_1 = 1.03803994$, $a_2 = 0.23238211$ and $a_3 = 0.05938718$. The statistic $n^{\frac{1}{2}}T_{n,1}/(\widehat{Var}T_{n,1})^{\frac{1}{2}}$ will be asymptotically normal under the normality assumption, and can be used to test for normality. The asymptotic approximation is good for $n$ as small as 12, at least in the tails. Similar results are true for $T_{n,2}$, which can be used in the same way. Locke and Spurrier also give a discussion of the relative powers of the statistics $T_{n,1}, T_{n,2}$ and the sample skewness for testing for normality against various types of non-symmetric alternatives.

For symmetric alternatives, they recommend using statistics of the form

$$D = U_n^{(1)}/s \quad \text{or} \quad T = U_n^{(2)}/s$$

where $U_n^{(1)}$ and $U_n^{(2)}$ are based on kernels $\phi^{(1)}(x_1, x_2) = |x_1 - x_2|$ and $\phi^{(2)}(x_1, x_2, x_3, x_4) = (x_{(3)} - x_{(2)})$ respectively. Details may be found in Locke and Spurrier (1977). For another approach, see Oja (1981,1983).

## 6.2.4 A test for independence

Let $(X, Y)$ be a bivariate r.v. with continuous joint and marginal densities and joint d.f. $F$. Then Hoeffding (1948b) introduces the parameter

$$\Delta = \int_{-\infty}^{\infty} \int_{-\infty}^{\infty} D^2(x, y) dF(x, y)$$

where $D(x, y) = F(x, y) - F(x, \infty) F(\infty, y)$. The parameter $\Delta$ has the property that $\Delta = 0$ if and only if $X$ and $Y$ are independent, provided we confine attention to continuous densities.

An alternative expression for $\Delta$ can be developed by introducing the functions

$$\psi(x_1, x_2, x_3) = \begin{cases} 1 & \text{if } x_2 \le x_1 < x_3, \\ 0 & \text{if } x_1 < x_2, x_3 \quad \text{or} \quad x_1 \ge x_2, x_3, \\ -1 & \text{if } x_3 \le x_1 < x_2 \end{cases}$$

and

$$\phi(x_1, y_1; \ldots; x_5, y_5) = \tfrac{1}{4} \psi(x_1, x_2, x_3) \psi(x_1, x_4, x_5) \psi(y_1, y_2, y_3) \psi(y_1, y_4, y_5).$$

The kernel $\phi$ can be interpreted in terms of the distribution of the four points $(x_2, y_2), \ldots, (x_5, y_5)$ in the four quadrants of a cartesian plane with origin $(x_1, y_1)$, and it can be shown that

$$\Delta = \int \cdots \int \phi(x_1, y_1; \ldots; x_5, y_5) dF(x_1, y_1) \ldots dF(x_5, y_5).$$

Hence $\Delta$ is a regular statistical functional, and is estimated by a $U$-statistic $D_n$ of degree 5 having kernel a symmetrised version of $\phi$. Note that $\phi$ depends only on the ranks of $X_1, \ldots, X_5$ and $Y_1, \ldots, Y_5$ and so by making a probability integral transformation we may assume that the marginal distributions of $X$ and $Y$ are uniform.

The sampling distribution of $D_n$ under independence can thus be completely worked out. The first two moments are $ED_n = 0$, and

$$Var\, D_n = \frac{2(n^2 + 5n - 32)}{9n(n-1)(n-3)(n-4)}.$$

261

The $U$-statistic has a degeneracy of order one under independence, and the asymptotic distribution of $nD_n$ is that of

$$\frac{10}{\pi^4} \sum_{k=1}^{\infty} (\chi^2_{r(k)} - r(k))/k^2$$

where $r(k)$ denotes the number of divisors of $k$, including 1 and $k$. A nonparametric test of independence may then be carried out by referrring $nD_n$ to the percentage points of this distribution, which may be calculated numerically.

### 6.2.5 Applications to the several-sample problem

Suppose we want to compare $m$ populations, and from each population have available independent random samples $X_{i,j}$, $j = 1, 2, \ldots n_i$. The r.v.s $X_{i,j}$ are independent and $X_{i,j}$ is assumed to have distribution function $F_i$ which we take to be absolutely continuous. Several authors have used $U$-statistic theory to derive non parametric tests for the hypothesis that the $F_i$'s are equal. For example, Bhapkar (1961) introduces generalised $U$-statistics $U^{(i)}$ defined by

$$U^{(i)} = U^{(i)}_{n_1,\ldots,n_m} = (n_1 \ldots n_m)^{-1} \sum_{j_1=1}^{n_1} \cdots \sum_{j_m=1}^{n_m} \phi^{(1)}(X_{1,j_1}; \ldots; X_{m,j_m})$$

for $i = 1, \ldots, m$, where

$$\phi^{(i)}(x_1; \ldots; x_m) = \begin{cases} 1 & \text{if } x_i \text{ is the smallest } x, \\ 0 & \text{otherwise.} \end{cases}$$

These $U$-statistics can be regarded as generalisations of the two-sample Wilcoxon statistic discussed in Example 1 of Section 2.2.

Let $\theta_i = EU^{(i)}$. Under the hypothesis of equal $F_i$'s, the quantities $\theta_i$ are all equal to $m^{-1}$ by symmetry, since in this case $\theta_i = Pr(X_i$ is the minimum of $X_1, \ldots, X_m)$ where $X_1, \ldots, X_m$ are i.i.d. Let $Z_i = U^{(i)} - m^{-1}$. A natural test statistic to measure departures from the null hypothesis of equal $F_i$'s is

$$V = N(2m - 1) \sum_{i=1}^{m} p_{i,N} (Z_i - \bar{Z})^2$$

where $N = n_1 + \cdots + n_m$, $\bar{Z} = \sum_{i=1}^{m} p_{i,N} Z_i$ and $p_{i,N} = n_i N^{-1}$. We will show that the asymptotic distribution of $V$ is $\chi^2_{m-1}$ when the null hypothesis is true, provided that $n_1, \ldots, n_m \to \infty$ in such a way that $p_{i,N} \to p_i > 0$ for each $i$.

By Theorem 2 of Section 3.7.1, it follows that the random vector $N^{\frac{1}{2}}(Z_1, \ldots, Z_m)$ converges in distribution to a multivariate normal distribution having mean zero and covariance matrix $\boldsymbol{\Sigma} = p_1^{-1} \boldsymbol{\Sigma}_1 + \ldots + p_m^{-1} \boldsymbol{\Sigma}_m$ where the covariance matrices $\boldsymbol{\Sigma}_i$ are defined in Theorem 2 of Section 3.7.1. Furthermore, we can write

$$ V = (2m - 1)(N^{\frac{1}{2}}\mathbf{Z})^T (\mathbf{I} - \mathbf{P}_N \mathbf{J})\mathbf{P}(\mathbf{I} - \mathbf{J}\mathbf{P}_N)(N^{\frac{1}{2}}\mathbf{Z}) $$

where $\mathbf{P}_N = \text{diag}\,(p_{1,N}, \ldots, p_{m,N})$ and $\mathbf{J}$ is a $m \times m$ matrix of ones. Thus the asymptotic distribution of $V$ will be that of

$$ (2m - 1)\mathbf{Z}_*^T (\mathbf{I} - \mathbf{P}\mathbf{J})\mathbf{P}(\mathbf{I} - \mathbf{J}\mathbf{P})\mathbf{Z}_* $$

where $\mathbf{Z}_*$ is $MN(0, \boldsymbol{\Sigma})$ and $\mathbf{P} = \text{diag}\,(p_1, \ldots, p_m)$. To demonstrate that this limit is $\chi^2_{m-1}$ we need to show that $(\mathbf{I} - \mathbf{P}\mathbf{J})\mathbf{P}(\mathbf{I} - \mathbf{J}\mathbf{P})\boldsymbol{\Sigma}$ is idempotent and of rank $m - 1$.

We first compute the elements of $\boldsymbol{\Sigma}$. Note that the $H$-decomposition functions (c.f. Theorem 2 of Section 3.7.1) are given by

$$ h^{(i)}(x) = \begin{cases} \dfrac{Pr(X_1 > x, \ldots, X_{i-1} > x, X_{i+1} > x, \ldots, X_m > x)}{\phantom{xx} - m^{-1},} & \text{if } i = l \\[2ex] \dfrac{Pr(X_1 > X_i, \ldots, X_{l-1} > X_i, x > X_i, X_{l+1} > X_i, \ldots)}{\phantom{xx} - m^{-1},} & \text{if } i \neq l \end{cases} $$

where $X_1, \ldots, X_m$ are i.i.d. $F$, and so if $f$ is the density of $F$,

$$ h_i^{(i)}(x) = (1 - F((x))^{m-1} - m^{-1} $$

and

$$ h_l^{(i)}(x) = \int_{-\infty}^{x} f(y)(1 - F(y))^{m-2} dy - m^{-1} $$
$$ = \frac{1 - (1 - F(x))^{m-2}}{m - 1} - m^{-1} $$

263

when $l \neq i$. Since $F(X_{ij})$ is uniformly distributed, it follows that if $\Sigma_l = (\sigma_{rs}^{(l)})$, we have

$$\sigma_{r,s}^{(l)} = Eh_l^{(r)}(X_{l,1})h_l^{(s)}(X_{l,1})$$

$$= \begin{cases} \frac{(m-1)^2}{m^2(2m-1)}, & \text{if } r = s = l, \\ \frac{-(m-1)}{m^2(2m-1)}, & \text{if } r = l \text{ or } s = l \text{ but } r \neq s, \\ \frac{1}{m^2(2m-1)}, & \text{when neither } r \text{ nor } s \text{ equals } l. \end{cases}$$

Writing $\Sigma = (\sigma_{r,s})$ we obtain

$$m^2(2m-1)\sigma_{rs} = \begin{cases} (m-1)^2 p_r^{-1} + \sum_{l \neq r} p_l^{-1}, & \text{when } r = s, \\ \sum_{l=1}^m p_l^{-1} - m(p_r^{-1} + p_s^{-1}), & \text{when } r \neq s. \end{cases}$$

Define $\mathbf{p} = (p_1, \ldots, p_m), \mathbf{q} = (p_1^{-1}, \ldots, p_m^{-1})$ and let $\mathbf{1}$ be an $m$-vector of ones. Then

$$\Sigma = \{m^2(2m-1)\}^{-1}\left\{\sum_{l=1}^m p_l^{-1}\mathbf{J} + m\mathbf{P}^{-1} - mq\mathbf{1}^T - m\mathbf{1}q^T\right\}$$

and using the relations $\mathbf{JPJ} = \mathbf{J}, \mathbf{Pq1}^T = \mathbf{J}, \mathbf{P1} = \mathbf{p}$ and $\mathbf{PJpq}^T = \mathbf{pq}^T$ we see that

$$(2m-1)(\mathbf{I} - \mathbf{PJ})\mathbf{P}(\mathbf{I} - \mathbf{JP})\Sigma = \mathbf{I} - \mathbf{J}/m$$

which is idempotent and has rank $m - 1$.

Another approach is taken by Quade (1965). He defines a score $Y_{ij}$ for each observation $X_{ij}$, and performs an analysis of variance of the scores. Under the null hypothesis, the scores if suitably defined will be exchangeable, and the $F$-test will be asymptotically valid. We illustrate the techniques used by considering the Kruskal-Wallis test, where the score $Y_{ij}$ is just the rank, $R_{ij}$ say, of $X_{ij}$ in the combined sample. Obviously, we may just as well take the score $Y_{ij} = R_{i,j} - 1$, since the $F$-statistic is invariant under shifts in the data.

Assuming no ties, we can write

$$R_{ij} - 1 = \sum_{r=1}^m \sum_{s=1}^{n_r} I\{X_{ij} > X_{rs}\}$$

so that

$$\frac{1}{n_i} \sum_{j=1}^{n_i} (R_{ij} - 1) = \sum_{r=1}^{m} \frac{1}{n_i} \sum_{j=1}^{n_i} \sum_{j=1}^{n_r} I\{X_{ij} > X_{rs}\}$$

$$= \sum_{r=1}^{m} n_r U_{n_i n_r}^{(i,r)}, \tag{1}$$

where $U_{n_1 n_r}^{(i,r)}$ is the generalised $U$-statistic based on the kernel $\psi(x_i; x_r) = I\{x_i > x_r\}$. The mean $\theta_{ir}$ of $U_{n_i n_r}^{(i,r)}$ is just $P_r(X_{i1} > X_{r1})$. Consider the averages $\bar{Y}_{i.} = \frac{1}{n_1} \sum_{j=1}^{n_i} (R_{ij} - 1)$. Using (1), we see that the expectation of $\bar{Y}_{i.}$ is $\sum_{r=1}^{m} n_r \theta_{ir} = \eta_i$ say, and because of the relationship

$$N^{-\frac{1}{2}} (\bar{Y}_{i.} - \eta_i) = \sum_{r=1}^{m} p_{r,N} N^{\frac{1}{2}} (U_{n_i n_r}^{(i,r)} - \theta_{ir})$$

$$= \sum_{r=1}^{m} p_r N^{\frac{1}{2}} (U_{n_i n_r}^{(i,r)} - \theta_{ir}) + o_p(1)$$

the joint asymptotic normality of the $U$'s (c.f .Theorem 2 of Section 3.7.1) entails that of the quantities $N^{-\frac{1}{2}} (\bar{Y}_{i.} - \eta_i)$.

Now suppose that the null hypothesis of equal $F_i$'s is true. Then $\theta_{i,r} = \frac{1}{2}$ and hence $\eta_i = \frac{N}{2}$. Let $Z_{i,N} = N^{-\frac{1}{2}} (\bar{Y}_{i.} - \frac{N}{2})$ and put $\mathbf{Z_N} = (Z_{1,N}, \ldots, Z_{m,N})$. Then $\mathbf{Z_N} \xrightarrow{D} MN(0, \Sigma)$ by the above reasoning, where $\Sigma$ is some covariance matrix to be determined.

The numerator of the usual $F$-ratio, computed from the scores $Y_{ij}$, is proportional to

$$N^{-2} \sum_{i=1}^{n} n_i (\bar{Y}_{i.} - \bar{Y}_{..})^2 = \sum_{i=1}^{n} p_{i,N} (Z_{1,N} - \bar{Z}_N)^2 \tag{2}$$

where $\bar{Z}_N = \sum_{i=1}^{m} p_{1,N} Z_{i,N}$. Using the same notation as in the Bhapkar example, we can write (2) as

$$\mathbf{Z}_N^T (\mathbf{I} - \mathbf{P}_N \mathbf{J}) \mathbf{P}_N (\mathbf{I} - \mathbf{J} \mathbf{P}_N) \mathbf{Z}_N$$

and hence (2) converges in distribution to $\mathbf{Z}_*^T (\mathbf{I} - \mathbf{PJ}) \mathbf{P} (\mathbf{I} - \mathbf{JP}) \mathbf{Z}_*$ where $\mathbf{Z}_*$ is $MN(0, \Sigma)$. We claim that the so-called Kruskal-Wallis statistic

$$H = \frac{12}{N(N+1)} \sum_{i=1}^{n_i} (\bar{Y}_{i.} - \bar{Y}_{..})^2 \tag{3}$$

265

is asymptotically $\chi^2_{m-1}$. It is enough to show that $12(\mathbf{I} - \mathbf{PJ})\mathbf{P}(\mathbf{I} - \mathbf{JP})\Sigma$ is idempotent and of rank $m - 1$. To show this, we need to evaluate $\Sigma = (\sigma_{ij})$ say. Let $\mathcal{D}(\mathbf{Z}_N)$ denote the covariance matrix of $\mathbf{Z}_N$. Then $\Sigma$ equals $\lim_N \mathcal{D}(\mathbf{Z}_N)$, so to evaluate $\Sigma$ we need to compute the quantities $Var(\sum_{r=1}^m n_r U^{(i,r)})$ and $Cov(\sum_{r=1}^m n_r U^{(i,r)}, \sum_{s=1}^m n_j U^{(j,s)})$. From Example 1 of Section 2.2, we can see that

$$Var\, U_{n_1 n_2}^{(i,r)} = \frac{1}{12n_i} + \frac{1}{12n_r},$$

and arguing similarly,

$$Cov(U_{n_i n_r}^{(i,r)}, U_{n_i n_s}^{(j,s)}) = \frac{1}{12n_i}.$$

Hence

$$Var\left(\sum_{r=1}^m n_r U_{n_i n_r}^{(i,r)}\right) = \sum_{r=1}^m n_r^2 \left(\frac{1}{12n_i} + \frac{1}{12n_r}\right) + \sum_{r \neq s} \frac{n_r n_s}{12n_i}$$

$$= \tfrac{1}{12}N(p_{i,N}^{-1} + 1),$$

so that $\sigma_{ii} = \frac{1}{12}(p_i^{-1} + 1)$. Similar arguments show that $\sigma_{rs} = \frac{1}{12}$ for $r \neq s$, so that $\Sigma = \frac{1}{12}(\mathbf{J} + \mathbf{P}^{-1})$. Hence, using the fact that $\mathbf{JPJ} = \mathbf{J}$, we see that

$$12(\mathbf{I} - \mathbf{PJ})\mathbf{P}(\mathbf{I} - \mathbf{JP})\Sigma = (\mathbf{I} - \mathbf{PJ})\mathbf{P}(\mathbf{I} - \mathbf{JP})(\mathbf{J} + \mathbf{P}^{-1})$$

$$= \mathbf{I} - \mathbf{PJ}$$

and so is idempotent and of rank $m - 1$. The proof is completed by using (3) to see that $(m - 1)$ times the $F$-ratio computed from the scores is just $(N+1)(N-m)H/(N^2 - 1 + (N+1)H)$ which is asymptotically equivalent to $H$, and hence is asymptotically $\chi^2_{m-1}$. Quade's paper has other examples using the same idea, where the numerator of the $F$-ratio computed from the scores is expressed as a function of generalised $U$-statistics.

### Non-parametric analysis of covariance

Suppose for each of $m$ samples selected at random for $m$ populations, we observe responses $Y_{ij}, j = 1, \ldots, n_i, \ i = 1, \ldots, n$ and in addition for each response we have available a covariate $x_{ij}$. Quade (1965), (1967),

(1982) has suggested techniques of nonparametric analysis of covariance based on both adjustment of the responses and matching the responses on the basis of the covariates.

In both cases we obtain a "score" for each response, and the hypothesis of identical conditional distributions (i.e. that the distribution of the response conditional on the covariate is the same for each population) is tested by performing an analysis of variance on the scores. Assume for simplicity that we have a single covariate $x$. Then Quade (1965) defines for each observation a score

$$Z_{ij} = R_{ij} - \frac{N+1}{2} - c(S_{ij} - \frac{N+1}{2})$$

where $c$ is a constant, $R_{ij}$ is the rank of $Y_{ij}$ in the (pooled) $Y$-sample, and $S_{ij}$ is the rank of $x_{ij}$ in the (pooled) sample of covariates.

Assume that the distribution of the covariate is the same in each population (Quade calls this the assumption of *concomitance*) so that equal conditional distributions imply equal joint distributions. An argument similar to that in the Kruskal-Wallis example above indicates that the $F$-statistic based on these scores is asymptotically $\chi^2_{m-1}$.

Alternatively, we may rely on matching to generate scores. For each response $Y_{ij}$, we can estimate $E(Y|x_{ij})$ by the average of all $Y's$ that have the corresponding $x$'s within $\epsilon$ of $x_{ij}$, i.e. we estimate the conditional expectation by

$$\hat{Y}_{ij} = \frac{\sum_{r=1}^{m} \sum_{s=1}^{n_r} Y_{rs} I\{|x_{ij} - x_{rs}| < \epsilon\}}{\sum_{r=1}^{m} \sum_{s=1}^{n_r} I\{|X_{ij} - X_{rs}| < \epsilon\}}.$$

The test then proceeds as usual given the scores $Z_{ij} = Y_{ij} - \hat{Y}_{ij}$. An alternative set of scores is

$$Z_{ij} = \frac{\sum_{r=1}^{n} \sum_{s=1}^{n_r} \text{sgn}(Y_{ij} - Y_{rs}) I\{|x_{ij} - x_{rs}| < \epsilon\}}{\sum_{r=1}^{n} \sum_{s=1}^{n_r} I\{|x_{ij} - x_{rs}| < \epsilon\}}.$$

Both sets of scores are exchangeable under the null hypothesis so that the $F$-test is asymptotically correct.

### 6.2.6 A test for "New better than used"

In the theory of reliability, a life distribution $F$ is the distribution of a non-negative random variable. Consider two independent units having lifetimes distributed as $F$, one of age at least $y$ and one brand new. The life distribution $F$ is said to be *new better than used* (NBU) if the new unit has the greater probability of not failing for an additional period of duration $x$ for all $x > 0$ and $y > 0$. In other words, if $X$ and $Y$ denote the lifetimes of these two units, then the NBU property is equivalent to

$$Pr(X > x) \geq Pr(Y > x + y | Y > y)$$

for all $x > 0$ and $y > 0$, or, in terms of $F$, assuming $F$ is absolutely continuous,

$$1 - F(x + y) \leq (1 - F(x))(1 - F(y)).$$

Hollander and Proschan (1972) introduce the parameter

$$\Delta(F) = E(1 - F(X + Y))$$

to measure the degree to which a life distribution $F$ has the NBU property. If in fact $F$ does have this property, then

$$\begin{aligned}
\Delta(F) &= \int \int (1 - F(x + y)) dF(x) dF(y) \\
&\leq \left( \int (1 - F(x)) dF \right)^2 \\
&= \tfrac{1}{4}
\end{aligned}$$

since $1 - F(X)$ is uniformly distributed. Note that the boundary value $\tfrac{1}{4}$ is attained when $F$ is exponential. It is natural to estimate $\Delta(F)$ by $\Delta(F_n)$, where $F_n$ is the empirical d.f. of $F$ (c.f. Section 4.2). We have

$$\begin{aligned}
\Delta(F_n) &= \frac{1}{n^2} \sum_{i=1}^{n} \sum_{j=1}^{n} (1 - F_n(X_i + X_j)) \\
&= \frac{1}{n^2} \sum_{i=1}^{n} \sum_{j=1}^{n} \sum_{k=1}^{n} I\{X_k > X_i + X_j\}.
\end{aligned}$$

268

An asymptotically equivalent $U$-statistic is

$$U_n = \binom{n}{3}^{-1} \sum_{(n,3)} \psi(X_i, X_j, X_k)$$

where $\psi(x_1, x_2, x_3) = \frac{1}{3}(I\{x_1 > x_1 + x_2\} + I\{x_2 > x_1 + x_3\} + I\{x_3 > x_1 + x_2\})$. To test the hypothesis that $F$ is exponential, versus an alternative that $F$ is NBU, we can use the statistic $n^{\frac{1}{2}}(U_n - \frac{1}{4})$, and reject the null hypothesis for small values of the statistic. Since $\sigma_1^2 = 5/3888$ when $F$ is exponential, the asymptotic distribution of this statistic under the null hypothesis is normal with mean zero and variance $45/3888$. For the small sample distribution, see the article by Hollander and Proschan cited above. For a modification of the test, see Ahmad (1975) and Deshpande and Kochar (1983). Deshpande (1983) and Bandyopadhyay and Basu (1989) consider tests of exponentiality against "increasing failure rate average" alternatives.

## 6.3 Applications of Poisson convergence

Several applications of Theorem 2 of Section 3.2.4 are to be found in the literature. Notable among these are applications to the multiple comparison of correlation coefficients and testing for randomness in planar point patterns.

### 6.3.1 Comparing correlations

Suppose we make measurements of $k$ characteristics on each of $n$ individuals, and want to decide which if any of the characteristics are associated. The common approach is to calculate a correlation matrix for the data and pick out the pairs of characteristics for which the correlations are significantly different from zero. The difficulty is that the $k(k-1)/2$ coefficients are strongly dependent, and this must be taken into account in the selection procedure.

In the case of normally distributed observations, Moran (1980) has developed a procedure which involves finding the distribution of a maximum of a set of dependent sample correlations, assuming the characteristics are actually independent. This has been extended to a nonparametric context by Eagleson (1983) and Best, Cameron and Eagleson (1983) using the Poisson limit theorem of Section 3.2.4.

269

Let $Y_{ij}$ denote the measurement of the $j$th characteristic for the $i$th individual, and assume that $Y^T = (Y_{i1}, \ldots, Y_{ik})$, $i = 1, \ldots, n$ are i.i.d. random vectors. The standardised measurements

$$X_{ij} = \frac{Y_{ij} - \overline{Y}_j}{\{\sum (Y_{ij} - \overline{Y}_j)^2\}^{\frac{1}{2}}}$$

form a matrix $X = (X_{ij})$, the columns of which we denote by $X_1, \ldots, X_k$. If in fact the $k$ characteristics are independent, then $X_1, \ldots, X_k$ are independent, and if we assume also that $Y_{11}, \ldots, Y_{1k}$ have distributions identical up to location and scale, the random vectors $X_1, \ldots, X_k$ will be identically distributed as well. Now let $\phi(x, y)$ denote the inner product between $n$-vectors $x$ and $y$, so that $\phi(X_i, X_j)$ denotes the (Pearson) sample correlation between characters $i$ and $j$. For constants $\beta_k > 0$, define kernels

$$\phi_k(X_i, X_j) = \begin{cases} 1 & \text{if } \phi(X_i, X_j) > \beta_k; \\ 0 & \text{otherwise.} \end{cases}$$

Then $T_k = \sum_{(k,2)} \phi_k(X_i, X_j)$ is the number of correlations exceeding $\beta_k$. The Poisson limit theorem is then used to prove the convergence of $T_k$ to a Poisson law as $k \to \infty$. Specifically, Eagleson (1983) proves

**Theorem 1.** *Suppose that the density of the random vectors $X_1, \ldots, X_k$ is bounded, and that the constants $\beta_k$ are chosen so that $\beta_k$ increases to unity and*

$$\lim_{k \to \infty} \binom{k}{2} Pr\left(\phi(X_1, X_2) > \beta_k\right) = \lambda.$$

*Then $T_k \xrightarrow{D} P(\lambda)$ and so the probability that the maximum correlation exceeds $\beta_k$ converges to $1 - e^{-\lambda}$.*

**Proof.** To prove the theorem, we need only verify that the condition (ii) of Theorem 2 of Section 3.2.4 is satisfied. In the present context, this amounts to proving that

$$\lim_{k \to \infty} k^{-3} Pr\left(\phi(X_1, X_2) > \beta_k \quad \text{and} \quad \phi(X_2, X_3) > \beta_k\right) = 0.$$

The random vector $X_1$ has length 1 and its elements sum to zero, so it lies on the intersection of an $n$-dimensional unit sphere and the hyperplane

270

$x_1 + x_2 + \cdots + x_n = 0$. This intersection may be identified with the unit sphere in $n - 1$ dimensional space, denoted conventionally by $S_{n-2}$. We may thus think of the density of $X_1$ as a (bounded) density on $S_{n-2}$. For a fixed vector $y$ in $S_{n-2}$, let $C(y, \rho)$ denote the set of points $z$ on $S_{n-2}$ such that the angle between $z$ and $y$ is less than $\rho$, so that $C(y, \rho)$ is a "cap" on $S_{n-2}$, centred on $y$, making an angle $2\rho$ at the centre of the hypersphere. Then the correlation between two vectors $X_i$ and $X_j$ exceeds $r$ if and only if $X_j$ is in $C(X_i, \cos^{-1}(r))$.

Now let $\nu$ denote the usual uniform measure on $S_{n-2}$, and let $\nu_k$ denote the $\nu$-measure (area) of the cap $C(x, \cos^{-1}\beta_k)$ which does not depend on $x$. Note that $\nu_k$ converges to zero as $k \to \infty$. We can write

$$Pr\left(\phi(X_1, X_2) > \beta_k\right) = \int_{S_{n-2}} f(x_1) \int_{C(x_1, \cos^{-1}\beta_k)} f(x_2)\nu(dx_2)\nu(dx_1)$$

and so

$$\lim_{k\to\infty} Pr\left(\phi(X_1, X_2) > \beta_k\right)/\nu_k$$

$$= \lim_{k\to\infty} \int_{S_{n-2}} f(x_1)\nu_k^{-1} \int_{C(x_1, \cos^{-1}(\beta_k))} f(x_2)\,\nu(dx_2)\,\nu(dx_1)$$

$$= \int_{S_{n-2}} f^2(x_1)\,\nu(dx_1)$$

since $\lim_{k\to\infty} \nu_k^{-1} \int_{c(x_1, \cos^{-1}\beta_k)} f(x_2)\,\nu(dx_2) = f(x_1)$. Hence if

$$\lim_{k\to\infty} \binom{k}{2} Pr\left(\phi(X_1, X_2) > \beta_k\right) = \lambda$$

in view of the fact that $f$ is bounded by $M$, say, we must have

$$\lim_{k\to\infty} \binom{k}{2}\nu_k \int_{S_{n-2}} f^2(x)\,\nu(dx) = \lambda$$

and in particular $\nu_k = O(k^{-2})$. To verify condition (ii) of Theorem 2 of Section 3.2.4, write

$$Pr\left(\phi(X_1, X_2) > \beta_k \text{ and } \phi(X_2, X_3) > \beta_k\right)$$

$$= \int_{S_{n-2}} \left\{ \int_{C(x_1, \cos^{-1} \beta_k)} f(x_2)\,\nu(dx_2) \right\}^2 f(x_1)\,\nu(dx_1)$$

$$\leq M^2 \nu_k^2 \int_{S_{n-2}} f(x_1)\nu(dx_1)$$

$$= O(k^{-4})$$

so that $\lim_{k \to \infty} k^3 Pr\left( \phi(X_1, X_2) > \beta_k \text{ and } \phi(X_2, X_3) > \beta_k \right) = 0$ and condition (ii) is verified. Thus $T_k = \sum_{(k,2)} I\{\phi(X_i, X_j) > \beta_k\}$ converges to a Poisson variate with parameter $\lambda$, and since $\max_{(k,2)} \phi(X_i, X_j) \leq \beta_k$ if and only if $T_k = 0$, we must have $\lim_{k \to \infty} Pr\left(\max_{(k,2)} \phi(X_i, X_j) > \beta_k\right) = 1 - e^{-\lambda}$.

Theorem 1 remains true if the Pearson correlation is replaced by either the Spearman or Kendall rank correlations. Replace the matrix $X$ by a matrix $R$ of ranks: $R = (R_{ij})$ where $R_{ij}$ is the rank of $X_{ij}$ among $X_{ij}, \ldots, X_{nj}$. Assuming no ties, the vector $R_j = (R_{ij}, \ldots, R_{nj})$ will have a uniform distribution on the set of all $n!$ permutations of $\{1, 2, \ldots, n\}$, and under the assumption of independence of the $k$ characteristics, the vectors $R_1, \ldots, R_k$ will be independent. Denote a typical permutation of $\{1, 2, \ldots, n\}$ by $\mathbf{r} = (r_1, \ldots, r_n)$ and let $\mathbf{r}^{-1}$ be the permutation inverse to $\mathbf{r}$. Then the Spearman correlation between two permutations $\mathbf{r}^{(1)}$ and $\mathbf{r}^{(2)}$ is

$$Corr_S(\mathbf{r}^{(1)}, \mathbf{r}^{(2)}) = \frac{n^{-1} \sum_{i=1}^n r_i^{(1)} r_i^{(2)} - \{\frac{1}{2}(n+1)\}^2}{\frac{1}{12}(n^2 - 1)}$$

and the Kendall correlation is

$$Corr_K(\mathbf{r}^{(1)}, \mathbf{r}^{(2)}) = \frac{1}{2} \binom{n}{2}^{-1} \sum_{i=1}^n \sum_{j=1}^n \operatorname{sgn}(r_i^{(1)} - r_j^{(1)})(r_i^{(2)} - r_j^{(2)})$$

From these representations it is clear that for any $\mathbf{r}$, these two correlations satisfy

$$Corr_S(\mathbf{r}\mathbf{r}^{(1)}, \mathbf{r}\mathbf{r}^{(2)}) = Corr_S(\mathbf{r}^{(1)}, \mathbf{r}^{(2)}) \qquad (1)$$

and

$$Corr_K(\mathbf{r}\mathbf{r}^{(1)}, \mathbf{r}\mathbf{r}^{(2)}) = Corr_k(\mathbf{r}^{(1)}, \mathbf{r}^{(2)}).$$

From (1), writing $Corr(.,.)$ for either $Corr_S(.,.)$ or $Corr_K(.,.)$, it follows that when $R_1$ and $R_2$ are independent,

$$Pr\left(Corr(R_1, R_2) > \beta_k \mid R_1 = \mathbf{r}^{(1)}\right) = Pr\left(Corr(\mathbf{r}^{(1)}, R_2) > \beta_k\right)$$

$$= (n!)^{-1} \sum_{(n)} I\{Corr(\mathbf{r}^{(1)}, \mathbf{r}) > \beta_k\}$$

$$= (n!)^{-1} \sum_{(n)} I\{Corr(\mathbf{r}^{(2)}, \mathbf{r}) > \beta_k\}$$

$$= Pr\left(Corr(R_1, R_2) > \beta_k | R_1 = \mathbf{r}^{(2)}\right)$$

where as usual, $\sum_{(n)}$ denotes summations over all permutations $\mathbf{r}$ of the set $\{1, 2, \ldots, n\}$. Thus $Pr\left(Corr(R_1, R_2) > \beta_k | R_1 = \mathbf{r}\right)$ does not depend on $\mathbf{r}$ and so is equal to the unconditional probability $Pr\left(Corr(R_1, R_2) > \beta_k\right)$. To check (ii) of Theorem 2 of Section 3.2.4, note that using the above we get

$$Pr\left(Corr(R_1, R_2) > \beta_k \text{ and } Corr(R_2, R_3) > \beta_k\right)$$

$$= \sum_{(n)} Pr^2\left(Corr(R_1, \mathbf{r}) > \beta_k\right) Pr(R_2 = \mathbf{r})$$

$$= (n!)^{-1} \sum_{(n)} Pr^2\left(Corr(R_1, R_2) > \beta_k\right)$$

$$= Pr^2\left(Corr(R_1, R_2) > \beta_k\right)$$

$$= O(k^{-4}).$$

The assertion $Pr\left(Corr(R_1, R_2) > \beta_k\right) = O(k^{-2})$ follows from the defining property of $\beta_k$, namely that $\lim_{k \to \infty} \binom{k}{2} Pr\left(Corr(R_1, R_2) > \beta_k\right) = \lambda$. Thus (ii) is satisfied and Theorem 1 remains true for both rank correlations. We note in passing that the approximation is quite satisfactory for $k$ as small as 5, as seen in Eagleson (1983) and Best et. al. (1983).

### 6.3.2 Applications to spatial statistics

Consider the problem of testing for randomness in spatial patterns. If $X_1, \ldots, X_n$ are random vectors on the plane, the point pattern formed by these vectors is random if the vectors are i.i.d., and a test for randomness may be based on consideration of the number of pairs of points that are less than some fixed distance apart. Under the randomness assumption, the number of such pairs is asymptotically Poisson. More precisely, Silverman and Brown (1978) define pairs of points $(X_i, X_j)$ to be "$n$-close" if $|X_i - X_j| < n^{-1}\mu$ when $\mu > 0$ is some fixed constant. Then provided the common density, $f$ say, of the $X_i$ is bounded, they prove that the number of $n$-close pairs converges in distribution to a Poisson law with parameter $\frac{1}{2}\pi\mu^2 \int f^2$. The proof is almost identical to that of Theorem 1 above, and is consequently omitted.

Another possibility is to test randomness against the alternative that points tend to lie along straight lines. Given any triple of points in the plane $X_1, X_2, X_3$, define

$$\theta(X_1, X_2, X_3) = \pi - \quad \text{the largest angle of triangle} \quad X_1 X_2 X_3.$$

Also define $N_n(\epsilon)$ to be the number of triples among the $n$ points for which $\theta(X_1, X_2, X_3)$ is less that $\epsilon$. Silverman and Brown also prove that $N_n(\epsilon)$ is asymptotically Poisson with parameter depending on $f$; details may be found in their paper.

Other applications of Poisson convegence are described by Babour and Eagleson (1983), who consider a statistic for testing association between time and position for spatial point patterns evolving through time, and a statistic for testing randomness versus clustering in one-dimensional point patterns made up of two different types of points. For other applications to spatial statistics, see Ripley (1981).

## 6.4    Sequential estimation

A considerable amount of research (for a summary see Sen (1981), (1985)) has recently focused on the problem of sequential point and interval estimation. We focus briefly on the latter.

Let $X_1, \ldots, X_n$ be i.i.d. with mean $\mu$ and variance $\sigma^2$. Suppose we want to construct a confidence interval for $\mu$ with fixed length $2d$ and specified coverage probability $1 - \alpha$. How big a sample should we take? Suppose $\sigma$ is known, and that $z_\alpha$ denotes the upper $\alpha$ percentage point of the standard normal distribution. Elementary theory says that if we chose

$n$ to be the smallest integer larger than $(z_{\alpha/2}\sigma/d)^2$ then

$$\lim_{d \to 0} Pr(\overline{X}_{n_d} - d < \mu < \overline{X} - n_d + d) = 1 - \alpha. \tag{1}$$

If $\sigma^2$ is unknown, the problem has no exact solution. However, we can adopt a sequential approach, and define a *stopping rule*. We draw observations $X_1, X_2, \ldots$ until the stopping rule tells us to stop sampling, resulting in a (random) sample size $N_d$. We then use $\overline{X}_{N_d} \pm d$ as our interval. In a seminal paper, Chow and Robbins (1965) introduced a stopping rule which gave asymptotically correct coverages, in that (1) is true using sample size $N_d$. Specifically, let $s_n^2 = \dfrac{1}{n}\sum_{i=1}^{n}(X_i - \overline{X})^2$, and continue sampling until $s_n^2 + n^{-1} \leq d^2 n/z_{\alpha/2}^2$, resulting in a random sample size $N_d$. Then Chow and Robbins prove that the interval $\overline{X}_{N_d} \pm d$ gives asymptotically the correct coverage i.e. that

$$\lim_{d \to 0} Pr(\overline{X}_{N_d} - d \leq \mu \leq \overline{X}_{N_d} + d) = \alpha.$$

Csenki (1980) obtains a related order of convergence result: if $\epsilon > 0$ and $E|X_1|^{3+5/(2\delta)+\epsilon} < \infty$ for some $\delta$ in $(0, \frac{1}{2})$ then

$$Pr(\overline{X}_{N_d} - d \leq \mu \leq \overline{X}_{N_d} + d) = 1 - \alpha + O(d^{\frac{1}{2}-\delta}).$$

These results have been generalised to $U$-statistics by Sproule (1969) and Mukhopadhyay (1981). Suppose we want a sequential fixed width confidence interval for $\theta = E\psi(X_1, \ldots, X_k)$. Sproule (1969) proposed the following generalization of the Chow and Robbins approach: Let $S_n^2 = n\widehat{Var}(JACK)$ where $\widehat{Var}(JACK)$ is the jackknife estimate of $VarU_n$ introduced in 5.1.1. Then $S_n^2$ estimates $k^2\sigma_1^2$, and we can define a stopping rule analogously to the Chow and Robbins rule : sample until $S_n^2 + n^{-1} \leq d^2 n/z_{\alpha/2}^2$, obtaining a sample size $N_d$. Then Sproule (1974) proves the asymptotic normality of $N_d^{\frac{1}{2}}(U_{N_d} - \theta)/k\sigma_1$ and Mukhopadhyay (1981) shows that the interval $U_{N_d} \pm d$ gives the correct coverage asymptotically, and proves a rate of convergence. Drawing on results on the rate of convergence for random $U$-statistics to normality discussed in 3.7.6 he shows that

$$Pr(U_{N_d} - d \leq \theta \leq U_{N_d} + d) = 1 - \alpha + O(d^{\frac{1}{2}-\delta})$$

whenever $E|\psi(X_1,\ldots,X_k)|^{\epsilon+5/(2\delta)-1}$ (Note that this is an improvement on the Csenki (1980) result in the case $U_n = \overline{X}_n$.) Csenki (1981) and Mukhopadhyay and Vik (1985) consider the rate of approach of $N_d$ to normality. Two stage procedures are treated in Mukhopadhyay and Vik (1987).

A parallel theory of sequential point estimation is available. We refer the reader in search of further details to Sen (1981), and also mention Sen (1986), Sen and Ghosh (1981), Ghosh (1981). Williams and Sen (1973, 1974) and Williams (1978) consider multivariate generalisations.

## 6.5   Other applications

The applications described above by no means exhaust the range of problems in which $U$-statistic theory has been found useful. An important application in the theory of non-parametric statistics is that of permutation tests for $U$-statistics, due to Sen. In a series of papers Sen (1965, 1966, 1967) extends the classic theory of permutation tests developed by Wald and Wolfowitz for linear rank statistics to the class of generalised $U$-statistics. A full account of this theory, with additional references, is given Puri and Sen (1971), Section 3.4, so we shall not describe it here. Additional applications include (in no particular order) Bhapkar and Gore (1971) who describe the application of $U$-statistic theory to multivariate selection procedures, and consider the problem and selecting the "best" population on the basis of some characterising parameter. Gore (1971) has a test based on $U$-statistics for interaction in a two-way layout. Chatterjee (1966) considers the problem of testing equality of scale parameters in several samples. Sen (1969) also considers robustness properties (against heterogeneity of the sample observations) of tests based on $U$-statistics.

Applications to more specific problems of a nonparametric nature are made by Veraverbecke (1985), who considers the problem of estimating the functional $\int f^2 dx$ of a probability density. His estimator is a $U$-statistic whose kernel depends on the sample size, and has an asymptotically normal distribution. O'Reilly and Mulke (1980) apply $U$-statistic theory to multi-response permutation procedures in both the i.i.d. and finite population sampling cases. Delong and Sen (1981) estimate the probability that

276

one random quantity is stochastically larger than another. Palachek and
Schucany (1983, 1984) devise a measure of association between the rankings
given by a group of judges and an external ranking.

# REFERENCES

Abramowitz, M. and Stegun, I. (1965). *Handbook of Mathematical Functions*, Dover, New York.

Aerts, M. and Callaert, H. (1986). The exact approximation order in the central limit theorem for random $U$-statistics. *Comm. Statist Ser. C - Sequential Analysis* **5**, 19-35.

Ahmad, I.A. (1975). A non-parametric test for the monotonicity of the failure rate function. *Commun. Statist.* **4**, 967-974.

Ahmad, I.A. (1980). On the Berry-Esseen Theorem for random $U$-statistics. *Ann. Statist.* **6**, 1395-1398.

Ahmad, I.A. (1980). A Hájek-Rényi inequality for $U$-statistics. *Acta Math. Acad. Sci. Hungar.* **36**, 267-269.

Ahmad, I.A. (1981). On some asymptotic properties of $U$-statistics. *Scand. J. Statist.* **8**, 175-182.

Anscombe, F. (1952). Large sample theory of sequential estimation. *Proc. Camb. Philos. Soc.* **48**, 600-607.

Arvesen, J.N. (1969). Jackknifing $U$-statistics. *Ann. Math. Statist.* **40**, 2076-2100.

Athreya, K.B, Ghosh, M., Low, L.Y. and Sen, P.K. (1984). Laws of large numbers for bootstrapped $U$-statistics. *J. Statist. Planning Inference* **9**, 185-194.

Baldi, P. and Rinott, Y. (1989). Asymptotic normality of some graph-related statistics. *J. Appl. Prob.* **26**, 117-175.

Bandyopadhyay, D. and Basu, A.P. (1989). A note on tests for exponentiality by Deshpande. *Biometrika* **76**, 403-405.

Barbour, A.D. and Eagleson, G.K. (1983). Poisson approximation for some statistics based on exchangeable trials. *Adv. Appl. Prob.* **15**, 585-600.

Barbour, A.D. and Eagleson, G.K. (1984). Poisson convergence for dissociated statistics. *J. R. Statist. Soc. Ser. B* **46**, 397-402.

Barbour, A.D. and Eagleson, G.K. (1985). Multiple comparisons and sums of dissociated random variables. *Adv. Appl. Prob.* **17**, 147-162.

Barbour, A.D. and Eagleson, G.K. (1987). An improved Poisson limit theorem for sums of dissociated random variables. *J. Appl. Prob.* **24**, 586-599.

Bell, C.B., Blackwell, D. and Breiman, L. (1960). On the completeness of order statistics. *Ann. Math. Statist.* **31**, 794-797.

Beran, R. (1984). Jackknife approximations to bootstrap estimates. *Ann. Statist.* **12**, 101-118.

Berk, R.H. (1966). Limiting behaviour of posterior distributions when the model is incorrect. *Ann. Math. Statist.* **37**, 51-58.

Berman, M. and Eagleson, G.K. (1983). A Poisson limit theorem for incomplete symmetric statistics. *J. Appl. Prob.* **20**, 47-60.

Best, S.J., Cameron, M.A. and Eagleson, G.K. (1983). A test for comparing large sets of tau values. *Biometrika* **70**, 447-453.

Bhapkar, V.P. (1961). A nonparametric test for the problem of several samples. *Ann. Math. Statist.* **32**, 1108-1117.

Bhapkar, V.P. and Gore, A.P. (1971). Some selection procedures based on $U$-statistics for the location and scale problems. *Ann. Inst. Stat. Math.* **23**, 375-386.

Bhattacharya, R.N. and Puri, M.L. (1983). On the order of magnitude of cumulants of von Mises functionals and related statistics. *Ann. Prob.* **11**, 346-354.

Bickel, P.J. (1974). Edgeworth expansions in nonparametric statistics. *Ann. Statist.* **2**, 1-20.

Bickel, P.J., Gotze, F. and van Zwet, W.R. (1986). The Edgeworth expansion for $U$-statistics of degree two. *Ann. Statist.* **14**, 1463-1484.

Billingsley, P. (1968). *Convergence of Probability Measures*, Wiley, New York.

Billingsley, P. (1979). *Probability and Measure*, Wiley, New York.

Blom, G. (1976). Some properties of incomplete $U$-statistics. *Biometrika* **63**, 573-580.

Blom, G. (1980). Extrapolation of linear estimates to larger sample sizes. *J. Amer. Statist. Assoc.* **63**, 573-580.

Boos, D.D. and Serfling, R.J. (1980). A note on differentials, and the CLT and LIL for statistical functions, with application to M-estimates. *Ann. Statist.* **8**, 618-624.

Borovskikh, Yu.V. (1984). Asymptotic properties of distributions of $U$-statistics and Von Mises functionals. *Theory Prob. Applns.* **28**, 195-197.

Breiman, L. (1968). *Probability*, Addison-Wesley, Reading, Massachusetts.

Brown, B.M. and Kildea, D.G. (1978). Reduced $U$-statistics and the Hodges-Lehmann estimator. *Ann. Statist.* **6**, 828-835.

Brown, T.C. and Silverman, B.W. (1979). Rates of Poisson convergence for $U$-statistics. *J. Appl. Prob.* **16**, 428-432.

Burrill, C. W. (1972). *Measure, Integration and Probability*, McGraw-Hill, New York.

Callaert, H. and Janssen, P. (1978). The Berry-Esseen theorem for $U$-statistics. *Ann. Statist.* **6**, 417-421.

Callaert, H., Janssen, P. and Veraverbeke, N. (1980). An Edgeworth expansion for $U$-statistics. *Ann. Statist.* **8**, 299-312.

Callaert, H. and Veraverbeke, N. (1981). The order of the normal approximation for a studentized $U$-statistic. *Ann. Statist.* **9**, 194-200.

Chan, Y.K. and Wierman, J. (1977). On the Berry-Esseen theorem for $U$-statistics. *Ann. Prob.* **5**, 136-139.

Chatterjee, S.K. (1966). A multi-sample non-parametric scale test based on $U$-statistics. *Cal. Statist. Assoc. Bull.* **15**, 109-119.

Chen, L.H.Y. (1975). Poisson approximation for dependent trials. *Ann. Prob.* **3**, 534-545.

Chen, X. (1980). On limiting properties of $U$-statistics and von Mises statistics. *Sci. Sinica* **23**, 1079-1091.

Cheng, K.F. (1981). On Berry-Esseen rates for jackknife estimators. *Ann. Statist.* **9**, 694-696.

Chernoff, H. and Teicher, H. (1958). A central limit theorem for sums of interchangeable random variables *Ann. Math. Statist.* **29**, 118-130.

Chow, Y.S. and Robbins, H. (1965). On the asymptotic theory of fixed width sequential confidence intervals for the mean. *Ann. Math. Statist.* **36**, 457-462.

Chow, Y.S. and Teicher, H. (1978). *Probability Theory – Independence, Interchangeability, Martingales*, Springer-Verlag, New York.

Csenki, A. (1980). On the convegence rate of fixed width sequential confidence intervals. *Scand. Actuarial J.* 107-111.

Csenki, A. (1981). A theorem on the departure of randomly indexed $U$-statistics from normality with an application in fixed width sequential interval estimation. *Sankhyā A* **43**, 84-99.

Csörgő, M and Horváth, L. (1988). Invariance principles for changepoint problems. *J. Multivariate Anal.* **27**, 151-168.

Dasgupta, R. (1984). On large deviation probabilities of $U$-statistics in non i.i.d. case. *Sankhyā A* **46**, 110-116.

David, H.A. and Rogers, M.P. (1983). Order statistics in overlapping samples, moving order statistics and $U$-statistics. *Biometrika* **70**, 245-9.

Davis, C.E. and Quade, D. (1978). $U$-statistics for skewness or symmetry. *Comm. Statist. A – Theory Methods* **7**, 413-418.

Dehling, H. (1989a). The functional law of the iterated logarithm for von Mises functionals and multiple Wiener integrals. *J. Multivariate Anal.* **28**, 177-189.

Dehling, H. (1989b). Complete convergence of triangular arrays and the law of the iterated logarithm for $U$-statistics. *Statistics and Prob. Letters* **7**, 319-321.

Dehling, H., Denker, M. and Phillip, W. (1984). Invariance principles for von Mises and $U$-statistics. *Z. Wahrsch. und Verw. Gebiete* **67**, 139-67.

Dehling, H., Denker, M. and Phillipp, W. (1986). A bounded law of the interated logarithm for Hilbert-space valued martingales and its application to $U$-statistics. *Prob. Th. Rel. Fields.* **72**, (1986). 111-131.

Delong, E.R. and Sen, P.K. (1981). Estimation of $P(X > Y)$ based on progressively truncated versions of the Wilcoxon-Mann-Whitney statistics. *Comm. Statist. A – Theory Methods* **10**, 963-981.

Denker, M., Grillenberger, C. and Keller, G. (1985). A note on invariance principles for von Mises statistics. *Metrika* **32**, 197-214.

Denker, M. and Keller, G. (1983). On $U$-statistics and v. Mises' statistics for weakly dependent processes. *Z. Wahrsch. und Verw. Gebiete* **64**, 505-522.

Deshpande, J.V. (1983). A class of tests for exponentiality against increasing failure rate average alternatives. *Biometrika* **70**, 514-518.

Deshpande, J.V. and Kochar, S.C. (1983). A linear combination of $U$-statistics for testing new better than used. *Comm. Statist. A – Theory Methods* **12**, 153-159.

de Wet, T. (1987). Degenerate $U$- and $V$-statistics. *South African Statist. J.* **21**, 99-129.

de Wet, T. and Randles, R.H. (1987). On the effect of substituting parameters in limiting $\chi^2$ $U$- and $V$-statistics. *Ann. Statist.* **15**, 398-412.

Dharmadhikari, S.W., Fabian, V. and Jogdeo, K. (1968). Bounds on the moments of martingales. *Ann. Math. Statist.* **39**, 1719-1723.

Donsker, M. (1951). An invariance principle for certain probability limit theorems. *Mem. Amer. Math. Soc.* **6**.

Doob, J.L. (1953). *Stochastic Processes*, Wiley, New York.

Dynkin, E.B. and Mandelbaum, A. (1983). Symmetric statistics, Poisson point processes and multiple Weiner integrals. *Ann. Statist.* **11**, 739-745.

Eagleson, G.K. (1979). Orthogonal expansions and $U$-statistics. *Austral. J. Statist.* **21**, 221-237.

Eagleson, G.K. (1982). A robust test for multiple comparisons of correlation coefficients. *Austral. J. Statist.* **25**, 256-263.

Efron, B. (1981). Nonparametric estimates of standard error: The jackknife, the bootstrap and other methods. *Biometrika* **68**, 589-599.

Efron, B. (1982). *The Jackknife, the Bootstrap and Other Resampling Plans*, SIAM, Philadelphia.

Efron, B. and Stein, C. (1981). The jackknife estimate of variance. *Ann. Statist.* **3**, 586-596.

Feller W. (1968). *An Introduction to Probability Theory and its Applications, Vol 1 (3rd Ed)*, Wiley, New York.

Feller W. (1971). *An Introduction to Probability Theory and its Applications, Vol 2 (2n d Ed)*, Wiley, New York.

Filippova, A.A. (1961). Mises' theorem on the asymptotic behaviour of functionals of empirical distribution functions and its statistical applications. *Theory Prob. Applns.* **17**, 24-57.

Fisher, N.I. (1982). Unbiased estimation for some non-parametric families of distributions. *Ann. Statist.* **10**, 603-615.

Fisher, N.I. and Lee, A.J. (1981). Nonparametric measures of angular-linear association. *Biometrika* **68**, 629-636.

Fisher, N.I. and Lee, A.J. (1982). Nonparametric measures of angular-angular association. *Biometrika* **69**, 315-321.

Fisher, N.I. and Lee, A.J. (1983). A correlation coefficient for circular data. *Biometrika* **70**, 327-332.

Fisher, N.I. and Lee, A.J. (1986). Correlation coefficients for random variables on the sphere and hypersphere. *Biometrika* **73**, 159-164.

Fraser, D.A.S. (1954). Completeness of order statistics. *Canad. J. Math.* **6**, 42-45.

Fraser, D.A.S. (1957). *Nonparametric Methods in Statistics*, Wiley, New York.

Frees, E.W. (1989). Infinite order $U$-statistics. *Scand. J. Statist.* **16**, 29-45.

Friedrich, K.O. (1989). A Berry-Esseen bound for functions of independent random variables. *Ann. Statist.* **17**, 170-183.

Ghosh, M. (1981). Sequential point estimation of the means of $U$-statistics in finite population sampling. *Comm. Statist. A – Theory Methods* **10**, 2215-2229.

Ghosh, M. (1985). Berry-Esseen bounds for functionals of $U$-statistics. *Sankhyā A* **47**, 255-270.

Ghosh, M. and Dasgupta, R. (1982). Berry-Esseen theorems for $U$-statistics in the non i.i.d. case. *Proc. Colloquia Math. Soc. Janos Bolyai on Nonparametric Inference.* Budapest, Hungary. 255-270.

Gore, A.P. (1971). A nonparametric test based on $U$-statistics for interaction in a two-way table (Abstract). *Ann. Math. Statist.* **41**, 1486.

Gotze, F. (1979). Asymptotic expansions for bivariate von Mises functionals. *Z. Wahrsch. und Verw. Gebiete* **50**, 333-335.

Gotze, F. (1987). Approximations for multivariate $U$-statistics. *J. Multivariate Anal.* **22**, 212-229.

Grams, W.F. and Serfling R.J. (1973). Convergence rates for $U$-statistics and related statistics. *Ann. Statist.* **1**, 153-160.

Gregory, G.G. (1977). Large sample theory for $U$-statistics and tests of fit. *Ann. Statist.* **5**, 110-123.

Grusho, A.A. (1986). On convergence of counting processes associated with $U$-statistics. *Theory Prob. Applns.* **30**, 626-630.

Grusho, A.A. (1987). On the problem of distributions of $U$-statistics. *Theory Prob. Applns.* **32**, 369-373.

Gupta, A.K. (1967). An asymptotically non-parametric test of symmetry. *Ann. Math. Statist.* **38**, 849-866.

Gupta, S.S. and Panchapakesan, S. (1982). Egeworth expansions in statistics: some recent developments. *Coll. Math. Soc. Janos Bolyai 36. Limit Theorems in Probability and Statistics.* Vezprém, Hungary.

Hall, P. (1979). An invariance theorem for $U$-statistics. *Stochastic Proc. Applns.* **9**, 163-174.

Halmos, P.R. (1946). The theory of unbiased estimation. *Ann. Math. Statist.* **17**, 34-43.

Halmos, P.R. (1950). *Measure Theory*, Van Nostrand, New York.

Hanson, D.L. (1970). Some results on convergence rates for weighted sums. *Lecture Notes in Mathematics* **160**, 53-63.

Hartman, P, and Winter, A. (1941). On the law of the iterated logarithm. *Amer. J. Math.* **63**, 169-176.

Helmers, R. (1985). The Berry-Esseen bound for studentised $U$-statistics. *Canadian J. Statist.* **13**, 79-82.

Helmers, R. and Van Zwet, W.R. (1982). The Berry-Esseen bound for $U$-statistics. *Statistical Decision Theory and Related Topics III, Vol. 1*, 497-512.

Heyde, C.C. (1981). Invariance principles in statistics. *International Statistical Review* **49**, 143-152.

Hoeffding, W. (1948a). A class of statistics with asymptotically normal distribution. *Ann. Math. Statist.* **19**, 293-325.

Hoeffding, W. (1948b). A non-parametric test of independence. *Ann. Math. Statist.* **19**, 546-337.

Hoeffding, W. (1961). The strong law of large numbers for $U$-statistics. *Institute of Statistics Mimeo Series No 302, University of North Carolina.*

Hoeffding, W. (1977). Some complete and boundedly complete families of distributions. *Ann. Statist.* **5**, 278-291.

Hoeffding, W. and Robbins,H. (1948). The central limit theorem for dependent random variables. *Duke Math. J.* **14**, 773-780.

Hollander, M. and Proschan, F. (1972). Testing whether new is better than used. *Ann. Math. Statist.* **43**, 1136-1146.

Horváth, L. (1985). Strong laws for randomly indexed $U$-statistics. *Math. Proc. Camb. Phil. Soc.* **98**, 559-567.

Ibragimov, I.A. and Linnik, Yu.V. (1971). *Independent and Stationary Sequences of Random Variables,* Walters-Noordhoff, Groningen.

Ibragimov, I.A. and Rozanov,Y.A. (1978). *Guassian Random Processes,* Springer-Verlag, New York.

Ito, K. (1951). Multiple Weiner integral. *J. Math. Soc. Japan* **3**, 157-169.

Janson, S. (1984). The asymptotic distributions of incomplete $U$-statistics. *Z. Wahrsch. und Verw. Gebiete* **66**, 495-505.

Janssen, P. (1981). Rate of convergence in the central limit theorem and in the strong law of large numbers for von Mises statistics. *Metrika* **28**, 35-46.

Janssen, P., Serfling, R.J. and Veraverbeke, N. (1987). Asymptotic normality of $U$-statistics based on trimmed samples. *J. Statist. Planning Inference* **16**, 63-74.

Johnson, R.A. and Wehrly, T.E. (1977). Measures and models for angular correlation and angular-linear correlation. *J.R.Statist. Soc. B* **39**, 222-229.

Jupp, P.E. (1987). A nonparametric correlation coefficient and a two-sample test for random vectors or directions. *Biometrika* **74**, 887-890.

Jupp, P.E. and Mardia, K.V. (1980). A general correlation coeficient for directional data and related regression problems. *Biometrika* **67**, 163-173.

Karlin, S. and Rinott, Y. (1982). Applications of ANOVA decompositions for comparison of conditional variance statistics including jackknife estimates. *Ann. Statist.* **10**, 485-501.

Kendall, M.G. and Stuart, A. (1963). *The Advanced Theory of Statistics, Vol 1 (2nd Ed, )*. Griffin, London.

Khashimov, Sh. A. (1987). On the asymptotic distribution of the generalised $U$-statistics for dependent variables. *Theory Prob. Applns.* **32**, 373-375.

Kokic, P.N. (1987). Rates of convergence in the strong law of large numbers for degenerate $U$-statistics. *Statistics and Prob. Letters* **5**, 371-374.

Korolyuk, V.S. and Borovskikh, Yu.V. (1986a). Approximation of non-degenerate $U$-statistics. *Theory Prob. Applns.* **30**, 439-450.

Korolyuk, V.S. and Borovskikh, Yu.V. (1986b). Expansions for $U$-statistics and Mises functionals *Ukranian Math.J.* **37**, 358-364.

Korolyuk, V.S. and Borovskikh, Yu.V. (1988). Convergence rate for degenerate von Mises functionals. *Theory Prob. Applns.* **33**, 125-135.

Krewski, D. (1978). Jackknifing $U$-statistics in finite populations. *Comm. Statist. A – Theory Methods* **7**, 1-12.

Landers, D. and Rogge, L. (1976). The exact approximation order in the Central Limit Theorem for random summation. *Z. Wahrsch. und Verw. Gebiete* **36**, 269-283.

Lee, A.J. (1982). On incomplete $U$-statistics having minimum variance. *Austral. J. Statist.* **24**, 275-282.

Lee, A.J. (1985). On estimating the variance of a $U$-statistic. *Comm. Statist. A – Theory Methods* **14**, 289-301.

Lehmann, E.L. (1951). Consistency and unbiasedness of certain nonparametric tests. *Ann. Math. Statist.* **22**, 165-179.

Lehmann, E.L. (1983). *Theory of Point Estimation*, Wiley, New York.

Lenth, R. (1983). Some properties of $U$-statistics. *American Statistician* **37**, 311-313.

Lin, K-H. (1981). Convergence rates and the first exit time for $U$-statistics. *Bull. Inst. Math. Acad. Sinica* **9**, 129-143.

Locke, C. and Spurrier, J.D. (1976). The use of $U$-statistics for testing normality against non-symmetric alternatives. *Biometrika* **63**, 143-147.

Locke, C. and Spurrier, J.D. (1977). The use of $U$-statistics for testing normality against alternatives with both tails heavy or light. *Biometrika* **64**, 638-640.

Loynes, R.M. (1969). The central limit theorem for backwards martingales. *Z. Wahrsch. und Verw. Gebiete.* **13**, 1-8.

Loynes, R.M. (1970). An invariance principle for reversed martingales. *Proc. Amer. Math. Soc.* **25**, 56-64.

Majumdar, H. and Sen, P.K. (1978). Invariance principles for jackknifing $U$-statistics for finite population sampling and some applications. *Comm. Statist. A – Theory Methods* **7**, 1007-1025.

Malevich, T.L. and Abdalimov, B. (1977). Stable limit distributions for $U$-statistics. *Theory Prob. Applns.* **22**, 370-377.

Malevich, T.L. and Abdalimov, B. (1979). Large deviation probabilities for $U$-statistics. *Theory Prob. Applns.* **24**, 215-219.

Mardia, K.V. (1972). *Statistics of Directional Data*, Acedemic Press, London.

Mandelbaum, A. and Taqqu, M.S. (1984). Invariance principle for symmetric statistics. *Ann. Statist.* **12**, 483-496.

Milbrodt, H. (1987). An invariance principle for $U$-statistics in simple random sampling without replacement. *Metrika* **34**, 195-200.

Miller, R.G. and Sen, P.K. (1972). Weak convergence of $U$-statistics and von Mises differentiable statistical functions. *Ann. Math. Statist.* **43**, 31-41.

Moran, P.A.P. (1980). Testing the largest of a set of correlation coefficients. *Austral. J. Statist.* **22**, 289-297.

Mukhopadhyay, N. (1981). Convergence rates of sequential confidence intervals and tests for the mean of a $U$-statistic. *Comm. Statist. A – Theory Methods* **10**, 2231-2244.

Mukhopadhyay, N. and Vik, G. (1985). Asymptotic results for stopping times based on $U$-statistics. *Comm. Statist. C – Sequential Analysis* **4**, 83-109.

Mukhopadhyay, N. and Vik, G. (1988). Convergence rates for two-stage confidence intervals based in $U$-statistics. *Ann. Inst. Statist. Math.* **40**, 111-117.

Nandi, H.K. and Sen, P.K. (1963). On the properties of $U$-statistics when the observations are not independent II. *Cal. Statist. Assoc. Bull.* **12**, 125-143.

Neuhaus, G. (1977). Functional limit theorems for $U$-statistics in the degenerate case. *J Multivariate Anal.* **7**, 424-439.

Nolan, D. and Pollard, D. (1987). U processes: Rates of convergence. *Ann. Statist.* **15**, 780-799.

Nolan, D. and Pollard, D. (1988). Functional limit theorems for U processes. *Ann.Prob.* **16**, 1291-1298.

Nowicki, E. (1988). Asymptotic Poisson distributions with applications to statistical analysis of graphs. *Adv. Appl. Prob.* **20**, 315-330.

Nowicki, E. (1989). Asymptotic normality of graph statistics. *J. Statist. Planning Inference* **21**, 209-222.

Nowicki, E. and Wierman, J.C. (1987). Subgraph counts in random graphs using incomplete $U$-statistic methods. Unpublished preprint.

Oja, H. (1981). Two location and scale-free goodness-of-fit tests. *Biometrika* **68**, 637-640.

Oja, H. (1983). New tests for normality. *Biometrika* **70**, 297-299.

O'Reilly F.J. (1980). Asymptotic normality of MRPP statistics from invariance principles of $U$-statistics. *Comm. Statist. A – Theory Methods* **9**, 629-637.

Palachek, A.P., and Schucany, W.R. (1983). On the correlation of a group of rankings with an external ordering relative to the internal concordance. *Statistics and Prob. Letters* **1**, 259-263.

Palachek, A.P., and Schucany, W.R. (1984). On approximate confidence intervals for measures of concordance. *Psychometrika* **49**, 133-141.

Puri, M.L. and Sen, P.K. (1971). *Nonparametric Methods in Multivariate Analysis*, Wiley, New York.

Quade, D. (1965). On analysis of variance for the K-sample problem. *Ann. Math. Statist.* **37**, 1747-1758.

Quade, D. (1967). Rank analysis of covariance. *J. Amer. Statist. Assoc.* **62**, 1187-1200.

Quade, D. (1982). Nonparametric analysis of covariance by matching. *Biometrics* **38**, 597-611.

Raghavarao, D. (1971). *Constructions and Combinatorial Problems in Design of Experiments*, Wiley, New York.

Randles, R.H. (1982). On the asymptotic normality of statistics with estimated parameters. *Ann. Statist.* **10**, 462-474.

Randles, R.H., Fligner, M.A., Policello, G.E. and Wolfe, D.A. (1981). An asymptotically distribution-free test for symmetry versus assymmetry. *J. Amer. Statist. Assoc.* **75**, 168-172.

Randles, R.H. and Wolfe, D.A. (1979). *Introduction to the Theory of Nonparametric Statistics*, Wiley, New York.

Rao Jammalamadaka, S. and Janson, S. (1986). Limit theorems for a triangular scheme of $U$-statistics with applications to interpoint distances. *Ann. Prob.* **14**, 1347-1358.

Riordan, J. (1958). *An Introduction to Combinatorial Analysis*, Wiley, New York.

Ripley, B.D. (1981). *Spatial Statistics*, Wiley, New York.

Rivest, L.-P. (1982). Some statistical methods for bivariate circular data. *J. R. Statist. Soc. B* **44**, 81-90.

Ronzhin, A.F. (1982). Asymptotic formulas for the moments of $U$-statistics with degenerate kernel. *Theory Prob. Applns.* **27**, 49-58.

Ronzhin, A.F. (1986). A functional limit theorem for homogeneous $U$-statistics with a degenerate kernel. *Theory Prob. Applns.* **30**, 806-810.

Rubin, H. and Vitale, R.A. (1980). Asymptotic distribution of symmetric statistics. *Ann. Statist.* **8**, 165-170.

Sen, P.K. (1960). On some convergence properties of $U$-statistics. *Cal. Statist. Assoc. Bull.* **10**, 1-18.

Sen, P.K. (1963). On the properties of $U$-statistics when the observations are not independent. *Cal. Statist. Assoc. Bull.* **12**, 69-92.

Sen, P.K. (1965). On some permutation tests based on $U$-statistics. *Cal. Statist. Assoc. Bull.* **14**, 106-126.

Sen, P.K. (1966). On a class of bivariate two-sample nonparametric tests. *Proc. Fifth Berkeley Symp. Math. Stat. Prob.* **1**, 638-656.

Sen, P.K. (1967). On a class of multisample permutation tests based on $U$-statistics. *J. Amer. Statist. Assoc.* **62**, 1200-1213.

Sen, P.K. (1969). On a robustness property of a class of non-parametric tests based on $U$-statistics. *Bull. Cal. Stat. Ass.* **18**, 51-60.

Sen, P.K. (1972). Limiting behaviour of functionals of empirical distributions for *-mixing processes. *Z. Wahrsch. und Verw. Gebiete* **25**, 71-82.

Sen, P.K. (1974a). Weak convergence of generalised $U$-statistics. *Ann. Prob.* **2**, 90-102.

Sen, P.K. (1974b). Almost sure behaviour of $U$-statistics and von Mises differentiable statistical functions. *Ann. Statist.* **2**, 387-395.

Sen, P.K. (1977a). Almost sure convergence of generalized $U$-statistics. *Ann. Prob.* **5**, 287-290.

Sen, P.K. (1977b). Some invariance principles relating to jackknifing and their role in sequential analysis. *Ann. Statist.* **5**, 316-329.

Sen, P.K. (1981). *Sequential Nonparametrics: Invariance Principles and Statistical Inference*, Wiley, New York.

Sen, P.K. (1984). Invariance principles for $U$-statistics and von Mises' functionals in the non-I.D. case. *Sankhyā A* **46**, 416-425.

Sen, P.K. (1985). *Theory and applications of sequential nonparametrics*, SIAM, Philadelphia.

Sen, P.K. and Ghosh, M. (1981). Sequential point estimation of estimable parameters based on $U$-statistics. *Sankhyā A* **43**, 331-344.

Serfling, R.J. (1971). The law of the iterated logarithm for $U$-statistics and related von Mises statistics (abstract). *Ann. Math. Statist.* **42**, 1974.

Serfling, R.J. (1980). *Approximation Theorems of Mathematical Statistics*, Wiley, New York.

Serfling, R.J. (1984). Generalized L-, M- and R-statistics. *Ann. Statist.* **12**, 76-86.

Silverman, B.W. (1978). Distances on circles, toruses and spheres. *J. Appl. Prob.* **15**, 136-143.

Silverman, B. and Brown, T. (1978). Short distances, flat triangles and Poisson limits. *J. Appl. Prob.* **15**, 815-825.

Singh, K. (1981). On the asymptotic accuracy of Efron's bootstrap. *Ann. Statist.* **9**, 1187-1195.

Sprott, D.A. (1954). A note on balanced incomplete block designs. *Canad. J. Math.* **6**, 341-346.

Sproule, R.N. (1969). *A Sequential Fixed Fidth Fonfidence Interval for the mean of a U-statistic*, **199**, 55-64.

Sproule, R.N. (1974). Some asymptotic properties of U-statistics. PhD dissertation, University of North Ccarolina, Chapel Hill.

Sproule, R.N. (1985). Sequential non-parametric fixed width confidence intervals for U-statistics. *Ann. Statist.* **13**, 228-235.

Strassen, V. (1967). Almost sure behaviour of sums of independent random variables and martingales. *Proc. Fifth Berkeley Symp. on Math. Stat. and Prob.* **2**, 315-343.

Sukhatme, B.V. (1957). On certain two sample nonparametric tests for comparing variances. *Ann. Math. Statist.* **28**, 189-194.

Sukhatme, B.V. (1958). Testing the hypothesis that two populations differ only in location. *Ann. Math. Statist.* **29**, 60-78.

Takahashi, H. (1988). A note on Edgeworth expansions for the von Mises functionals. *J. Multivariate Anal.* **24**, 56-65.

Vandemaele, M. (1982). On large deviation probabilities for U-statistics. *Theory Prob. Applns.* **27**, 614.

Vandemaele, M. and Veraverbeke, N. (1985). Cramer-type large deviations for studentised U-statistics. *Metrika* **32**, 165-180.

van Zwet, W.R. (1984). A Berry-Esseen bound for symmetric statistics. *Z. Wahrsch. und Verw. Gebiete* **66**, 425-440.

Veraverbeke, N. (1985). Studentised estimation of a certain functional of a probability density function. *Scand. Actuarial J.* 131-147.

Von Mises, R. (1947). On the asymptotic distribution of differentiable statistical functions. *Ann. Math. Statist.* **18**, 301-348.

Weber, N.C. (1980). Rates of convergence for U-statistics with varying kernels. *Bull. Austral. Math. Soc.* **21**, 1-5.

Weber, N.C. (1981). Incomplete degenerate $U$-statistics. *Scand. J. Statist.* **8**, 120-123.

Weber, N.C. (1983). Central limit theorems for a class of symmetric statistics. *Math. Proc. Camb. Phil. Soc.* **94**, 307-313.

Williams, G.W. (1978). A pilot Monte Carlo study of two sequential estimation procedures based on $U$-statistics. *Comm. Statist. B – Simula. Computa.* **7**, 129-149.

Williams, G.W. and Sen, P.K. (1973). Asymptotically optimal sequential estimation of regular functionals of several distributions based on generalised $U$-statistics. *J. Multivariate Anal.* **3**, 469-482.

Williams, G.W. and Sen, P.K. (1974). On bounded maximum width sequential confidence elpsoids based on generalised $U$-statistics. *J.Multivariate Anal.* **4**, 453-468.

Withers, C.S. (1988). Some asymptotics for $U$-statistics. *Comm. Statist. A – Theory Methods* **17**, 3269-3276.

Yamato, H. and Maesono, Y. (1986). Invariant $U$-statistics. *Comm. Statist. A – Theory Methods* **15**, 3253-3263.

Yamato, H. and Maesono, Y. (1989). Deficiencies of $U$-statistics of degree 2 under symmetric distributions. *Comm. Statist. A – Theory Methods* **18**, 53-66.

Yoshihara, K. (1976). Limiting behaviour of $U$-statistics for stationary, absolutely regular processes. *Z. Wahrsch. und Verw. Gebiete* **35**, 237-252.

Yoshihara, K. (1984). The Berry-Esseen theorems for $U$-statistics generated by absolutely regular processes. *Yokohama Math. J.* **32**, 89-111.

Zhao, L. and Chen, X. (1987). Berry-Esseen bounds for finite population $U$-statistics. *Sci. Sinica* **30**, 113-127.

# INDEX

299

300

Printed in the United States
by Baker & Taylor Publisher Services